An Introduction to Ethical, Safety and Intellectual Property Rights Issues in Biotechnology

An Introduction to Ethical, Safety and Intellectual Property Rights Issues in Biotechnology

Padma Nambisan

Cochin University of Science & Technology, Cochin, Kerala, India

ACADEMIC PRESS

An imprint of Elsevier

Academic Press is an imprint of Elsevier
125 London Wall, London EC2Y 5AS, United Kingdom
525 B Street, Suite 1800, San Diego, CA 92101-4495, United States
50 Hampshire Street, 5th Floor, Cambridge, MA 02139, United States
The Boulevard, Langford Lane, Kidlington, Oxford OX5 1GB, United Kingdom

British Library Cataloguing-in-Publication Data
A catalogue record for this book is available from the British Library

Library of Congress Cataloging-in-Publication Data
A catalog record for this book is available from the Library of Congress

ISBN: 978-0-12-809231-6

For Information on all Academic Press publications
visit our website at https://www.elsevier.com/books-and-journals

Working together
to grow libraries in
developing countries

www.elsevier.com • www.bookaid.org

Publisher: Mica Haley
Acquisition Editor: Kristine Jones and Erin Hill-Parks
Editorial Project Manager: Molly McLaughlin and Tracy Tufaga
Production Project Manager: Sue Jakeman
Cover Designer: Mark Rogers

Typeset by MPS Limited, Chennai, India

For

Dr. P.N. Narayanan Nambisan

Mohan Panampilly

Gaurav Mohan Panampilly

Contents

Foreword xvii
Preface xix

CHAPTER 1   Genes, Genomes, and Genomics 1
    1.1 Introduction 2
    1.2 The Development of Gene Concept 2
    1.3 Genes 3
    1.4 Genomes 8
    1.5 Genomics 9
    1.6 The Human Genome Project (HGP).. 10
    1.7 Applications of Genomics.. 13
        1.7.1 Phylogenetic Analysis 13
        1.7.2 Genetic Testing and Diagnostics 13
        1.7.3 Pharmacogenomics 16
        1.7.4 Forensics... 16
        1.7.5 Recombinant DNA Technology 17
        1.7.6 Gene Therapy 17
        1.7.7 Genome Editing 18
        1.7.8 Synthetic Biology 18
    1.8 Ethical, Legal, and Social Implication 19
    1.9 Genetic Reductionism 19
  1.10 Summary.. 20
        References.. 21
        Further Reading 23

CHAPTER 2   Cloning 25
    2.1 Introduction 26
    2.2 Cloning Animals 26
        2.2.1 Dolly... 28
        2.2.2 Progress in Cloning After Dolly... 29
        2.2.3 Limitations of Nuclear Transfer 30
        2.2.4 Applications of Cloning 30
        2.2.5 Ethical Issues in Cloning Animals 36
        2.2.6 Laws and Public Policy on Reproductive Cloning in Animals ... 38
    2.3 Human Cloning 41
        2.3.1 Ethical Considerations 42
        2.3.2 Laws and Public Policy on Reproductive Cloning in Humans ... 45

2.4 Summary..50
References...50
Further Reading 53

CHAPTER 3   Stem Cell Research **55**
3.1 Introduction 56
3.2 Sources of Stem Cells...57
3.2.1 Embryonic Stem Cells...58
3.2.2 Nuclear Transfer—Embryonic Stem Cells.................................58
3.2.3 Fetal Stem Cells..59
3.2.4 Cord Blood Stem Cells or Neonatal Stem Cells........................59
3.2.5 Adult Stem Cells...59
3.2.6 Induced Pluripotent Stem Cells 60
3.2.7 Stimulus-Triggered Acquisition of Pluripotency (STAP) Cells 60
3.3 Benefit to Society...60
3.3.1 Understanding Cellular Differentiation 64
3.3.2 Study Disease Progression..65
3.3.3 Regenerative Medicine 65
3.3.4 Tissue Engineering..66
3.3.5 Grow Organs for Transplantation...66
3.3.6 Alter Current Biomedical Practices for Treatment of Cancer....66
3.3.7 Identify Drug Targets and Test Potential Therapeutics 67
3.3.8 Toxicity Testing 67
3.4 Ethical Issues in Stem Cell Research 68
3.4.1 Moral Status of Embryos—Fetalistic Viewpoint 68
3.4.2 Exploitation of Women—Feministic Viewpoint........................70
3.5 Ethical Issues in Stem Cell Translation..71
3.5.1 Clinical Trials Using Stem Cells..71
3.5.2 Guidelines for the Clinical Translation of Stem Cells...............71
3.6 Stem Cell Research Policy...73
3.7 The Role of Politics and Public Opinion in Shaping Stem Cell Policy 76
3.8 Summary...77
References...78
Further Reading 80

**CHAPTER 4   Recombinant DNA Technology and Genetically Modified
            Organisms**..**83**
4.1 Introduction 84
4.2 Gene Therapy...86
4.2.1 Technique 88
4.2.2 Application of Gene Therapy 88

4.2.3 Severe Combined Immune Deficiency (SCID) 89
4.2.4 Challenges in Gene Therapy 89
4.3 Gene Editing Therapy 90
4.4 GMOs and the New Biotech Industry 91
4.4.1 Recombinant Proteins 91
4.4.2 Recombinant Antibodies 91
4.4.3 Pharming 97
4.5 Genetically Modified Crops (GM Crops) 100
4.5.1 Health and Safety Concerns 103
4.5.2 Environmental Concerns 104
4.5.3 Ethical and Socioeconomic Concerns 105
4.6 Transgenic Animals 105
4.6.1 Disease Models 106
4.6.2 Increasing Food Production 106
4.6.3 Producing Proteins for Industry 108
4.6.4 Vector Control 109
4.6.5 Pets 109
4.6.6 Hypoallergenic Pets 111
4.7 Synthetic Organisms 111
4.8 Challenges in Applications of rDNA Technology and GMO Release 112
4.8.1 Effect on Environment 112
4.8.2 Effect on Biodiversity 112
4.8.3 Effect on Sociocultural Norms 113
4.8.4 Effect on Socioeconomic Status of Farmers 113
4.9 Objections to Genetic Engineering and GMOs 114
4.9.1 Proportionality 114
4.9.2 Slippery Slope 114
4.9.3 Subsidiarity 114
4.10 GMOs and "Biopolitics" 117
4.11 Summary 121
References 122
Further Reading 126

CHAPTER 5   Relevance of Ethics in Biotechnology **127**
5.1 Introduction 128
5.1.1 Ethics and Bioethics 129
5.1.2 "Values," "Morals," and "Ethics" 129
5.2 Theories in Ethics 129
5.3 Promoting Ethically Sound Science 131
5.4 Ethical Issues in Biotechnology 131
5.4.1 Environmental Ethics 131

        5.4.2 Ethical Issues in Plant Biotechnology 132
        5.4.3 Animal Rights and the Use of Animals in Medical Research 134
        5.4.4 Ethical Issues in Human Clinical Trials 138
    5.5 Medical Ethics to Biomedical Ethics 140
    5.6 Good Clinical Practices 141
    5.7 Bioethics 143
    5.8 Summary 146
        References 146
        Further Reading 148

CHAPTER 6   Bioterrorism and Dual-Use Research of Concern 149
    6.1 Introduction 150
    6.2 Bioterrorism 151
        6.2.1 Weaponizing Microbes 151
        6.2.2 Biological Weapons in History 152
    6.3 The 1972 Biological Weapons Convention 153
    6.4 Biosecurity Measures for Preventing Bioterrorism 155
    6.5 United States Approach to Bioterrorism 157
    6.6 European Union Approach to Bioterrorism 161
    6.7 Biodefense Programs 162
    6.8 Dual-Use Research of Concern 163
        6.8.1 National Science Advisory Board for Biosecurity 163
        6.8.2 Core Responsibilities of Life Scientists in Regard to Dual-Use
               Research of Concern 165
    6.9 Summary 165
        References 167
        Further Reading 169

CHAPTER 7   Genetic Testing, Genetic Discrimination and Human Rights 171
    7.1 Introduction 172
    7.2 Genetic Testing 172
    7.3 Genetic Exceptionalism 175
    7.4 Genetic Discrimination 175
    7.5 Ethical, Legal, Social Implication (ELSI) 177
        7.5.1 The ELSI of Human Genome Project 178
        7.5.2 International and National Programs on ELSI 179
    7.6 Mechanisms for Preventing Genetic Discrimination 180
        7.6.1 Rights-Based Advocacy 180
        7.6.2 Rights-Based Policy and Tools 182
        7.6.3 National Legislation 183
    7.7 Summary 186

References.. 186
Further Reading 187

CHAPTER 8 Biodiversity and Sharing of Biological Resources......................... **189**
8.1 Introduction 190
8.2 Concept of Biodiversity 191
8.2.1 The Functions of Biodiversity 191
8.2.2 Bioprospecting 194
8.2.3 Concerns Regarding Bioprospecting 194
8.3 The United Nations Convention on Biological Diversity 195
8.3.1 Conference of the Parties.. 196
8.3.2 Bonn Guidelines on Access to Genetic Resources and Fair and
        Equitable Sharing of the Benefits Arising Out of Their Utilization
        (Bonn Guidelines on ABS, 2001)..................................... 198
8.3.3 The Nagoya Protocol on Access and Benefit-Sharing...... 199
8.4 The United Nations Convention on the Law of the Sea.................. 201
8.5 FAO International Treaty on Plant Genetic Resources for Food and
        Agriculture.. 202
8.5.1 Access and Benefit Sharing (ABS) Obligations 203
8.5.2 Intellectual Property and Farmer's Rights 203
8.5.3 Ex Situ Conservation Centers... 203
8.6 Regional and National Significance and ABS Frameworks 204
8.6.1 Category 1: Countries With No National Laws Specifically
        Devoted to ABS 205
8.6.2 Category 2: Countries With a Biodiversity or Environmental
        Law With Provisions on ABS 206
8.6.3 Category 3: Countries With National Laws Specifically Devoted
        to ABS... 206
8.7 Summary.. 209
References.. 209
Further Reading 209

CHAPTER 9 Ensuring Safety in Biotechnology **211**
9.1 Introduction 212
9.2 Safety Issues in Recombinant DNA Technology............................. 213
9.2.1 Effect of Genetically Modified Organisms on the Environment............... 213
9.2.2 Foods Derived from Genetically Modified Organisms............ 214
9.2.3 Safety Issues in the Use of Recombinant DNA Technology
        in Medicine 214
9.3 International Organizations, Treaties, and Conventions Addressing Biosafety ... 215
9.3.1 Cartagena Protocol on Biosafety to the Convention on Biological
        Diversity 217

      9.3.2 United Nations Food and Agricultural Organization (FAO) Instruments That Deal With Issues Pertaining to Biosafety 221

      9.3.3 Codex Alimentarius Instruments That Deal With Issues Pertaining to Food Safety 222

9.4 Regulatory Oversight for Handling of Genetically Modified Organisms 223

      9.4.1 Organization for Economic Cooperation and Development (OECD) Guidelines 223

      9.4.2 European Union Regulatory Approach 224

      9.4.3 National Institutes of Health (NIH) Guidelines and Regulation of GMOs in the United States 225

      9.4.4 Other National Frameworks for Regulation of GMOs 228

9.5 Summary 231

      References 231

      Further Reading 232

CHAPTER 10 Risk Analysis **233**

10.1 Introduction 234

10.2 Risk Assessment 234

10.3 Risk Management 237

10.4 Risk Communication 238

10.5 Risk Assessment for Genetically Modified Microorganisms 238

10.6 Risk Assessment for Genetically Modified Crops 240

      10.6.1 Stacked Events 241

      10.6.2 Pest-Risk Analysis for Quarantine Pests Including Analysis of Environmental Risks and Living Modified Organisms 241

10.7 Risk Assessment for Transgenic Animals 243

      10.7.1 World Organization for Animal Health 243

10.8 Safety Assessment of Foods Derived from Genetically Modified Organisms 247

      10.8.1 Codex Alimentarius Commission 247

      10.8.2 Postrelease Monitoring 249

10.9 Safety Assessment in Clinical Trials 249

10.10 Precautionary Principle in GMO Regulation 250

10.11 Summary 251

      References 251

      Further Reading 252

CHAPTER 11 Laboratory Biosafety and Good Laboratory Practices **253**

11.1 Introduction 254

11.2 Risk Categories of Microorganisms 255

11.3 Biosafety Levels 256

      11.3.1 Physical Containment 257

      11.3.2 Biological Containment 263

11.4 Physical and Biological Containment for Research Involving Plants 263
11.5 Physical and Biological Containment for Research Involving Animals 267
11.6 Good Laboratory Practice 270
11.7 Summary... 271
        References.. 271

CHAPTER 12 Recombinant DNA Safety Considerations in Large-Scale Applications and Good Manufacturing Practice **273**
12.1 Introduction 274
12.2 Safety Considerations for Industrial Applications of Organisms Derived by Recombinant DNA Techniques 274
12.3 Good Industrial Large-Scale Practice 275
12.4 Good Developmental Principles 278
12.5 Safety Considerations for Field/Market Release of GMOs and/or Their Products 279
        12.5.1 Safety Considerations for Field Release of Genetically Modified (GM) Crops 279
        12.5.2 Safety Considerations for Field Release of Genetically Modified Animals 282
        12.5.3 Safety Considerations for Marketing of Foods from Genetically Modified Organisms 283
        12.5.4 Safety Considerations for Market Approval of Biopharmaceuticals 284
        12.5.5 Safety Considerations for Market Approval of Biosimilars 286
12.6 Good Manufacturing Practices.. 286
12.7 Summary... 289
        References.. 289
        Further Reading 290

CHAPTER 13 Relevance of Intellectual Property Rights in Biotechnology **291**
13.1 Introduction 292
13.2 Intellectual Property Rights 293
13.3 Role of Intellectual Property Rights in Trade 293
13.4 International Conventions and Treaties for Protection of IPRs 295
        13.4.1 Paris Convention 296
        13.4.2 World Intellectual Property Organization 296
13.5 Intellectual Property Rights Issues in Biotechnology 297
        13.5.1 Issues of Patentability 298
        13.5.2 Issues of Ownership 302
        13.5.3 Issues of Enforcement... 302
        13.5.4 Issues of Sharing of Costs and Benefits....................................... 306
        13.5.5 Issues of Ethics 307

13.6 Summary...308
References...308

CHAPTER 14 Patenting of Life Forms.....................................**311**
    14.1 Introduction 312
    14.2 Criteria for Award of Patent 312
        14.2.1 Prerequisites for Patentability.................................313
        14.2.2 Essential Requirements for Patentability................314
    14.3 Patenting Cells and Cell Lines...316
    14.4 Patenting Genes and DNA Sequences..................................318
        14.4.1 Sequences Used for Diagnostic Testing..................319
        14.4.2 Sequences Used as Research Tools.........................319
        14.4.3 Sequences Used in Gene Therapy...........................321
        14.4.4 Sequences Used in Production of Therapeutic Proteins...........321
    14.5 Patenting of Animals..321
    14.6 Protection of Plant Varieties..323
        14.6.1 Patents 323
        14.6.2 Sui Generis Forms of Protection of Plant Varieties: International Union for the Protection of New Varieties of Plants (*Union internationale pour la protection des obtentions végétales*, UPOV).......324
        14.6.3 Geographical Indications 325
    14.7 Summary..326
References...326
Further Reading 327

CHAPTER 15 Patents in Biopharma...**329**
    15.1 Introduction 330
    15.2 Pharmaceuticals and Biopharmaceuticals............................331
        15.2.1 Differences Between Biologics and Conventional Drugs........331
        15.2.2 Biosimilars and "Interchangeable" Biologics 333
    15.3 The Need for Intellectual Property Protection of Biopharmaceuticals.................334
    15.4 Patent Protection for Biologics..335
        15.4.1 Incremental Innovation 336
        15.4.2 "Evergreening"...336
    15.5 Patent Protection for Diagnostics 338
    15.6 The Impact of Patent Protection in Genetic Testing............340
    15.7 International Trade Agreements in Medicines 342
    15.8 Summary..343
References...344

CHAPTER 16 Protection of Traditional Knowledge Associated With Genetic Resources **345**

16.1 Introduction 346

16.2 The Importance and Need for Protection 346

    16.2.1 The Convention on Biological Diversity and Traditional Knowledge Associated With Genetic Resources 347

    16.2.2 The World Intellectual Property Organization and Traditional Knowledge 348

    16.2.3 The World Trade Organization and Traditional Knowledge 350

    16.2.4 The Food and Agriculture Organization and Traditional Knowledge 352

16.3 Legal Protection for Traditional Knowledge 352

    16.3.1 Positive Protection Strategies 353

    16.3.2 Defensive Protection Strategies 353

    16.3.3 Problems With Implementation 354

16.4 Summary 354

    References 356

Index 357

Foreword

Discoveries and inventions made in the 21st century have led to a deeper understanding of biological systems, which in turn have empowered us with tools for manipulating life itself. We have entered an era in which novel life forms are being created through recombinant DNA technology, genome editing, and synthetic biology. The question being posed to biologists today in the context of manipulating living organisms is no longer "*can we*" but "*should we.*" Now, more than ever before, the implications of current policies are being perceived to have the potential of affecting future generations of scientists. Increasingly, awareness and participation of public in influencing policy decisions are impacting the contours of scientific research and its downstream translation. Examples that immediately come to mind are policies regarding stem cell research, and those with respect to the adoption of genetically modified (GM) crops. In both instances, arguments put forward by scientists in support of their adoption have been overwhelmed by public perceptions of ethics and safety. It is therefore important that students of biology, as also the general public interested in developments in science, have access to information that is unbiased and helps the reader to better understand the advantages and possible pitfalls in the adoption of different biotechnologies. This book aims to do just that: it introduces the reader the issues in biotechnology that pertain to the ethics, safety, and intellectual property rights involved in the technology.

The field of biotechnology is vast and encompasses varied applications in ***healthcare***, both for diagnosis and treatment of diseases; ***agriculture***, for the production of food, feed, and fiber; and ***environment***, especially the conservation of biological diversity and genetic resources, and practices that sustain livelihood. To do justice to all aspects of the translation of technologies in multiple areas is challenging. Nevertheless, this book attempts to collate and organize information on current attitudes and policies in several emerging areas of biotechnology. The first four chapters of this book introduce four important technologies in biology such as genome analysis, cloning, stem cell research, and recombinant DNA technology, and sensitize the reader to some of the ethical and safety issues associated with each. Chapters 5—8 highlight the importance of ethics in biotechnology and discuss issues such as unethical use of biological agents in terrorism, concerns regarding the handling of genetic information, and fair use of biological resources. In Chapters 9—12, current mechanisms for ensuring safety in biotechnology are discussed with reference to analysis of risk associated with genetically modified organisms and how they are to be handled in laboratories as well as in large-scale manufacturing units such that they do not cause harm to laboratory workers, or affect the health of humans, animals, or the environment if released from laboratory settings for large-scale use. Chapters 13—16 deal with the relevance of Intellectual Property Rights (IPR) in biotechnology and the use of IPR instruments such as patents for protecting IP in the context of novel life forms, pharmaceuticals, and traditional knowledge associated with genetic resources. By including references to the original sources and web links wherever possible, the author has ensured a rich learning experience for the interested students. I trust that this book would serve to be a valuable resource to students in interdisciplinary areas of study to better understand the nuances of applications emerging in the field of biology. I am deeply impressed by the approach of Prof. Padma Nambisan in handling a complex and difficult to deal with area in this book. I am confident that both students of biotechnology and public interest groups interested in downstream descent of this powerful technology will find this book of immense value.

V.L. Chopra

Preface

The motivation for writing a textbook comes from many sources. For me it came from a need to present my students with a source of unbiased information that would help them form their own opinions on the use of biotechnology. Faced with having to teach a newly introduced paper "Bioethics, Biosafety, and Intellectual Property Rights" to students in the second year of a post-graduate course in Biotechnology in 2005, I was hampered with not having suitable textbooks; the available textbooks were too technical to sustain the interest of most students and the World Wide Web was inundated with articles and papers which were often biased and left the student confused. By 2010, I had compiled sufficient content to attempt an e-learning course running on software developed by Mohan Panampilly, my husband. It ran for 3 years on the university website before it was hacked and became unavailable to users. Unfortunately, I lost my husband to cancer in 2013, and not knowing how to maintain the software, the e-learning course was abandoned, to be resurrected as a book project. I hope by this effort to help graduate students in biotechnology and allied fields to put into perspective some of the recent developments in the application of technology using biological systems. The goal is to empower the student to taking a proactive role in shaping the future of technologies that are perceived to be capable of altering the fabric of human society. Off the cuff examples of such technologies include genome editing in humans, the adoption of genetically modified (GM) foods and perceptions of its safety, the future of stem cell therapy, the sharing of biological resources, patenting of biopharmaceuticals, and pharmacogenomics. As it happens, many of these issues are increasingly being discussed not only in classrooms but by the public. Many countries are in the process of shaping science policies to deal with regulatory concerns regarding the use of biotechnology.

One of the major challenges I faced while writing the book was keeping abreast of current developments. The rapid pace at which technologies are developing and the issues that are constantly emerging in their adoption threatens to render the content outdated even before it is published. I hope that the book would still be of use to students albeit to understand the evolution of technologies from a historical perspective. A few illustrative examples are given below:

- In the United Kingdom, laws for human reproduction have been modified in January 2017 to permit genomic editing for certain rare genetic disorders and techniques such as the mitochondrial replacement therapy which combines genetic material from three parents. Already, in Mexico, in September 2016, a healthy boy was born to a mother carrying fatal genetic defects in her mitochondria using this therapy, while a second child was born in Russia in January 2017. While being heralded as a technological leap for in vitro fertilization, concerns are being expressed on its safety (the fatally flawed mitochondria could reappear or increase the vulnerability of the baby to new diseases), and that this may be only a short step away from the development of genetically enhanced "designer" babies. Will human genome editing be permitted in other nations or regions?
- The world is divided with regard to the adoption of GM crops (and the safety of GM foods) but a change in perception is seen even in regions which have embraced the technology. For instance, the United States has been a supporter of GM crops, and GM foods have been available in the US markets since the early eighties. Labeling of GM foods was not considered

necessary as the USFDA did not consider food from GM crops to be substantially different from that from unmodified crops. Yet in 2016, President Obama signed into law labeling laws that required food containing GM ingredients to carry a text label, a symbol or an electronic code readable by smartphone. In India, despite the success of GM cotton in propelling the country to becoming the highest producer of cotton, the cultivation of GM food crops is banned due to safety concerns. GM mustard developed by Deepak Pental's laboratory in Delhi University had received clearance from the GEAC (the regulatory agency in India) by November 2016, but its release to farmer's fields has been put on hold by the Supreme Court in January 2017 pending admission of a public petition raising doubts on its safety. A new Directive in European Union in 2015 gave member nations more freedom to make decisions regarding growing GM crops resulting in 19 of the 28 countries banning GM crops on the grounds of safety. In an effort to assure the public that fears regarding GM crops have no scientific basis, 133 Nobel laureates have signed a letter on June 29, 2016, soliciting more support for the development of humanitarian efforts such as the development of GM rice, the Golden Rice. Although many agree that the scientific basis of anti-GM policies is specious, negative public perceptions regarding GM crops currently overwhelm scientific opinion in several countries. It is thus uncertain whether GM crops and GM foods will ever gain universal acceptability.

- Biotechnology provides several new techniques for the diagnosis and treatment of human diseases, but many of the options unfortunately are expensive. Keeping healthcare affordable is a challenge facing most governments. For innovators in the pharma industry, patents and exclusive marketing rights are crucial for retaining market edge and protecting profits. International trade negotiations such as the Trade Related Intellectual Property Rights (TRIPS) and Agreement and Trans Pacific Partnership (TPP) have therefore included provisions that influence the cost of medicines. The TPP for instance has been a source of concern to organizations such as the *Medecins Sans Frontieres*, as the provisions were considered to possibly raise the cost of healthcare. With the new US President Donald Trump having signaled the withdrawal of the United States from the TPP in his first week in office, while also signing an executive order to begin unraveling the Affordable Care Act (ACA, or "Obamacare") and ease the "burdens of Obamacare," the impact on pharma companies and the biotech sector remains to be seen.

ACKNOWLEDGMENTS

In the development of this book project, I have depended on the help and support of several who in this inter-connected world have been only a mouse-click away, ready to share material and words of encouragement. This book would not have been possible without the faith that my Elsevier Editor Kristine Jones had in me and the able hand-holding provided by her and Molly McLaughlin (Editorial Project Manager) to a first-time author. Always ready with crisp and insightful comments and suggestions, it has been pure pleasure to work with them and the rest of Kristine Jones' (later Erin Hill-Parks') team including Sue Jakeman (Production Manager) and Jyotsna Gopichandran. For helping me ensure the veracity of the content, imperative for a text-book, I have depended on the expert advice of Dr. Balakrishnan Rajan, Professor and Head, Department of Radiation Oncology, National Oncology Center, The Royal Hospital, Muscat, Sultanate of Oman, (former Director, Regional Cancer Centre, Trivandrum, India); Professor Eunjoo Huising Pacifici, Associate Director,

Graduate Programs, International Center for Regulatory Science, Assistant Professor of Clinical Pharmacy, Los Angeles, CA; and for the chapters on intellectual property rights (IPR), Professor N.S. Gopalakrishnan, Director, Inter University Center for IPR Studies, Cochin University of Science and Technology, Cochin, India. They have individually reviewed the work, pointed out errors, suggested better ways of presentation, and in some cases contributed content to improve the textbook, for all of which I am grateful. Any remaining errors, of course, are entirely mine. I am deeply indebted to Professor V.L. Chopra, former Director-General, Indian Council of Agricultural Research and member of the Planning Commission of India, instrumental in introducing me to the fascinating realm of biotechnology, for providing the Foreword to this book, a token of his abiding support for this endeavor. Writing a book is time-consuming and would not have been possible without the academic freedom that I enjoyed and the cooperation of my colleagues in Cochin University of Science and Technology. To them as to my students, I am indebted also for countless discussions and interactive sessions which helped mature opinions and views regarding many technologies. For images used in the presentation of the content, I have corresponded with and depended on the kindness and generosity of several including Hillary S. Kativa, Curator of Photographic and Moving Image Collections, Othmer Library of Chemical History, Chemical Heritage Foundation; Margaret Engel of The Alicia Patterson Foundation; Janice F. Goldblum of the National Academy of Sciences Archives; Marin P. Allen, Deputy Associate Director for Communications and Public Liaison and Director of Public Information, National Institutes of Health; Adam Freestone, Director of Communications & Digital Media, Immune Deficiency Foundation; Daniel Hartwig, University Archivist at Stanford; and the WHO Permissions team. I thank each of them for their patience and cooperation. I am especially grateful to Adrian Dubock of the Golden Rice Humanitarian Board; Mathew Warren of Oxitec; Rhodora R. Aldemita of the ISAAA SE Asia Center; and Dave Conley of AquaBounty Technologies, Inc., for having reviewed the contents of specific Boxes for accuracy and relevance, and for providing images I could use to supplement the text.

This book is dedicated to the memory of my father, Dr.P.N.N. Nambisan, botanist and plant breeder, responsible for igniting in me a love for science and biology at a very early age; and to the memory of my husband, Mohan Panampilly, a software engineer who developed a keen interest in biotechnology (he was especially drawn to the information storage and retrieval properties of nucleic acids) and encouraged me to write a book he could understand. Finally, I would like to thank my son, Gaurav Mohan Panampilly, without whose encouragement and unwavering support I would not have been able to write this book.

HOW TO USE THIS BOOK

The book is divided loosely into four sections of four chapters each: The first section describes four areas crucial to modern biotechnology—genomics, cloning, stem cell research, and recombinant DNA technology. The subsequent sections deal respectively with issues relevant to ethics, safety, and IPR in biotechnology. Needless to say these issues are not mutually exclusive and are in reality highly interlinked. For instance, ensuring safety is of paramount importance in normative ethical practices in biotechnology, and patents may be foregone for ethical reasons and collective good. Each chapter is followed by a list of references that is rather more extensive than that usually found in textbooks. For most, links to online copies of the article have also been provided. In order to derive the full benefit of the textbook, students are strongly urged to read the original articles. The need for brevity severely limits a detailed discussion of several viewpoints in the textbook, and the references are therefore provided to complement the text. Additional material and examples are provided in Boxes, while key concepts are summarized in "Key Takeaways" sections.

It is hoped that this introductory book would be of use to students as well as the general reading public to gain a better understanding of the future implications of modern biotechnology.

GENES, GENOMES, AND GENOMICS

The greatest history book ever written is the one hidden in our DNA.
-Spencer Wells, geneticist, anthropologist and author of "The Journey of Man"

CHAPTER OUTLINE

1.1 Introduction 2
1.2 The Development of Gene Concept 2
1.3 Genes 3
1.4 Genomes 8
1.5 Genomics 9
1.6 The Human Genome Project (HGP) 10
1.7 Applications of Genomics 13
    1.7.1 Phylogenetic Analysis 13
    1.7.2 Genetic Testing and Diagnostics 13
    1.7.3 Pharmacogenomics.. 16
    1.7.4 Forensics 16
    1.7.5 Recombinant DNA Technology 17
    1.7.6 Gene Therapy... 17
    1.7.7 Genome Editing 18
    1.7.8 Synthetic Biology 18
1.8 Ethical, Legal, and Social Implication 19
1.9 Genetic Reductionism... 19
1.10 Summary 20
References 21
Further Reading 23

An Introduction to Ethical, Safety and Intellectual Property Rights Issues in Biotechnology.
DOI: http://dx.doi.org/10.1016/B978-0-12-809231-6.00001-6

On hand at a press conference that followed the White House genome announcement are (from l) Dr. Craig Venter, Celera; Dr. Ari Patrinos, US Department of Energy, and Dr. Francis Collins, director, NHGRI. DOE and NIH were the two federal agencies involved in the Human Genome Project.

Photo courtesy: National Institutes of Health.

1.1 INTRODUCTION

The 21st century has been referred to as the "Century of Biology" (Venter & Cohen, 2004) primarily because of the explosion of information regarding the manner in which living organisms are structurally and functionally organized and their characteristics passed down from one generation to the next. Starting with a very basic premise at the turn of the century that the genetic information is of a particulate nature, the end of the 20th century saw the construction of detailed genetic maps, including sequence information, from taxonomically diverse organisms. From **Mendel to the Human Genome Project** (**HGP**), the events in the 20th century have changed the way we view life and living things. It has created new ways of using life forms, of manipulating genetic information in organisms to perform new tasks or make new products, but it has also made it possible for this information to be misused in ways that threaten global safety.

This chapter discusses the various milestones in our understanding of genetic information that paved the way for the technology to manipulate genes to obtain new genetically modified organisms (GMOs). It will also discuss the development of genomics and cover some of the current and potential applications of this technology.

1.2 THE DEVELOPMENT OF GENE CONCEPT

In 1900, the **principles of inheritance of characters** as proposed by **Gregor Mendel** in 1865 were rediscovered independently by three botanists. These principles intuitively implied the **involvement**

of particulate matter, which Mendel termed as "factors" (and later came to be known as "**genes**") in the transmission of traits from parent to offspring. The behavior of these "factors" was **predictable** and outcomes of crosses between parents differing in characters could therefore be estimated on the basis of **simple probability rules** (see Fig. 1.1). Although at that time the physical or chemical nature of these factors was not known, very soon it was established that the behavior of the factors was similar to the behavior of "thread-like" structures, the **chromosomes**, observed in the nuclei of cells. Whether it was the proteins or the nucleic acid in chromosomes that served as the genetic material remained a mystery till 1952 when Alfred Hershey and Martha Chase performed experiments with differentially radioactively labeled bacteriophages (Fig. 1.2) which proved unequivocally that it was the **nucleic acid which encoded genetic information** (Hershey & Chase, 1952).

Studies on the chemical structure of nucleic acids culminated in the proposal of the **double helical model for DNA** by James Watson and Francis Crick in 1953 (Watson & Crick, 1953). The model (Fig. 1.3) fascinated biologists for several reasons: for one, the model **accounted for the stability of DNA** and provided a plausible reason for its ubiquity—DNA had been observed in almost all living organisms across all taxonomic classes and did not seem to be easily destroyed by most environmental factors. For another, being double helical, made up of two antiparallel strands, it was easy to imagine **fidelity of transmission** from parent to offspring if one assumed that each strand would serve as a **template** for synthesis of the new strand. The structure had the added advantage that it could **account for mutations** in genetic traits seen in different organisms in nature if one assumed that **changes in base sequence could arise and persist**. The molecule innately had an **infinite capacity to store information** in the form of **different sequence combinations of the four bases**—Adenine, Thymine, Cytosine, and Guanine, abbreviated as A, T, C, and G, respectively. **Information retrieval involved the synthesis of another nucleic acid, RNA**, using one of the two DNA strands, as a template. This **RNA directs the synthesis of proteins** which either form the structural components of the cells, or act as enzymes bringing about other metabolic reactions responsible for all the characteristic features of the cell. This essentially was what was summed up by Crick as the "**Central Dogma of Biology**" (Crick, 1970).

1.3 GENES

In classical genetics, a **gene** came to be **defined on the basis of its function;** in other words, the **ability to direct the expression of a character, and to undergo mutations** which altered the expression of the character. Genes were assumed to be strung together on chromosomes much like beads on a string. Genes had fixed positions on the chromosomes but could be shuffled by **recombination** which could happen by a physical exchange between paired chromosomes during a specific stage in cell division. Recombination was thought to occur only between genes, so the **gene was considered to be indivisible and a unit of function, recombination, and mutation**. The concept was altered in the sixties with the experiments of Seymour Benzer and the **gene as a unit of function** was termed the "cistron," as a unit of mutation, a "muton," and as a unit of recombination, a "recon" (Benzer, 1955, 1961). It is now generally accepted that a **gene is a sequence of nucleotides in DNA which performs a function, such as encoding the synthesis of a protein, or binding of factors that regulate gene expression.**

I: Law of Segregation:

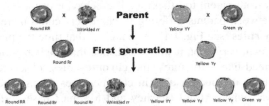

II: Law of Independent Assortment:

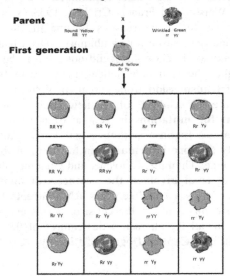

Second generation: 9 : 3 : 3 : 1

FIGURE 1.1

Mendel's Laws:

I: Law of Segregation based on monohybrid crosses. The first generation consistently showed only one of the two alternative traits, known as the dominant trait. (In this example, round is dominant over wrinkled and yellow dominant over green peas.) The recessive trait reappeared in the second generation in a consistent ratio of 3 dominant:1 recessive.

II: Law of Independent Assortment based on dihybrid crosses. The first generation showed only the two dominant traits (round and yellow), but the second generation showed a phenotypic ratio of **9** (round, yellow):**3** (round, green):**3** (wrinkled, yellow):**1** (wrinkled, green). This was the ratio that could be expected from the simultaneous occurrence of two independent events—since each character segregated with a probability of 3:1, and the probability of the simultaneous occurrence of two independent events is the product of their individual probabilities: that is—(3:1)(3:1) = 9:3:3:1.

Structurally, the gene can be divided into three regions—the ***promoter region*** to which the RNA polymerase binds, the ***coding region*** which is transcribed into RNA and specifies the sequence of amino acids in the polypeptide, and the ***terminator region*** that indicates the end of the reading frame and causes the RNA polymerase to dissociate from the DNA. In prokaryotes which are unicellular, rapid response to environmental cues (such as the presence or absence of a nutrient in the substrate, or response to a stressor) is often crucial to survival. Genes specifying enzymes in the same metabolic pathway are often **clustered and coordinately expressed**. Known as *operons*, these gene assemblies have several *structural genes* specifying the various enzymes, expressed using a common ***promoter gene*** (Fig. 1.4). Switching on or off of the operons is achieved by regulatory factors (the "**repressor**" or the "**inducer**") binding to the *operator gene*.

In eukaryotes, the coding region is not continuous, but "split" into coding and noncoding regions (Fig. 1.5). The **noncoding stretches** (known as the *introns*) are **excised from the RNA transcript** and the **coding regions** (known as the *exons*) are "**spliced**" together during maturation of the RNA transcript. An evolutionary advantage in having split genes lies in the possibility to generate different transcripts (and hence proteins) by **alternate splicing**, including or excluding some of the exons in the final transcript. Another major difference between genes in prokaryotes and eukaryotes is that in eukaryotes, the **promoter region is rather large**, and in addition to the RNA polymerase, **binds several transcription factors** which serve to influence the expression of the gene.

Expression of genes in both prokaryotes and eukaryotes involves synthesis of a **RNA intermediate**, a process referred to as "*transcription.*" The nucleotide bases in this RNA known as the "**messenger RNA**" (**mRNA**) are read by the protein synthesizing machine, the ribosomes, in groups of three bases (**triplet codons**). During the process of protein synthesis, referred to as "*translation*," amino acids are incorporated as specified by each codon in the **mRNA reading frame** into the growing polypeptide chain that eventually matures into a protein. Codon usage appears to be **universal**; almost all organisms use the same **genetic code** to specify the

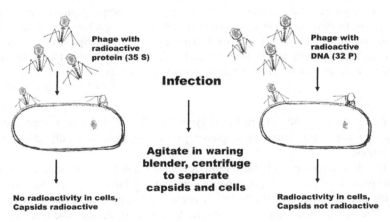

FIGURE 1.2 Hershey and Chase Experiments

Infecting *Escherichia coli* cells with differentially labeled T2 phages established that DNA and not protein is the genetic material.

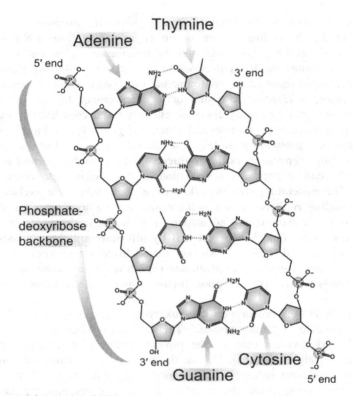

FIGURE 1.3 Watson and Crick Model of DNA

The model consists of two "backbones" made of the deoxyribose sugar linked by phosphodiester bonds. Bridging the two backbones in the manner of the rungs of a ladder are the complementarily paired nitrogenous bases—each "rung" has a purine (A or G) linked to a pyrimidine (T or C) by hydrogen bonds resulting in a uniform diameter of approximately 2 nm. The two strands run in opposite directions ("antiparallel") and are twisted into a right-handed helix making a complete turn in about 3.4 nm with about 10 bases per turn. The actual dimensions of the DNA strand are dependent on the level of hydration and the ionic concentration.

By Madeleine Price Ball via Wikimedia Commons.

incorporation of a specific amino acid in the polypeptide. The code is also degenerate; more than one codon specifies the same amino acid, possibly because there are only 20 amino acids, but $4^3 = 64$ **triplet codes**.

In prokaryotes, transcription and translation follow in quick succession, the nascent mRNA is translated into protein even as it itself is being synthesized. However, in eukaryotes, the DNA is sequestered in the nucleus and only transcription occurs in the nucleus. Protein synthesis occurs in the cytoplasm. The RNA transcript therefore has to move from the nucleus to the cytoplasm in order to be translated. Several modifications are made in the RNA before it is translated, such as addition of a "**cap**" and a poly(A) "**tail**" at the ends of the RNA as well as cutting and removing noncoding intron regions and **splicing** together the exon regions of the reading frame.

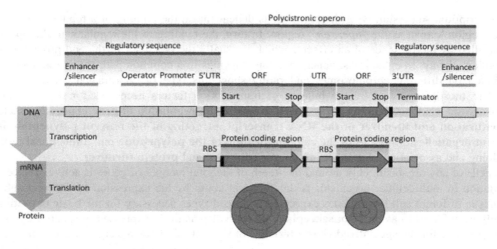

FIGURE 1.4 Organization of genetic elements in prokaryotes

[ORF: Open Reading Frame, UTR: Untranslated regions, RBS: Ribosome Binding Sites.]

By Thomas Shafee, via Wikimedia Commons.

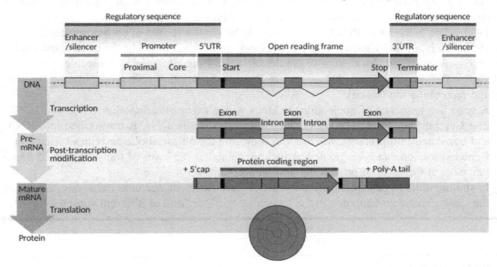

FIGURE 1.5 Organization of genetic elements in eukaryotes

[UTR: Untranslated regions.]

By Thomas Shafee, via Wikimedia Commons.

   Regulation of gene expression is effected at several points during transcription and translation. In prokaryotes, genes encoding enzymes required in a particular metabolic pathway are often clustered and coordinately regulated, referred to as "*operons*." Switching "on" or "off" of an operon hinges on the "*promoter*" region of the operon being accessible to an RNA polymerase required for transcription. Accessibility of the promoter is dependent on the binding of a regulator to a contiguous region known as the *"operator"* region. In turn, the binding of the regulator to the operator is controlled by the presence of an effector which is usually the precursor or the end product of the metabolic pathway. In eukaryotes, genes encoding enzymes in a common pathway are more widely distributed on different chromosomes and rarely show this type of coordinated regulation. The genes are mostly regulated by the binding of transcription factors necessary for the binding of the eukaryotic RNA polymerase. Other **check points in the regulation of gene expression** include the **maturation and turnover of the RNA transcript, targeting of the nascent polypeptide into different organelles** or regions of the cell, **maturation of the polypeptide** into a functional protein by folding and association with other proteins or cofactors, and **protein turnover**.

In cells of any organism, only around one-tenth of the total number of genes is active. Tissue differentiation in multicellular organisms is thus brought about by the expression of different subsets of genes in different cell lines. Genes expressed in all cell types necessary for the **basic activities of the cell** are referred to as the **"housekeeping" genes**, whereas those expressed in specific cell lines and responsible for the **specialized characteristics** of the cells are known as the "luxury" genes.

KEY TAKEAWAYS

- 1865—Gregor Mendel proposes "factors" as units of inheritance with predictable behavior based on simple laws of probability in controlled crosses
- 1903—The behavior of the "factors" found to be similar to that of chromosomes
- 1952—Hershey and Chase identify DNA (and not protein) as being the genetic material
- 1953—Watson and Crick propose the double helical model for DNA molecule
- 1961—Benzer establishes that genes are the unit of function (cistron), mutation (muton) and recombination (recon)
- A gene is a sequence of nucleotides in DNA which performs a function
- The Central Dogma of Biology is that the information in DNA is transcribed into RNA, and translated into proteins, which are responsible for all metabolic activities of the cell
- Gene expression can be regulated at various points: at the level of transcription, of translation, or at the level of protein maturation and turnover
- In multicellular organisms, different tissues have different characteristics despite having the same genetic makeup due to the switching "off" or "on" of different subsets of genes

---

1.4 GENOMES

A **"genome"** is the entire DNA in an organism, including its genes. Although simple unicellular organisms have smaller genomes than higher eukaryotes, the amount of DNA is not proportional to the morphological complexity of the organism or the probable number of genes. It is estimated that the **human genome has around three billion base pairs of DNA** and has between 20,000 and 25,000 genes. The length of gene sequences is between 2000 and 20,000 base pairs which accounts for only a fraction of the total DNA in the genome. Apparently, in most eukaryotes, a large proportion of DNA sequences in the genome have no known function. This fraction has often been referred to as "selfish DNA" or "junk DNA." Putative functions ascribed to this fraction are that these sequences are a reservoir for generating new genes often necessary for the evolution of the species, or that the sequences act as spacers positioning active genes in locations in the nucleus where access to the enzymes and factors for gene expression is improved. Roughly 75% of DNA that is transcribed into RNA does not code for proteins but for short or long noncoding RNAs, such as endogenous small interfering RNAs (endo-siRNAs), PIWI-interacting RNAs (piRNAs), small nucleoli RNAs (snoRNAs), natural antisense transcripts (NATs), circularRNAs (circRNAs), long intergenic noncoding RNAs (link RNAs), enhancer noncoding RNAs, and transcribed ultraconserved regions (T-UCRs) which have important functions in the body.

Sequences in the genome fall into one of **three classes**, the **"unique"** sequences which has one to few copies (and invariably all gene sequences fall in this class), the **"moderately repeated"** sequences which exist in 50 to a few hundred copies, and **"highly repetitive"** sequences which exists in thousands of copies. Sequencing genomes have given us insights into evolutionary relationships between organisms. All organisms despite the diversity of structure and organization appear to be connected by common DNA sequences. One advantage is that sequence information from nonhuman genomes too could further enhance our understanding of human biology.

1.5 GENOMICS

The HGP and subsequent projects have resulted in the genomes of more than 180 organisms being sequenced since 1995. Computational techniques/bioinformatics have served to make sense of the sequence data generated and provided support to the study of genetics. **"Omics"** is a generic term for a discipline in science and engineering for **analyzing the interactions between biological information elements** in diverse "omes," for example genome, proteome, metabolome, expressome, and interactome (see Table 1.1). The main focus of omics is on **mapping elements**, understanding the **interaction** between them, and **engineering networks to manipulate the elements and their interactions**. Mapping involves the identification of biological information elements like genes, proteins, and ligands.

The term "genomics" was suggested by T. H. Roderick of the Jackson Laboratory, Bar Harbor, Maine, and is considered to be a new discipline born *"from a marriage of molecular and cell biology with classical genetics, fostered by computational science"* (McKusick & Ruddle, 1987). Genomics includes:

- *Functional genomics*—understanding the structure of genes and their functions in terms of synthesis of mRNA and proteins.
- *Structural genomics*—studying the spatial arrangement of genes and sequences on chromosomes.
- *Comparative genomics*—understanding the evolutionary relationships between the different species on the basis of conservation of genes and proteins.
- *Epigenomics* (epigenetics)—looking at the effect of DNA methylation patterns, imprinting, and DNA packaging on gene function.
- *Pharmacogenomics*—using sequence based information for designing drugs and vaccines as well as effecting personalized medicine.

Table 1.1 "Omics"	
Genomics:	The genome is the complete set of DNA within a single cell of an organism. Genomics is that branch of genetics that seeks to understand the structure and function of genes based on the DNA sequence information.
Proteomics:	Proteins are vital to cells as they make up the structural components of cells, and as enzymes, catalyze all metabolic reactions in the cell. Proteomics is the study of proteins with respect to their structure and function.
Metabolomics:	Cellular processes vary and are characteristic of different cells resulting in unique chemical profiles. Metabolomics is a systematic study of the chemical profiles of cells resulting from metabolic pathways operating in them.
Transcriptomics/ Expressomics:	In living cells, only subsets of genes are "active" that is transcribed into mRNA which would be translated into protein (known as "gene expression"). Understanding gene expression patterns helps to decipher the function of specific genes, for example, in diseases, or in behavior. The field of transcriptomics studies the expression patterns of genes and provides a link between the genome, the proteome, and the cellular phenotype.
Phenomics:	Study of growth, performance, and composition helps to understand physiological and biochemical processes and linking it to the gene(s).
Interactomics:	This is a discipline at the intersection of bioinformatics and biology that deals with studying both the interactions and the consequences of those interactions between and among proteins, and other molecules within a cell.

1.6 THE HUMAN GENOME PROJECT (HGP)

The HGP initiated in 1990 was coordinated by the US Department of Energy (DOE) and National Institute of Health (NIH). In all, 18 countries have participated in the worldwide effort at different times and stages of the project. The most significant contributions came from the Sanger Center in the United Kingdom and research centers in Germany, France, and Japan. **The primary objective of this initiative was to determine the sequence of the approximately three billion base pairs of the human genome and to archive it in online databases.** The purpose of the exercise was to disseminate this information to educational, research, and private sector institutions so as to enable rapid commercialization of these findings primarily in human healthcare. The HGP also provided a framework for analysis of ethical, legal and social implications (ELSI) of the application of genome based technologies especially in diagnostics and genetic screening.

DNA sequencing is the process of determining the exact order of the three billion chemical bases that make up the DNA of the 24 different human chromosomes. The DNA sequence maps that were produced from the research provide a comprehensive understanding of human biology and other complex phenomena. The **reference sequences** built during the HGP served as a starting point for comparisons across all of humanity. The first reference genome did not represent any one person's genome; it was **generated from five anonymous donors**, two males and three females, from different racial and ethnic groups including two Caucasians, one Hispanic, one African, and one Asian (see Box 1.1 for more information on how the genome was sequenced). The knowledge

BOX 1.1 SEQUENCING AND MAPPING THE HUMAN GENOME

Automatic sequencing machines.

Photo courtesy: By Flickr user jurvetson (Flickr) [CC BY 2.0 (http://creativecommons.org/licenses/by/2.0)],
via Wikimedia Commons.

A biochemical technique for determining the sequence of nucleotides that makes up DNA was first developed by **Frederick Sanger** and colleagues in 1977. Known as the **dideoxy sequencing technique**, it relies on the generation of fragments of DNA resulting from the incorporation of a dideoxy ribonucleoside triphosphates into a growing DNA strand during replication under in vitro (cell free) conditions. Arthur Kornberg had earlier demonstrated DNA replication in a cell free bacterial extract which contained a DNA polymerase (which he discovered in 1955), nucleotide building blocks and single stranded DNA (ssDNA). By inclusion of short nucleotide fragments (primers) which bind to the ssDNA and provide a free 3′ hydroxyl end, nucleotides could be incorporated by DNA polymerase in the growing DNA stand as specified by the ssDNA template. **In the dideoxy sequencing technique, chain termination occurs upon the incorporation of a dideoxy nucleotide due to the absence of a free 3′hydroxyl end. These fragments could be size separated by electrophoresis in polyacrylamide gels.** The earliest sequencing experiments had four reaction mixtures, each containing one of the four dideoxy nucleotides and radiolabelled nucleotides. Contents from each of the four vials were run in different lanes in a sequencing gel, and the autoradiograms were "read" to determine the sequence. The process was soon **automated using fluorescently labeled dideoxy nucleotides and capillary gel electrophoresis.** Samples passing a detection window are excited by laser, the emitted fluorescence is read by a CCD camera, and the signals are converted into base calls. The advantage was that the resolution was improved as was the read length to up to 1000 bases. Further sophistication has led to the **Next Gen Sequencers** capable of analyzing 384 sequencing reactions in parallel ushering in an era of high-throughput whole genome sequencing. While Sanger sequencing techniques rely on chain termination with dideoxy nucleotides, in 1996, a new technique based on the **detection of pyrophosphate release on nucleotide incorporation** was developed in the Royal Institute of Technology, Stockholm. Known as "pyrosequencing," the activity of DNA polymerase is **detected with another chemiluminescent enzyme (luciferase).** The ssDNA template is hybridized to a sequencing primer in a reaction mixture that contains DNA polymerase, ATP sulfurylase, luciferase, apyrase, and the substrates luciferin and adenosine 5′ phosphosulfate.

(Continued)

BOX 1.1 (CONTINUED)

The priority of the HGP was not in sequencing, as much as in **mapping sequences** in human and nonhuman ("model") organisms. Two types of maps were attempted: *genetic maps* ordering polymorphic markers (easily characterized small regions of DNA that vary among individuals) on chromosomes with a resolution of 2 to 5 centimorgans (cM), and *physical maps* of unique 100−200 basepair (bp) sequences (known as Sequence-Tagged-Site (STS) markers) spaced approximately 100 kilobases (kb) apart, and contiguous overlapping clones (known as "**contigs**") of 2 megabases (Mb). A crude physical map of the whole genome was first constructed by restriction mapping (using a class of enzymes known as restriction endonucleases which cut DNA at specific target sequences). The Sanger genome sequencing technique adopted by the HGP involved breaking the genome into overlapping fragments and cloning the fragments in bacteria using unique vectors known as Bacterial Artificial Chromosomes which could clone large fragments of DNA. The DNA from each clone was cut into smaller fragments and subcloned for sequencing. The process was tedious and time consuming but necessary as the DNA sequence of flanking regions was required in order to anneal primers for sequencing. Primary sequence reads were fed to computers and aligned based on homologies in overlapping fragments to assemble the contigs. An alternate method to this "**clone by clone**" method was the "**shotgun sequencing**" method which involved breaking the genome into random fragments using sound waves, inserting the fragments into vectors, sequencing each fragment, and assembling the fragments based on overlaps. This technique had been considered suitable for only small regions of the genome, but in a significant difference of opinion, Craig Venter the founder of The Institute for Genomic Research, Maryland, advocated the use of the technique for the whole genome. In 1998, Craig Venter formed a new private company (which eventually became Celera Genomics) to put his ideas to test. The move sparked off **a rivalry between the publicly funded and privately funded effort**. Ideological differences extended also to the manner in which the data was to be used. While HGP's viewpoint was to ensure that information generated by the project would be **made freely available to all** within 24 hours, **Craig Venter planned to file for preliminary patents on over 6000 genes and full patents on a few hundred genes before releasing the sequence.** His view was that it was not appropriate to give the sequence away for free and that patenting the sequence would ensure some degree of control on how the sequences were going to be used. One effect of the rivalry was that the **work progressed at an unprecedented pace**. Both groups were ready with a working draft of the human genome by **2000**. Under pressure from the White House to arrange a truce, the leaders of the HGP, Francis Collins (NIH) and Ari Patrinos (DOE), and Craig Venter joined the US President Bill Clinton (and the British Prime Minister Tony Blair by satellite link) on June 26, 2000, to announce to the world that the human genome had been sequenced.

Reference

National Human Genome Research Institute (2012). *An over view of the Human Genome Project: A brief history of the Human Genome Project.* Retrieved from http://www.genome.gov/12011239.

obtained from the sequences is generic and is applicable to all humans because all humans share the same basic set of genes and regions that regulate and control the development and maintenance of biological structures and processes. Other **"model"** organisms (representative of a taxonomic group or well-studied by geneticists) were also sequenced in the HGP initiative. These included baker's yeast (*Saccharomyces cerevisiae*), nematode worm (*Caenorhabditis elegans*), Arabidopsis (*Arabidopsis thaliana*), the fruit fly (*Drosophila melanogaster*), mouse (*Mus musculus*), pufferfish (*Takifugu rubripes*), and rice (*Oryza sativa*). Sequence data from these organisms are available in freely accessible online databases such as the Saccharomyces Genome Database (http://www.yeast-genome.org/), Worm Base (http://www.wormbase.org/), FlyBase (http://flybase.org/), Mouse Genome Database (http://www.informatics.jax.org/), and rice (http://rice.plantbiology.msu.edu/). The website maintained by the **National Center for Biotechnology Center (NCBI)** (http://www.ncbi.nlm.nih.gov/genome/guide/human/) to quote their website, serves as *"an integrated, one-stop,* *genomic information infrastructure for biomedical researchers from around the world so that they may use these data in their research efforts."*

Coinciding with the 50th anniversary of the discovery of the Watson and Crick's model of the DNA, on April 14, **2003**, the International Human Genome Sequencing Consortium **announced the completion of the HGP** (National Human Genome Research Institute, 2003). The implications and the proposed applications of the sequence of the human DNA was published in the journals *Science* (Collins, Morgan, & Patrinos, 2003b) and *Nature* (Collins et al., 2003a). By completely identifying the Human Genome Sequence the HGP revealed the estimated 20,000–25,000 human genes within our DNA as well as the regions controlling them. Every human being has a unique DNA sequence and it differs from that of any other human being by about 0.1% and this small difference affects characteristics such as how individuals look. Slight variations in DNA sequences can have a major impact on development of diseases and on responses to such factors as infectious microbes, toxins, drugs, and donated organs.

KEY TAKEAWAYS

- The Human Genome Project (HGP) was a collaborative project involving 18 countries coordinated by the US DOE and NIH.
- Sequencing of the three billion base pairs that make up the human genome was completed in 2003 and entered into a public database maintained at the NCBI.
- Sequencing of other "model" organisms including yeast, fruit fly, nematode worm, mouse, and rice was also completed.

1.7 APPLICATIONS OF GENOMICS

Some of the current and potential applications of genomics are described in the following sections.

1.7.1 PHYLOGENETIC ANALYSIS

Sequence information has been used to establish **evolutionary relationships between taxonomic groups** as it is generally understood that genomes evolve by the gradual accumulations of mutations resulting in species showing divergence from a common ancestor. By estimating the degree of conservation of sequences, **phylogenetic trees** are constructed. A typical example of sequence information being used to understand evolution is that of the Genographic Project that sought to trace the **evolutionary history of humans** (*Homo sapiens sapiens*) (Box 1.2).

1.7.2 GENETIC TESTING AND DIAGNOSTICS

Genetic analysis in humans has over time revealed a number of disorders that are inherited. Documentation of this information was published between 1966 and 1998 in 12 editions of

BOX 1.2 GENOGRAPHIC PROJECT

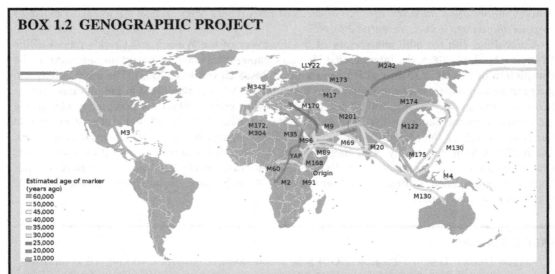

"Out of Africa" migration patterns of the human species.

By KVDP, via Wikimedia Commons.

The Genographic Project (https://genographic.nationalgeographic.com/) launched in April 2005 was a collaboration between IBM and National Geographic Society, funded by IBM, National Geographic, and the Waitt Family Foundation. The Director of the project, **Spencer Wells, a population geneticist**, came up with a theory that the DNA in every human being carries the **record of his/her descent** from the earliest human beings on earth (Wells, 2007). The theory was based on his postdoctoral research conducted in 1994 at the School of Medicine, Stanford University, under the guidance of Luca Cavelli-Sforza (often called the "father of anthropogenic genetics"). The investigative team in the Genographic Project had 13 population geneticists from the United States, Brazil, the United Kingdom, Spain, France, Lebanon, South Africa, Russia, India, China, and Australia who coordinated the collection and analysis of DNA from indigenous people from around the world. The National Geographic set up 11 regional collection centers worldwide and sought 100,000 participants. The project was also opened to the public who could request a kit and send in a sample to be analyzed, the results of which would be anonymously posted online. **Data from more than 395,000 individuals** is thus available in a global database. The project also serves as a powerful example of the use of computers in understanding biology and human evolution. As lead partner in the project, the IBM's Computational Biology Center led by Ajay Royyuru provided **life-science-oriented datamining tools and algorithms to analyze the massive amounts of data generated from the project.** The project itself was to be a five-year project but is now extended beyond 2011 on a yearly basis. Funds generated by the project from the sale of kits are being used to support indigenous communities and preservation of languages and customs.

The Genographic Project provides evidence for the "**out of Africa**" theory of human evolution. By using small variations of DNA sequence as "**markers of descent**," the Genographic Project was able to **trace the migrations to all the other continents**, of the human species (*Homo sapiens*) that **evolved in Africa some 60,000 years ago.** What did surprise many however was that despite the apparent diversity of color, creed, customs, and languages, **all humans were linked ("connected") by a common descent.**

References

Wells, S. (2007). A family tree for humanity [TED Global 2007, Video file]. Retrieved from https://www.ted.com/talks/spencer_wells_is_building_a_family_tree_for_all_humanity?language=en.

The Mapping of Humanity's Family Tree [IBM 100 Icons of Progress]. Retrieved from http://www-03.ibm.com/ibm/history/ibm100/us/en/icons/mappinghumanity/.

Mendelian Inheritance in Man by Victor McKusick of Johns Hopkins University School of Medicine (McKusick, 1966, 1994). Since 1987, the information became available online under the direction of the Welch Medical Library with financial support from the NCBI (McKusick, 2007). Since 2010, the website of *Online Mendelian Inheritance in Man* (http://www.omim.org/) is authored and edited at the McKusick-Nathans Institute of Genetic Medicine, Johns Hopkins University School of Medicine, with financial support from the National Human Genome Research Institute (NHGRI).

Changes in chromosomes, genes, or proteins correlated to disease conditions have been used to confirm or rule out a genetic condition. Since the 1950s, amniocentesis (studying free floating cells isolated from the amniotic fluid that surrounds the fetus in the womb) has been used by physicians to diagnose chromosomal aberrations, such as Down's syndrome, in the unborn fetus. More than thousand genetic tests have been developed and can be broadly classified into three categories:

- Molecular genetic tests (or gene tests),
- Chromosomal genetic tests, and
- Biochemical genetic tests that study the amount or activity level of proteins.

In addition to **diagnosis**, genetic testing can be used to **identify carriers** [individuals who have only one copy of the mutated gene hence do not show symptoms (phenotype) of the genetic disorder], and to determine if an individual has an **inherited predisposition** to a disease. The advantages of genetic testing are:

- *To provide an accurate diagnosis of a disease condition and to make informed decisions regarding the treatment:* Genetic testing can help to identify the cause of the diseased condition and to select the appropriate treatment regimen (including whether a patient would respond to a particular treatment option). This has been found to be especially useful in the treatment of certain cancers as the therapy is expensive (thereby posing a financial burden on the patient), and could cause adverse effects (thereby further reducing the quality of life of the patient).
- *To make life choices:* Amniocentesis has helped to make decisions regarding the medical termination of pregnancy in cases of numerical/structural abnormalities in chromosomes known to cause disease syndromes in humans. With the development of In Vitro Fertilization techniques, in which the embryo is created in a petri dish and later implanted into the womb, genetic testing is being increasingly used to select "healthy" embryos. **Preimplantation Genetic Testing** is especially advocated in instances when one of the biological parents is known to be affected by (or is a carrier of) a genetic disorder.
- *To make life-style choices:* Several genetic disorders are known to be age-dependent and the disease is often manifested only after the reproductive age of the patient, hence is transmitted to the progeny (example, Huntington's disease). Genetic testing could help make informed decisions regarding reproductive choices. It could also help to plan for future healthcare and end-of-life decisions for debilitating diseases such as Alzheimer's, Parkinson's, and some forms of dementia.

Due to the rather sensitive nature of the information revealed by the tests, genetic testing is invariably coupled with **genetic counseling** to help patients understand how the test results would affect them and to guide them to make **informed choices**. The field however is mired in ethical conundrums. The ELSI initiative of the HGP is an attempt to address issues of privacy and psychological impacts of genome information revealed by genetic testing. Some of the ethical aspects associated with genetic testing are discussed in Chapter 7, Genetic Testing, Genetic Discrimination, and Human Rights of this book.

Several privately held personal genomics companies have been established, for example, the California based company 23andMe (https://www.23andme.com/en-int/) founded by Linda Avey, Paul Cusenza, and Anne Wojcicki in 2006 that has been offering direct to customer genetic testing since November 2007. (See International Society of Genetic Genealogy Wiki (2016) for a country-wise listing of personal genomics companies.)

1.7.3 PHARMACOGENOMICS

A major objective of the HGP was **to use genome information to identify genes and pathways and their interaction with environment to determine the health of an individual**. The primary goal was to use this information for (1) **predicting disease susceptibility** or **drug response** and (2) **designing new drugs**. Recent analysis of approved medicines indicates that using human genetic data to support the selection of drug targets and indications can roughly double the chances of the clinical success of the drug (Nelson , Tipney, Painter, Shen, & Sanseau, 2015).

In 2005, the NHGRI initiated the project **The Cancer Genome Atlas** (http://cancergenome.nih.gov/) to map the genetic changes that are associated with cancer. Because there may be multiple causes to the same type of cancer, **more effective, individualized treatments** may be possible by connecting specific genomic changes with specific outcomes. In January 2015, President Obama announced the US$215 million **Precision Medicine Initiative** (White House, 2015) that proposes to assemble a cohort of 1 million US volunteers for longitudinal research, including genetic studies (Collins & Varmus, 2015).

The importance of pharmacogenomics lies in the observation that patients who have the same disease symptoms show differences in their response to medication. Advances in genetic research has helped to develop **"companion diagnostic tools"** that target specific alterations in diseased cells and **allow therapy to be individualized**, enabling patients and their physicians to choose treatments that can provide the highest likelihood of success. As of January 2016, there are 20 FDA-approved companion diagnostic tests, with many more in the pipeline, for the selection of drugs to treat various diseases and conditions on a more individualized basis. It has been most effective in the treatment of cancer. For example, Myriad's BRACAnalysis Cdx is an FDA-approved companion diagnostic test proven to effectively identify BRCA-positive patients with advanced ovarian cancer who are eligible for treatment with Lynparza (olaparib), a first-in-class drug for treating ovarian cancer (Lancaster, 2015). Understanding genes and their pathways can also help design drugs that specifically target and modify disease causing metabolic processes. An example is the development of imatinib mesylate (*Gleevec*) used in the treatment of chronic myelogenous leukemia. The molecule is an inhibitor of a specific enzyme (the BCR−ABL tyrosine kinase) and its use in therapy came from a detailed understanding of the disease's genetic cause.

1.7.4 FORENSICS

Modern technologies, such as the Polymerase Chain Reaction used to amplify (make millions of copies of) DNA, and sequencing of DNA fragments, are proving to be more powerful than tissue and fingerprints analysis and eye-witness accounts for catching criminals and solving crimes. Analysis methods include Short Tandem Repeats, Single Nucleotide Polymorphism, mitochondrial DNA, and Y-chromosome analysis. DNA evidence is generally linked in DNA databases to offender profiles. In the United States, in the late 1980s, the federal government created a system called **Combined DNA Index System (CODIS)**, which linked national, state, and local databases for the **storage and exchange of DNA profiles**, available to law enforcement agencies across the country. In the late 1980s and early 1990s, states began to pass laws requiring offenders convicted of certain crimes to provide DNA samples. **DNA evidence** is also proving to be a powerful tool to establish the innocence of prisoners who were wrongly convicted before DNA testing was available. DNA analysis has been used to solve hitherto "unsolvable" mysteries, for example, the fate of the last Russian monarchy, the Romanov family (Coble et al., 2009).

1.7.5 RECOMBINANT DNA TECHNOLOGY

Conventional breeding of livestock and crop plants involves making crosses between two parents selected for superior characteristics and screening the second and further generations of their progeny for individuals that combine the desirable characteristics of both parents. The process is imprecise and time consuming (typically 6−8 generations need to be raised in order to get stable inheritance of the combined characters, taking several years), and relies on chance and the skill of the breeder. Breeding has often been considered to be an "art" rather than a "science." Genomics has revolutionized the process of breeding. Desirable genes can be identified, and using the tools of recombinant DNA technology, teased out of donor species, spliced onto vectors, and mobilized into recipient species. As DNA is ubiquitous, with small modifications, **genes from taxonomically unrelated donor sources can be expressed in recipients** (something that could never be done by conventional methods). Chapter 4, Recombinant DNA Technology and Genetically Modified Organisms in this book, discusses the use of recombinant DNA technology in the **creation of GMOs** and the concerns regarding the use of GMOs in different industries, including biopharmaceuticals and food. Issues regarding safety and regulations to ensure ethical and safe use of the technology are discussed in Chapter 9, Ensuring Safety in Biotechnology.

1.7.6 GENE THERAPY

One of the major objectives of the HGP was to use genomic information in healthcare. Genetic studies had established that **many diseases were due to mutations in single genes** that render them nonfunctional ("inborn errors of metabolism"). For example, cystic fibrosis is caused by a mutation (deletion of three nucleotides) in the gene *cystic fibrosis transmembrane conductance regulator (CFTR)*, whereas phenylketonuria is due to impaired activity of an enzyme *phenylalanine hydroxylase (PAH)* responsible for the metabolism of the amino acid phenylalanine. Using genomics and recombinant DNA technology, **identification of defective genes and replacement with normal genes**, known as "gene therapy," could serve as a permanent cure in affected patients (see Chapter 4: Recombinant DNA Technology and Genetically Modified Organisms). Although mooted since 1972, the first successful clinical trial was in 1990, in which 4-year-old Ashanti DeSilva was treated for Severe Combined Immune Deficiency (SCID) by inserting a healthy adenosine deaminase (ADA) gene in her white blood cells. Subsequent gene therapy trials for ADA-SCID had to be stopped due to patients developing leukemia possibly due to the retroviral vector used inserting near oncogenes. Adenoviral vectors have been successfully used in a gene therapy approved by European Medicines Agency called Glybera (see Chapter 4: Recombinant DNA Technology and Genetically Modified Organisms).

1.7.7 GENOME EDITING

Genome editing technology involves making changes to the **genetic information in germ line cells so that the changes can be inherited**. The technology aims to treat human diseases by deleting harmful mutations. Although the technology could help cure several forms of cancers and blood disorders such as sickle-cell anemia and hemophilia, the possibility of the technology being used for **nontherapeutic modifications** and **eugenics** raises ethical, legal, and health concerns. Around 40 countries worldwide, especially in western Europe with 15 of 22 countries **discourage or ban germ line editing**. Several groups of scientists have called for a voluntary moratorium on experiments on germ line alterations (Lanphier, Urnov, Haecker, Werner, & Smolenski, 2015).

Genome editing (also known as **Genome Editing with Engineered Nuclease, GEEN**) involves a precise modification in the DNA sequence such as a deletion, insertion, or substitution effected by enzymes that cause double-strand breaks at the desired location of the genome, which are then repaired by homologous recombination or nonhomologous end joining. Currently four families of engineered nucleases have been developed for this purpose. These include the zinc finger nucleases, meganucleases, Transcription Activator-Like Effector-based Nucleases (TALEN), and the Clustered Regularly-Interspaced Short Palindromic Repeats (CRISPR)—Cas system. In January 2016, in the United Kingdom, representing a historical first, the Human Fertilization and Embryology Authority has **approved** the application of Kathy Niakan of the Francis Crick Institute in London to **conduct gene editing experiments**. The approval permits the experiments to be conducted in the first seven days after fertilization, although it would still be illegal for scientists to implant the embryos in a woman (Gallagher, 2016).

1.7.8 SYNTHETIC BIOLOGY

Synthetic biology was pioneered by electrical engineers Tom Knight and Randy Rettberg of Massachusetts Institute of Technology (MIT) who attempted to apply engineering principles to biology. Knight was keen to see if biology could be "**modular**" so that cells could be treated like living circuit boards with genes substituting for electrical components such as resistors and capacitors. If successful, **cells could be programed to make unusual products** such as plastics, fuels, or drugs. Difficulties they encountered in this attempt included being able to snip and tease out gene sequences and being able to test them in a modified organism. In 2003, Knight proposed the concept of standardized pieces of DNA that he called a **BioBrick**, each of which includes a gene associated with a specific trait, cataloged in a **Registry of Standard Biological Parts**. BioBricks are made by entering the sequence of a desired gene into a DNA synthesizer, and could be connected to other parts as they are capped at both ends with DNA sequences for this purpose (Trafton, 2011). Since 2004, MIT has been hosting an annual competition, the **International Genetically Engineered Machine**, in which undergraduate student teams present projects to build unusual organisms for various applications (http://igem.org/Main_Page or http://ung.igem.org/About).

Significant progress in the understanding of synthetic life forms has come from the work of Craig Venter and his team. In 2003, researchers at the institute founded by him, the J. Craig Venter Institute (JCVI), created a **synthetic version** of the bacteriophage PhiX174. From his work on a simple bacterium *Mycoplasma genitalium*, it was possible to determine a minimalistic genome, the bare minimum number of genes required for a living cell. By splicing together these genes, it was possible to assemble an entire **synthetic genome**, and by transferring this to a cell devoid of DNA, to create a **synthetic cell**. Announcement of the first synthetic cell whose "**parent is a computer**" was made by Craig Venter in May 2010 (Venter, 2010). Major **applications** envisaged by the JCVI include **alternate fuels** to substitute for fossil fuels and production of **new vaccines**.

Synthetic biology could expedite vaccine development, which normally takes over six months, to a few days. By t**reating viral sequences like open source software**, even **without shipping of viruses between laboratories**, vaccine seed stock could be prepared within days. The sequence could be **downloaded from the Internet and synthesized**; the artificial genes validated using the published sequence, inserted into preexisting viral backbones, and viruses harvested from infected eukaryotic cells in culture. Synthetic biology is being used to develop vaccines for chlamydia scheduled for clinical trials in 2017 (Synbicite, 2015). In trying to expedite vaccine development for the Zika virus implicated in cases of microcephaly and Guillain—Barre Syndrome, declared as a global public health emergency by World Health Organization in February 2016 (WHO Media Center, 2016), researchers at the University of Wisconsin-Madison in collaboration with scientists in Brazil have started uploading real time study results of their experiments on virus dynamics in rhesus macaques at https://goo.gl/IXF79u (Prasad, 2016).

1.8 ETHICAL, LEGAL, AND SOCIAL IMPLICATION

The ELSI program was established in 1990 by the NHGRI acknowledging the importance of the ELSI of the genetics and genomics research conducted in the HGP. It was legislatively initiated by the NIH Revitalization Act of 1993 under which "not less than" 5% of the NIH budget for the HGP was to be set aside for research on these aspects. The DOE which also funded the HGP set aside 3% of its funds for this purpose. The ELSI program is **essentially a research program**, but in the initial days, the term was used more broadly, and the ELSI program was understood to be responsible for development of policy solutions to issues raised by genomics research. The ELSI Program is currently the only dedicated extramural bioethics research program at the NIH and is the largest US funder of research focused on ELSI in genetics and genomics (McEwen et al., 2014). Topics addressed by the research include issues related to **informed consent, privacy, and psychological impacts of the research and test results** (see also Chapter 7: Genetic Testing, Genetic Discrimination, and Human Rights).

1.9 GENETIC REDUCTIONISM

A better understanding of the influence of genes on characters resulted in a "**gene centric**" view of biology in the late 20th century and early 21st century. An extreme form of such reductionist thinking is seen in the writings of the famous author, Richard Dawkins. In his books, Dawkins opines that all of evolution could be explained on the basis of the gene being a self-replicating and self-sufficient hereditary particle, and that organisms were merely vessels (or in his words "gene machines") for survival of genes.

Establishing the **role of genes** in determining human **behavioral traits such as altruism, sexual orientation, and criminality**, and the influence of environmental factors on gene expression (**"nature v nurture"**) remain challenging fields of study for molecular biologists. For example, a systematic study involving 900 Finnish prisoners, two genes **MAO A** (coding for monoamine oxidase A necessary for the turnover of the neurotransmitter dopamine) and **CDH 13** (coding for cadherin 13 a neuronal membrane adhesion protein, also associated with substance abuse and Attention Deficit Hyperactivity Disorder) were **found to be associated with extremely violent behavior** (Hogenboom, 2014). The lead author of the study Jari Tiihonen of the Karolinska Institutet in Sweden, however, emphasized that merely possessing the genes should not influence trial outcomes in criminal courts. Lighter sentences for criminal behavior have been awarded in a few cases in which the MAO A variant has been detected in the accused. For instance, in 2009, an Italian court reduced the sentence given to a convicted murderer by a year because he had genes linked to violent behavior (Feresin, 2009). Similarly, in 2010, a US court awarded a lighter sentence of voluntary manslaughter instead of murder based on genetic evidence that the murder accused Bradley Waldroup had the MAO A mutation (Hagerty, 2010). Most sociological criminologists argue that these defenses merely serve to "beat the system," and that crime should be punished with the same severity irrespective of who commits it (Wilson, 2011).

Current understanding is that **genes do not act alone** and that the expression of genes (as a manifestation of a character) is **dependent on its interaction with other genes and with the environment**. For instance, it is now fairly well-documented that in women, having mutations in two genes BRCA-1 and BRCA-2, the probability of developing breast cancer increases to 85%, but evidently the mutations alone are not causative, as in 15% no disease symptoms are seen. Considerable research is now directed toward analyzing "**interactomes**," the **protein-protein interactions** and **protein-nucleic acid interactions** in genetic diseases, and **Genome Wide Association Studies**.

1.10 SUMMARY

This chapter tracks the evolution of gene concept in the 20th century from the discovery of Mendel's work on inheritance to the sequencing of the human genome. Virtually, all biological systems known today have DNA (or RNA) as the genetic material. The central dogma of biology is that the information encoded by the DNA is transcribed into an RNA intermediate and translated into proteins. The characteristic features of an organism are a result of various metabolic pathways mediated by proteinaceous enzymes. Determining genome sequences have improved our understanding of how genes work and equipped us with the tools to manipulate characters, enabling genetic engineering of crops and livestock, or precise diagnosis and treatment of diseases. Washington health policy consultant and ethicist Kathi Hanna writes in the closing essay of *The Genomic Revolution*:

> *"The knowledge gained [about the human genome] could cure cancer, prevent heart disease, and feed millions. . .. At the same time, its improper use can discriminate, stigmatize, and cheapen life through frivolous enhancement technologies. Because of the promise for great good, we all need to understand more about the science and application of human genomics to ensure that the harms do not materialize."*
>
> **Hanna (2002)**

REFERENCES

Benzer, S. (1955). Fine structure of a genetic region in bacteriophage. *Proceedings of the National Academy of Sciences of the United States of America, 41*, 344−354. Retrieved from http://ttk.pte.hu/biologia/genetika/gen/Benzer.pdf.

Benzer, S. (1961). On the topography of the genetic fine structure. *Proceedings of the National Academy of Sciences of the United States of America, 47*(3), 403−415. Retrieved from https://www.ncbi.nlm.nih.gov/pmc/articles/PMC221592/.

Coble, M. D., Loreille, O. M., Wadhams, M. J., Edson, S. M., Maynard, K., Meyer, C. E., . . . Finelli, L. N. (2009). Mystery solved: The identification of the two missing Romanov children using DNA analysis. *PLoS ONE, 4*(3), e4838, 2009 Published online Mar 11, 2009. doi:10.1371/journal.pone.0004838.

Collins, F. S., & Varmus, H. (2015). A new initiative on precision medicine. *New England Journal of Medicine, 374*(9), 793−795.

Collins, F. S., Green, E. D., Guttmacher, A. E., & Guyer, M. S. (2003a). A vision for the future of genomic research: A blueprint for the genomic era. *Nature, 422*(6934), 835−847.

Collins, F. S., Morgan, M., & Patrinos, A. (2003b). The Human Genome Project: Lessons from large-scale biology. *Science, 300*, 286.

Crick, F. (1970). Central dogma of molecular biology. *Nature, 227*(5258), 561−563. Retrieved from http://www.nature.com/nature/focus/crick/pdf/crick227.pdf.

Feresin, E. (Reporter) (October 30, 2009). Lighter sentence for murderer with "bad genes". Italian court reduces jail term after tests identify genes linked to violent behavior. Published online 30 October 2009 | Nature | doi:10.1038/news.2009.1050. Retrieved from http://www.nature.com/news/2009/091030/full/news.2009.1050.html.

Gallagher, J. (Reporter) (February 1, 2016). Scientists get "gene editing" go-ahead. [BBC News website]. Retrieved from http://www.bbc.com/news/health-35459054.

Hagerty, B.B. (Reporter) (July1, 2010). Can your genes make you murder? [npr]. Retrieved from http://www.npr.org/templates/story/story.php?storyId = 128043329.

Hanna, K. E. (2002). Summing up: finding our way through the revolution. In Rob DeSalle, & Michael Yudell (Eds.), *The Genomic Revolution* (pp. 199−208). Washington, D.C.: Joseph Henry Press.

Hershey, A., & Chase, M. (1952). Independent functions of viral protein and nucleic acid in growth of bacteriophage. *Journal of General Physiology, 36*(1), 39−56. Retrieved from http://www.ncbi.nlm.nih.gov/pmc/articles/PMC2147348/.

Hogenboom, M. (Reporter) (October 28, 2014). Two genes linked with violent crime [BBC News]. Retrieved from http://www.bbc.com/news/science-environment-29760212.

International Society of Genetic Genealogy Wiki (2016). List of personal genomics companies. Retrieved from http://isogg.org/wiki/List_of_personal_genomics_companies.

Lancaster, J. (June 1, 2015) The importance of companion diagnostics in cancer and beyond [Myriad Matters Blog]. Retrieved from https://www.myriad.com/importance-companion-diagnostics-cancer-beyond/.

Lanphier, E., Urnov, F., Haecker, S. E., Werner, M., & Smolenski, J. (2015). Don't edit the human germ line. *Nature*, *519*, 410−411. (26 March 2015) doi:10.1038/519410a. Retrieved from http://www.nature.com/news/don-t-edit-the-human-germ-line-1.17111.

McEwen, J.E., Boyer, J.T., Sun, K.Y., Rothenberg, K.H., Lockhart, N.C., & Guyer, M.S. (2014). The ethical, legal and social implications program of the National Human Genome Research Institute: Reflections on an ongoing experiment. Annual Review of Genomics and Human Genetics. 15: 481−505. First published online as a Review in Advance on April 24, 2014 doi: 10.1146/annurev-genom-090413-025327. Retrieved from http://www.annualreviews.org/eprint/eDSR5xjQy7XjwMQ9VDXs/full/10.1146/annurev-genom-090413-025327.

McKusick, V. A. (2007). Mendelian Inheritance in Man and its online version, OMIM. *American Journal of Human Genetics*, *80*(4), 588−604. Available from http://dx.doi.org/10.1086/514346, PMC 1852721. PMID 17357067.

McKusick, V. A. (1994). *Mendelian inheritance in man. A catalog of human genes and genetic disorders* (11th ed, 12th ed, 1998). Baltimore, MD: Johns Hopkins University Press.

McKusick, V. A. (1966). *Mendelian inheritance in man. Catalogs of autosomal dominant, autosomal recessive and X-linked phenotypes* (1st ed, 2nd ed, 1969; 3rd ed, 1971; 4th ed, 1975; 5th ed, 1978; 6th ed, 1983; 7th ed, 1986; 8th ed, 1988; 9th ed, 1990; 10th ed, 1992). Baltimore, MD: Johns Hopkins University Press.

McKusick, V. A., & Ruddle, F. H. (1987). A new discipline, a new name, a new journal [editorial] *Genomics*. *September*, *1*, 1−2.

National Human Genome Research Institute (2003) International consortium completes Human Genome Project, all goals achieved; New vision for genome research unveiled. Retrieved from https://www.genome.gov/11006929.

Nelson, M. R., Tipney, H., Painter, J. L., Shen, J., ... Sanseau, P. (2015). The support of human genetic evidence for approved drug indications. *Nature Genetics*, *47*(8), 856−860.

Prasad, R. (Reporter) (February 26, 2016). *The benefits of open science* [The Hindu]. Retrieved from http://www.thehindu.com/opinion/op-ed/zika-virus-the-benefits-of-open-science/article8281301.ece.

Synbicite (August 3, 2015). *New vaccine for Chlamydia to use synthetic biology*. Retrieved from http://www.synbicite.com/news-events/2015/aug/3/new-vaccine-chlamydia-use-synthetic-biology/.

Trafton, A. (Reporter) (April 19, 2011) Rewiring cells: How a handful of MIT electrical engineers pioneered synthetic biology [MIT Technology Review]. Retrieved from https://www.technologyreview.com/s/423703/rewiring-cells/.

Venter, C. (2010, May) Watch me unveil "synthetic life" [TED in the Field, Video file]. Retrieved from https://www.ted.com/talks/craig_venter_unveils_synthetic_life?language=en.

Venter, C., & Cohen, D. (2004). The century of biology. *New Perspectives Quarterly*, *21*(4), 73−77. Retrieved from http://onlinelibrary.wiley.com/doi/10.1111/j.1540-5842.2004.00701.x/abstract.

Watson, J. D., & Crick, F. H. C. (1953). A structure for deoxyribose nucleic acid. *Nature*, *171*, 737−738. Retrieved from http://www.nature.com/scitable/content/Molecular-Structure-of-Nucleic-Acids-16331.

White House (January 30, 2015). President Obama's Precision Medicine Initiative. Retrieved from https://www.whitehouse.gov/the-press-office/2015/01/30/fact-sheet-president-obama-s-precision-medicine-initiative.

WHO Media Centre (February 1, 2016). WHO statement on the first meeting of the International Health Regulations (2005) (IHR 2005) Emergency Committee on Zika virus and observed increase in neurological disorders and neonatal malformations. Retrieved from http://www.who.int/mediacentre/news/statements/2016/1st-emergency-committee-zika/en/.

Wilson, J. W. (2011). Debating genetics as a predictor of criminal offending and sentencing. *Student Pulse*, *3*(11). Retrieved from http://www.studentpulse.com/a?id=593.

FURTHER READING

Reviews

Collins, F. S., Green, E. D., Guttmacher, A. E., & Guyer, M. S. (2003). A vision for the future of genomics research. *Nature, 422*, 835−847. (24 April 2003). Retrieved from http://www.nature.com/nature/journal/v422/n6934/pdf/nature01626.pdf.

Gannett, L. (2014). The Human Genome Project, *the Stanford encyclopedia of philosophy* (Winter 2014 Edition), Edward N. Zalta (ed.). Retrieved from http://plato.stanford.edu/archives/win2014/entries/human-genome/.

Wheeler, D. A., & Wang, L. (2013). From human genome to cancer genome: The first decade. *Genome Research, 23*, 1054−1062 . Retrieved from http://genome.cshlp.org/content/23/7/1054.full.pdf + html. Published online Mar 11, 2009. doi:10.1371/journal.pone.0004838.

Web resources
Genes and Genetics

Cold Spring Harbor Laboratory Dolan DNA learning center website: https://www.dnalc.org/.

Virginia Commonwealth University Secrets of the Sequence—video series on the life sciences website: http://www.sosq.vcu.edu/videos.aspx.

Human genome project

Cold Spring Harbor Oral History Collection- Genome Research website: http://library.cshl.edu/oral-history/category/genome-research/.

Human Genome Project Information Archive 1990−2003 website: http://web.ornl.gov/sci/techresources/Human_Genome/project/index.shtml National Human Genome Research Institute website: https://www.genome.gov/.

ELSI

McEwen, J.E., Boyer, J.T., Sun, K.Y., Rothenberg, K.H., Lockhart, N.C., & Guyer, M.S. (2014). The ethical, legal and social implications program of the National Human Genome Research Institute: Reflections on an ongoing experiment. Annual Review of Genomics and Human Genetics. 15: 481−505. First published online as a Review in Advance on April 24, 2014 doi: 10.1146/annurev-genom-090413-025327. Retrieved from http://www.annualreviews.org/eprint/eDSR5xjQy7XjwMQ9VDXs/full/10.1146/annurev-genom-090413-025327.

Synthetic Biology

J. Craig Venter Institute. Synthetic biology and bioenergy website: http://jcvi.org/cms/research/groups/synthetic-biology-bioenergy/.

CLONING

Cloning represents a very clear, powerful, and immediate example in which we are in danger of turning procreation into manufacture.

-Leon Kass, American physician, scientist, educator

CHAPTER OUTLINE

2.1 Introduction 26
2.2 Cloning Animals 26
    2.2.1 Dolly 28
    2.2.2 Progress in Cloning After Dolly.. 29
    2.2.3 Limitations of Nuclear Transfer 30
    2.2.4 Applications of Cloning 30
    2.2.5 Ethical Issues in Cloning Animals 36
    2.2.6 Laws and Public Policy on Reproductive Cloning in Animals................ 38
2.3 Human Cloning 41
    2.3.1 Ethical Considerations 42
    2.3.2 Laws and Public Policy on Reproductive Cloning in Humans 45
2.4 Summary 50
References 50
Further Reading 53

Dolly

(By Toni Barros from São Paulo, Brasil (Hello, Dolly!) [CC BY-SA 2.0 (http://creativecommons.org/licenses/by-sa/2.0)], via Wikimedia Commons).

An Introduction to Ethical, Safety and Intellectual Property Rights Issues in Biotechnology.
DOI: http://dx.doi.org/10.1016/B978-0-12-809231-6.00002-8

2.1 **INTRODUCTION**

The development of a multicellular organism from a zygote formed by the fusion of egg and sperm is complex and tightly orchestrated by environmental and hormonal cues. The process depends on the expression of several subsets of genes in different cell lines formed by mitotic cell division of the zygote. Genes are expressed sequentially, whereby products of the first set of genes are required for the expression of the next set of genes. Very early in development, cells become *committed* to developmental pathways that result in the acquisition of specialized functions, a process known as *differentiation*. In most higher plants, differentiation is reversible; cells may be induced to lose specialized functions and behave similar to the zygote through hormonal and environmental manipulation. Plant cells thus have the potential to develop into all of the cell types that make up the whole plant, a phenomenon known as *totipotency*. However, in animals, irrevocable changes occur in gene expression during differentiation, and until very recently, cells could at best be induced only to modify the developmental pathway and give rise to other cell types (referred to as becoming *pluripotent*), but not to revert to a zygotic state. Totipotency of higher plants has enabled plant biotechnologists to excise tissues ("explants") from mature plants, to grow them in culture media, and to generate new plants from them. This "tissue culture" of plants allows scientists to make several copies, each genetically identical to the mother plant from which the explants were sourced. This process is called "cloning" (a "**clone**" is by definition, a **population of genetically identical individuals**). This chapter examines application of the process in animals such as rodents, pets, and livestock and discusses some of the ethical issues surrounding the extension of the technology to human beings.

2.2 **CLONING ANIMALS**

Cloning of animals was not thought to be possible, until it was found that in amphibians the first few cell divisions after fertilization produce cells that are totipotent. These cells could be induced to develop into new animals if the nuclei from the cells were isolated and injected into enucleated eggs, which provides the nutritional milieu for the introduced nucleus to divide and differentiate into an embryo. The embryo is implanted in a womb in order to raise a young one that is genetically identical to the donor of the nucleus. This technique of cloning by "**somatic cell nuclear transfer (SCNT)**" was **first reported in 1952 in frogs** (Briggs & King, 1952) and has since been used widely to study early development in amphibians. SCNT involves several steps, some of which require a microscope and micromanipulator (Fig. 2.1). As the embryo develops further, the cells lose totipotency, and the success of nuclear transfer rapidly declines. Some nuclear transfer experiments using cells from adult frogs produced viable embryos, but these never developed beyond the tadpole stage.

In mammals, cloning by nuclear transfer was first demonstrated in **mice in 1977**, but was not found to be repeatable. Despite this, interest in the area was sustained because of potential applications in cattle breeding and the prospect of large commercial benefits from multiplying elite embryos. The first report of cloning by transfer of nuclei from early embryos (partially

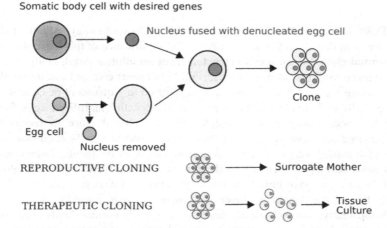

FIGURE 2.1 Cloning by Somatic Cell Nuclear Transfer (SCNT)

(By en: converted to SVG by Belkorin, modified and translated by Wikibob [GFDL (http://www.gnu.org/copyleft/fdl.html) or CC-BY-SA-3.0 (http://creativecommons.org/licenses/by-sa/3.0/)], via Wikimedia Commons).

differentiated cells at the 64- and 128-cell stage of embryo development) **in sheep** was made by Willadsen (1986). The next major breakthrough came in **1995 from the Roslin Institute, Edinburgh, Scotland, when Keith Campbell, Ian Wilmut et al. produced live lambs—Megan and Morag**—by nuclear transfer from cells of early embryos that had been cultured for several months in the laboratory (Campbell, McWhir, Ritchie, & Wilmut, 1994). The success of this experiment was attributed to the induction of "quiescence," the suspension of cell division in the donor cells. Further studies were conducted in the Roslin Institute at Edinburg to test if successful nuclear transfer was restricted to embryo-derived cells or could be carried out with a wider range of cell types, such as fetal fibroblast cells and adult cells. **Four lambs were born from embryo cells, three from the fetal cells and one, subsequently named Dolly, from an adult cell**. Their identity was confirmed by DNA testing, and the results were published in *Nature* on **February 27, 1997** (Wilmut et al., 1997). Further, in July 1997, Roslin Institute and a private company, PPL Therapeutics, announced the production of Polly, the first transgenic lamb produced by nuclear transfer (Schnieke et al., 1997). In this case, the donor cell was a fetal fibroblast that had been transfected with the gene coding for human blood clotting factor IX.

Although it was Dolly that attracted the media attention, the key practical discovery in this sequence of events was made earlier in 1995 with the production of live lambs from an established cell line. The Roslin Institute has two patent applications to cover its invention filed with a priority date of August 31, 1995. The first, PCT/GB96/02099, is entitled "Quiescent cell populations for nuclear transfer," and the second, PCT/GB96/ 02098, is entitled "Unactivated oocytes as cytoplast recipients for nuclear transfer." These cover use of the technology in most of the countries of the world and in all animals.

2.2.1 DOLLY

The creation of Dolly was a landmark event in the history of biotechnology. Although cloning in animals had previously been demonstrated in amphibians, the significance of this event was that this sheep was **the first mammal cloned from a cell extracted from an adult animal**. Dolly was derived from cells that had been taken from the udder of a 6-year-old Finn Dorset ewe and cultured in the laboratory. By nutritionally starving the cells, the cell nuclei were reduced to a quiescent or G_0 state, in which the chromosomes are inactive. Individual cells were then fused with unfertilized eggs from which the genetic material had been removed and subjected to an electrical pulse to reactivate the nuclei (Fig. 2.2). Two hundred and seventy-seven of these "reconstructed eggs"—each now with a diploid nucleus from the adult animal were cultured for 6 days in temporary recipients. Twenty-nine of the eggs that appeared to have developed normally to the blastocyst stage were implanted into 13 surrogate Scottish Blackface ewes. One gave birth to a live lamb, some 148 days later, on **July 5, 1996**, named Dolly (after Dolly Parton, the American singer, song-writer, actress, legendary for her well-endowed breast!) Although apparently, morphologically, and physiologically normal (Dolly even produced lambs for 3 years), **DNA analysis of her telomeres confirmed that they were shorter than age-matched sheep**. (The telomere is the tip of the chromosome and is of variable length being made up of repeat sequences. The length of the telomere has been known to decrease with increasing age of the cell attributable to the reduced activity of the telomere synthesizing enzyme, telomerase, in older cells.)

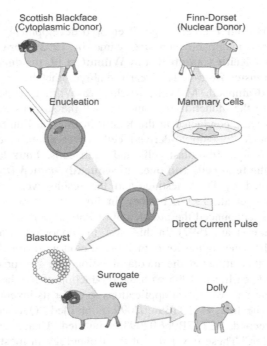

FIGURE 2.2 The Creation of Dolly

(By Squidonius (Own work (Original text: self-made)) [Public domain], via Wikimedia Commons.)

Dolly was clinically diagnosed to be arthritic in autumn 2001. Lesions in both left and right stifle joints were confirmed by radiography, and later pathological investigation confirmed extensive changes such as erosions and new bone production in the left stifle and less-advanced changes in the right joint. Anti-inflammatory drugs were used to treat the arthritis. On February 10, 2003, Dolly showed raised respiratory effort and coughing. Computer tomography examination conducted 4 days later showed increased density suggesting consolidation in the ventral lung lobe, and Dolly was clinically diagnosed with ovine pulmonary adenocarcinoma (OPA). Dolly was euthanized on Friday, February 14, 2003, at an age of 6 years and 7 months.

For scientists, the creation of Dolly challenged one of the fundamental tenets of developmental biology—that differentiation in animal cells is irreversible. However, the identity of the cell type that donated its nucleus to produce Dolly was not known, and the possibility that she derived from a mammary stem cell rather than a terminally differentiated epithelial cell could not be ruled out. The other question that remained unanswered was whether the arthritis and OPA that Dolly manifested were linked to the symptoms of aging (for instance, the shorter telomeres in Dolly compared to other age-matched sheep). The scientists at Roslin Institute were convinced that there was no reason to believe that Dolly was more vulnerable because she was a clone—arthritis had been reported in commercial sheep of similar age and other noncloned animals in the barn had developed OPA. It remains to be unequivocally established that clones would not have accelerated aging and that the technology can be applied without compromising animal health and welfare.

2.2.2 PROGRESS IN CLONING AFTER DOLLY

The nuclear transfer technique used to create Dolly was used in 1998 to clone over 50 mice by a group in Hawaii (Altonn, 1998) and subsequently in cattle, sheep, mice, goats, pigs, and pet cats and dogs by several groups around the world. In all cases, **success rates were low, only about 1% of cloned embryos resulted in live births**. Various reasons have been attributed for the low success rates including differences in early development between species and the cell division stage until which cells retain developmental plasticity. Many cloned offsprings die in the womb or shortly after birth probably because of developmental abnormalities. Normal fetal development is dependent on the methylation state of the DNA of the maternal and paternal chromosomes (referred to as *genetic imprinting*) and the alteration in the methylation pattern after fertilization. As somatic nuclei differ in chromatin structure from the nucleus of the zygote, **a number of abnormalities possibly arise during the "reprogramming" of the transferred nucleus**. Efforts to improve success rates include the use of genomic analysis to understand differences in gene expression and in imprinting patterns in transferred nuclei compared to nuclei arising from in vitro or in vivo fertilization.

KEY TAKEAWAYS
- Birth of Dolly in 1996—a landmark in biology, representing the first mammal to be cloned from an adult cell
- Derived from cell taken from udder of Finn Dorset ewe, fused by an electric pulse to an enucleated egg, and implanted in a Scottish Blackface surrogate ewe

- Morphologically and physiologically normal, produced lambs for 3 years
- However, DNA analysis of telomeres indicated that they were shorter than those of age--matched sheep
- Developed arthritis at age 5 years, and was diagnosed with ovine pulmonary carcinoma at 6 years and 7 months
- Technique has since been successfully used in mice, goats, pigs, cattle, horses, pet cats, and dogs
- Success rates are low, with only about 1% of cloned embryos resulting in live births

2.2.3 LIMITATIONS OF NUCLEAR TRANSFER

One important limitation of the technique is the requirement of an **intact nucleus with functioning chromosomes**. DNA on its own is not enough. The technology therefore does not as such support resurrection of extinct organisms from fossils or preserved samples that usually have fragmented genomic DNA. Another serious limitation is the **requirement of egg cells and surrogate mothers**. It is possible to clone only in closely related species as the chance of carrying a pregnancy to term would be increasingly unlikely if the eggs and surrogate mothers are from distantly related species. This again is a serious limitation in the conservation efforts of many rare and endangered species with no close relatives. Despite these drawbacks, attempts are being made globally to use the technique with modifications for "de-extinction" or resurrection of extinct animals. Animals targeted for de-extinction include the woolly mammoth, the gastric brooding frog, the passenger pigeon, the dodo, the Pyrenean ibex, the Carolina parakeet, the woolly rhino, the moa, and the Tasmanian Tiger (thylacine) (see more in Box 2.1).

2.2.4 APPLICATIONS OF CLONING

The primary use of clones is to produce **breeding stock**. The clones are copies of the best animals in the herd and are used for conventional breeding. The sexually reproduced offspring of the animal clones become the food-producing animals. SCNT has the potential to create unlimited copies of an individual with desirable traits and represents an advancement over other forms of **assisted reproductive technologies (ART)** used in cattle breeding, such as **embryo splitting** and **nuclear transfer of cells from blastomeres (blastomere nuclear transfer, BNT)**. In embryo splitting, in most cases, cells retain the ability to give rise to a young one only till the 8-celled stage of the embryo. Consequently, only eight cloned individuals can be produced. In BNT, the nuclei for cloning are obtained from embryos with more than eight cells (the blastomeres), and therefore clone sizes may be higher, but as some reprograming of nuclei is necessary, gestational and postnatal abnormalities sometimes occur. One major **drawback** of both embryo splitting and BNT is that it **is impossible to generate clones genetically identical to the elite animal**: the embryo for cloning is derived by fusing the egg (or sperm) of the elite animal with that of another parent and can thus have **only 50% genetic identity with the elite parent**.

BOX 2.1 DE-EXTINCTION

Woolly Mammoth.
(By Flying Puffin (Mammut Uploaded by FunkMonk) [CC BY-SA 2.0 (http://creativecommons.org/licenses/by-sa/2.0)], via Wikimedia Commons.)

Compelling reasons to bring back species that are extinct or on the brink of extinction include: (1) the preservation of biodiversity on earth, (2) to restore diminished ecosystems, and (3) an attempt to undo the harm to nature that human beings have knowingly or unknowingly caused in the past. Many species, now extinct, were *keystone* species in their habitats and resurrecting them could possibly restore the biodiversity in several regions of the globe. This process of "genetic rescue" depends on several biotechnologies. The first level of genetic rescue involves an understanding of the genetic make-up of the species achieved through sequencing of DNA and comparing gene sequences with other related living or extinct species. Clues from comparative genomics serve as leads on the genetic diversity of the species, and of past events such as the scale and timing of major changes in population sizes. Samples for genomic study often come from fossils and preserved remains in museums and conservatories, DNA extracted from some of which may be degraded (fragmented). Genomic information could be used in the second level for genome editing in living, endangered species to restore genetic diversity and remove disease susceptibilities threatening the survival of the species. At a third level, genomic information could be used for genome editing to revive extinct species. This involves transferring synthetic DNA sequences or amplified gene sequences from the fossil or preserved specimens (Ancient Genome Assemblies) into genomes of closely related species. Nuclei with ancient DNA could be cloned by SCNT into eggs from closely related species and the developing embryos inserted into surrogate mothers of related species (cross species cloning). Although cloning is not possible in the absence of DNA, it is considered plausible for ancient creatures whose genomes can be reconstructed. Examples of such efforts include the resurrection of the Tasmanian tiger which went extinct due to hunting and trapping by humans in 1936. Palaeontologist Michael Archer used samples of teeth, skins, and dried tissues of the animal from the Australian Museum to isolate DNA, fragments of which could be successfully spliced into mouse. Possibly the genome of the animal could be assembled and transplanted into an

(Continued)

BOX 2.1 (CONTINUED)

enucleated egg of a close relative, the Tasmanian devil. The resulting embryo implanted in a Tasmanian devil could potentially resurrect the Tasmanian tiger. Michael Archer led researchers at the University of Newcastle Australian in science project called the Lazarus Project which has been successful in bringing back to life the gastric brooding frog (Australian Associated Press, 2013).

Comparative genomics of the woolly mammoth with the modern day Asian and African elephants has helped to identify more than 400,000 sites where elephant genomes differ from that of the mammoth. Technically, the elephant genome could thus be modified (by allele replacement) and injected into an enucleated elephant egg cell, cloned in an elephant surrogate mother to give birth to a woolly mammoth. Considerable work in this area has been done by evolutionary geneticist Hendrik Poinar of the Ancient DNA Centre at McMaster University, and geneticist George Church of Harvard University, in association with the Swedish Museum of Natural History.

The concept of de-extinction has been controversial and severely criticized for diverting funds from conserving and protecting the existing biodiversity for the expensive sequencing and cloning efforts involved. Also in question is the habitat to which the resurrected animal could be released as it is impossible to recreate the original habitat and the ramifications involved with release into the existing ones are not known. De-extinction has also been criticized on account of the ethics involved—bringing a mammoth to life could be cruel to both the mammoth and its elephant mother without offering any benefit.

Reference

Australian Associated Press. (2013, November 22). *Extinct frog resurrected with 'de-extinction technology.* Retrieved from http://www.theguardian.com/environment/2013/nov/22/extinct-frog-resurrected-with-de-extinction-technology.

Further Reading

The Long Now Foundation website: http://longnow.org/revive/what-we-do/extinction-continuum/.
Switek, B. (2013). (Reporter). How to resurrect lost species. *National Geographic.* Retrieved from http://news.nationalgeographic.com/news/2013/13/130310-extinct-species-cloning-deextinction-genetics-science/.
Shapiro, B. (2015). *How to clone a mammoth: The science of de-extinction* (p. 240). Princeton, NJ: Princeton University Press.

2.2.4.1 Multiplication of elite strains of livestock for meat and milk

The technique could be used to create infinite number of identical copies of farm animals such as pigs and cattle reared primarily for food. **Clones could directly be used as food or used as breeding stock** integrated into conventional breeding programs. The only drawback to widespread application is that it is not currently established that clones would consistently deliver the commercial performance.

One issue of concern is whether the meat or milk produced from cloned animal would be **safe for human consumption**. In the 1980s, cloning technologies (embryo splitting and blastomere nuclear transfer) were introduced into cattle breeding. Holstein cattle were first cloned from somatic cells in 1997, but food and feed from animal clones and their progeny were not introduced in the market due to doubts regarding the safety of the products. In 1999, the **Centre for Veterinary Medicine (CVM)** at the US Food and Drug Administration (FDA) began an extensive review of food products from nongenetically engineered cloned animals including cattle, swine, and goat clones (Rudenko, Matheson, & Sundlof, 2007). The comprehensive risk assessment conducted by the CVM had two objectives: (1) to determine whether cloning introduced any unique hazards to animals involved in the cloning process, and (2) whether the meat and milk derived from them posed different risks from those posed by foods produced by current agricultural practices. The risk assessment concluded that cloning was essentially only an extension of ARTs and **poses no unique**

risks either to animal health or to food consumption. The CVM also opined that **no special labeling of food products derived from clones was necessary** < https://www.fda.gov/AnimalVeterinary/SafetyHealth/AnimalCloning/ucm055516.htm#labeling >. Based on its risk assessment, the CVM issued a **Guidance for Industry CVM GFI #179** *Use of Edible Products from Animal Clones/Progeny for Human Food/Animal Feed*—draft in 2006 (FDA, 2006). The guidance however **excludes sheep clones for food uses** due to insufficient data on safety.

2.2.4.2 Creation of transgenic livestock for production of pharmaceuticals ("Pharming")

Transgenic or genetically modified (GM) organisms are those in which "foreign" genes have been integrated in the genome using the tools of recombinant DNA technology (see Chapter 4: Recombinant DNA Technology and Genetically Modified Organisms). Several such organisms have been created in order to express genes encoding proteins of therapeutic use. Although several human therapeutic proteins have been produced in bioreactors by transgenic bacteria, problems with the technology include difficulty in purification and absence of appropriate post-translational modifications associated with the production of a protein of eukaryotic origin in a prokaryotic system. Alternatives have been to produce the proteins in human cell cultures (too expensive) and purification from blood (risk of contamination with viruses such as the AIDS or hepatitis C virus). A viable alternative appears to be to **produce human proteins** that have appropriate post-translational modifications **in the milk of transgenic sheep, goats, and cattle**. Output can be as high as 35−40 g/L milk, and costs are relatively low.

Transgenic livestock can be produced by pronuclear injection by which method 200−300 copies of the transgene are transferred into a recently fertilized egg that is then implanted in a surrogate mother (see Fig. 4.4 in Chapter 4: Recombinant DNA Technology and Genetically Modified Organisms). The drawback with the technique is that only 2%−3% of eggs injected give rise to transgenic offsprings, and only a fraction of these express the added gene at sufficiently high levels to be of commercial interest. If animals can be derived from cells in culture, then it is possible to carry out much more precise genetic modifications, including the removal or substitution of specific genes, and to ensure that the human genes are inserted in specific points in the genome. Producing transgenic animals by nuclear transfer has immediate practical advantages; it uses less than half of the experimental animals to generate founder animals than does pronuclear injection. By specifying the sex of the offspring and generating a small number of identical clones, it is possible to generate a flock that is sufficiently large to produce the drug for the conduct of clinical trials within one and half years. Chapter 4: Recombinant DNA Technology and Genetically Modified Organisms (Section 4.4.3), discusses a few examples of cloned transgenic livestock created for the production of pharmaceuticals.

2.2.4.3 Creation of transgenic livestock for production of nutraceuticals

Gene targeting and transgenesis could be used to **improve the nutritional quality of cow's milk**. Problems associated with the use of cow's milk by human beings, especially premature babies, include lactose intolerance and allergenicity to certain proteins present in cow's milk. By replacing these with human proteins, special herds of cattle could be created that produce milk similar to human milk. In 2012, AgResearch laboratories in Hamilton, New Zealand, created a calf capable of producing **milk free of the allergen beta lactoglobin (BLG) protein** by genetically modifying skin cells of a cow to block synthesis of BLG and transplanting the nucleus from these cells into enucleated eggs. Of 100 attempts at transplantation into surrogates, one live birth of a calf named Daisy was reported (Sample, 2012). Hormonally induced lactation two years later has proved that Daisy produces BLG-free milk, but establishing a herd is still several years away.

2.2.4.4 Derivation of stem cells for cell-based therapies (therapeutic cloning)

Replacement of damaged or diseased cells with cells from normal individuals has been attempted in the treatment of several diseases like diabetes, stroke, heart disease, and Parkinson's disease. However, transplanted cells cause immunological responses in the recipient similar to organ transplants and are as likely to be rejected. It is envisaged that the ability to clone adult cells may give rise to opportunities to grow healthy tissue from a patient's own cells, thereby circumventing problems of rejection. This technique is referred to as **_therapeutic cloning_** in which cloning is done in order to derive stem cells (see Chapter 3: Stem Cell Research for more information on this), rather than to create a copy of the individual (**_reproductive cloning_**). Although stem cells occur in several organs, almost all early reports of the occurrence of these cells were confined to the embryo, known as **_embryonic stem cells_**. **Cloning by nuclear transplantation has the potential to convert differentiated adult cells to embryonic stem cells.** In 2005, Professor Woo-Suk Hwang in Korea garnered media and scientists' attention with reports that he had derived stem cells from a cloned human embryo, demonstrating for the first time that it was possible to clone human cells, and that it was possible to obtain embryonic stem cells from the resulting clone. Although this work remained controversial with his paper being retracted in 2006, it appears that he had made some headway in creating stem cell lines through "parthenogenesis" (creation of embryos without fertilization) in human beings (Minkel, 2007).

2.2.4.5 Production of transgenic animals for xenotransplantation

In order to overcome the **shortage of organs for transplantation**, transgenic animals, such as pigs, are being developed which express human proteins (such as complement inhibitory factor) that coats the pig tissues and is intended to prevent immediate rejection of the transplanted heart or kidney. Also, by nuclear transfer, genes may be deleted from pigs so that the **organs no longer elicit an immune response when transplanted into a human patient**. Xenotransplantation products are subject to regulation by the FDA under section 351 of the US Public Health Service Act [42 U.S.C. 262] and the Federal Food, Drug and Cosmetic Act [21 U.S.C. 321 et seq]. In accordance with the statutory provisions governing premarket development, xenotransplantation products are subject to FDA review and approval (Samdani, 2014).

2.2.4.6 Models for understanding aging and cancer

In all multicellular organisms, several rounds of cell division occur, first to generate all the tissues and organs that make the organism, and later to replace dead or damaged cells and tissues. It is known that with time, the process of cell division becomes less precise, resulting in the **accumulation of several genetic and chromosomal mutation in cells**. Morphological and physiological manifestation of aging, including the development of several age-related diseases and increased incidence of certain cancers, is attributed to this accumulation of cellular or somatic mutations. It may be possible to follow the changes that occur in gene expression and trace the biochemical and physiological changes occurring in diseased cells from start to finish, if cells from a diseased patient could be cloned (by inserting the nucleus from affected cells of a diseased individual into an enucleated egg cell).

2.2.4.7 Conservation of rare breeds

Cryopreservation of plant cells and tissues is a viable method of preserving rare or endangered species. The technique involves freezing and storing cells, tissues or plant organs in liquid nitrogen. Once thawed, and transferred to suitable culture media, plants could be regenerated. The technique was not possible in animal systems. At best, sperm or eggs could be cryopreserved and used for creating new individuals. The drawback with this strategy is that **sperm and egg can each only provide half the genes** and also, if implantation of an embryo fails, the embryo is lost forever. In contrast, **cloning provides an unlimited supply of cells with which to keep trying for a successful pregnancy**. Scientists have suggested that cloning could be used to "save" rare or endangered species such as the Giant Panda, perhaps by using female bears or related species as a source of eggs and surrogate mothers (see also Box 2.2)

2.2.4.8 Cloning for companionship and entertainment

Among animals which have been successfully cloned are equines—the first male mule named Idaho Gem was born on May 4, 2003 in Idaho, America (Horse and Hound, 2003a), and first

---

BOX 2.2 MISSIPLICITY PROJECT

In 1997, media reports of the success of Dolly inspired a millionaire businessman, John Sperling to fund the Missiplicity project aimed at cloning his pet dog, Missy. The project was managed by Genetic Saving and Clone (GSC), which Sperling founded when he found that there was a popular demand from others who wanted to bank and clone their pets. The company operated in association with the Texas A&M University, USA. In 2001, GSC was able to clone a cat in Operation CopyCat—Carbon Copy (CC) was born on December 22, 2001, at Texas A&M College of Veterinary Medicine. However, cloning dogs proved to be more complex and after Missy died in 2002, the funding for the project to the tune of US $3.7 million was withdrawn in November 2002 (AgBiotechNet, 2002).

Five years after Missy died, three clones of the dog were produced in South Korea from her conserved cells. This was made possible because one of the key scientists in the Missiplicity project was Dr. Taeyoung Shin, a student of Dr. Hwang Woo-Suk of Seoul National University. Dr. Shin prompted work in Dr. Hwang's laboratory to try and overcome the problems encountered in the Missiplicity project for cloning dogs. Dr. Hwang and team were successful in cloning the world's first dog called Snuppy in 2005. On a request from the former CEO of GSC (now CEO of BioArts International), Lou Hawthorne, conserved cells of Missy were flown to Seoul, and Dr. Hwang's team at Sooam Biotech Research Foundation succeeded in producing three Missy clones—Mira (born December 5, 2007), Chingu, and Sarang (born February 15 and 19, 2008). Sooam (http://en.sooam.com/main.html) and BioArts International (http://www.bioartsinternational.com/) floated a program in 2008 called "Best Friends Again" for cloning dogs, although the program ended in 2009 due to disputes on patent infringements (Carlson, 2009). Sooam has since cloned over 400 dogs as well as hundreds of other animals such as cows, pigs, and a coyote.

References

AgBiotechNet. (2002, November 12). *Missiplicity project funding ends*. Retrieved from http://www.cabi.org/agbiotechnet/news/1988.

Carlson, B. (2009, September 10). BioArts International ends cloning service; blasts black market cloners. Retrieved from https://web.archive.org/web/20091009064712/http://www.bioarts.com/press_release/ba09_10_09.htm. Also, Watch video: [Missiplicity Proj] Dead dog Missy Cloned out by HWS. Retrieved from https://www.youtube.com/watch?v=xE3X5Tu5j2Y.

cloned foal, Prometea, on May 28, 2003, in Italy (Horse and Hound, 2003b). Although initially cloned horses were barred from competing in equestrian events, the Federation Equestre Internationale in 2012 announced that it would not forbid clones or their progenies from participating in international events. Efforts have been made to clone other pets too, such as dogs and cats (see Box 2.2 on the Missiplicity Project).

2.2.5 ETHICAL ISSUES IN CLONING ANIMALS

Given that, animals have been cloned for diverse applications mostly (though not solely) aimed at human ends such as better food production, understanding/treating diseases, or entertainment, ethical issues in cloning animals are complex. Although policy makers have often reduced the issue to a discussion of whether the food produced from cloned animals is safe to eat, ethical concerns go beyond such reductionalistic thinking. In a survey conducted in 2002 by the Pew Initiative on Food and Biotechnology and the Center for Veterinary Medicine of the FDA, **64% of Americans consider cloning "morally wrong"** (Tucker-Foreman, 2002). In another survey, **4.2% of people believed that meat and milk from clones and their offspring is unsafe to eat** (Brooks & Lusk, 2011). Some of the ethical issues arising from cloning animals are discussed below.

2.2.5.1 Pain and suffering experienced by the animals during the cloning process

This includes the suffering endured by the donor mother for surgical removal of eggs, the surrogate mother (as surgery is required for implantation), the health of the clone, and the suffering that it endures if cloned for the study of diseases. With cloning success rates being currently low (in the order of 1%−2% in almost all the animal species studied), the process is error prone with high fatality rates. Cloned animals are also reported to have more physical defects such as contracted tendons, heart and kidney problems, respiratory failure, head and limb defects, and shortened life span. This has prompted animal welfare groups to oppose this technology. For instance, in 2003, the **US Humane Society had advocated a ban on products coming from cloned animals or their offspring on grounds that the process was ethically unacceptable** (Humane Society of the United States, 2003). Proponents of the technology however argue that the benefits in terms of improved food production, pharmaceuticals, and even animal health, far outweigh the drawbacks of the technology, and that the issues faced by cloned animals are not quintessentially different from those of other experimental animals.

2.2.5.2 The effect of the cloned animals on other populations of animals

For example, opponents of pet cloning question the need to create copies of deceased pets when there are large numbers of pets available in animal shelters awaiting adoption.

2.2.5.3 Compromising the safety of food from livestock

With consumers already suspicious of GM foods, safety concerns exist with respect to meat and products from cloned livestock.

2.2.5.4 Effect on the environment

Negative outcomes could arise from interbreeding of cloned animals with others in the same ecological niche resulting in unintended consequences to biodiversity.

2.2.5.5 Creation of unwanted animals

Opponents of cloning have raised doubts as to whether there is indeed a need to use the technology to revive extinct animals such as the woolly mammoth and Tasmanian tiger, since even finding suitable habitats for them would be a challenge.

2.2.5.6 Unnatural modification of animal species—"playing God"

Although the world's spiritual leaders advocate caution, there is generally no consensus with regard to cloning, and the religious debate remains open-ended (Sullivan, 2004).

2.2.5.7 Potential for misleading bereaved pet owners regarding resurrection of their beloved pets

It is often difficult for people to accept that a cloned pet would not necessarily be the same as the loved pet.

KEY TAKEAWAYS

Applications of cloning:

- Multiplication of elite strains of livestock for milk and meat: clones could be used directly as food, or used as breeding stock to improve herds
- Creation of transgenic livestock for production of pharmaceuticals ("Pharming"): for example, human antithrombin, ATryn from cloned goats, or albumin from cows
- Creation of transgenic livestock for production of nutraceuticals: for example, creating herds of transgenic cattle that produce milk similar to human milk or free of the allergen BLG
- Derivation of stem cells for cell-based therapies or therapeutic cloning
- Production of transgenic animals for xenotransplantation: for example, transgenic pigs that express human proteins that allow the organs to be used for human transplantation
- Creation of models for understanding aging and cancer
- Conservation of rare breeds and "de-extinction"
- Cloning for entertainment: for example, cloning of prize horses and beloved pets such as dogs and cats

Ethical issues in cloning animals:

- Pain and suffering experienced by the animals during the cloning process
- The effect of the cloned animals on other populations of animals
- Compromising the safety of food from livestock
- Effect on the environment
- Creation of unwanted animals
- Unnatural modification of animal species—"playing God"
- Potential for misleading bereaved pet owners regarding resurrection of their beloved pets

---

2.2.6 LAWS AND PUBLIC POLICY ON REPRODUCTIVE CLONING IN ANIMALS

In March 2012, the European Commission (EC) conducted a survey in connection with a proposal for a Council Directive on placing on the market food from animal clones, on the cloning of animals of the bovine, porcine, ovine, caprine, and equine species kept and reproduced for farming purposes (European Commission, 2013). The survey was conducted in view of the **vast majority of EU citizens (above 80%) expressing negative perception of the cloning technique if used for food production**, and the European Food Safety Authority (EFSA) finding that **surrogate dams (carrying the clones) and the clones themselves suffer in the application of the technique**. In the survey, questionnaires were sent to 15 countries with major trade in meat, milk products, and reproductive materials with the EU, and revealed that cloning for food production takes place in the United States, Canada, Argentina, Brazil, and Australia. The same survey indicated that cloning for breeding purposes is practiced in Australia, Argentina, Brazil, and the United States (European Commission, 2013, p. 9). Based on the status of commercial cloning, legislation in different regions and countries is summarized in Table 2.1.

2.2.6.1 European Union

In Europe, the EC in 2008 asked the European Group on Ethics to issue an expert opinion on the ethical implications of cloning animals for food (The European Group on Ethics in Science and New Technologies to the European Commission, 2008). The group concluded that it could find **no ethical justification for cloning animals due to the animal suffering involved**. On December 18, 2013, the EC adopted two proposals (European Commission Press Release Database, 2013): one, on the cloning of animals kept and reproduced for farming purposes, and the other on the placing on the market of food from cloned animals. The key elements on cloning were that there would be **no cloning for farming purposes carried out in the European Union (EU) and no such clone would be imported**. Cloning would however **not be prohibited for research, production of pharmaceuticals or medical devices, and for conservation of rare breeds**. The draft directives also foresaw a temporary ban on placing on the market live clones and embryos, and food products such as milk or meat from animal clones (EUR-Lex-52013PC0892, 2013). Under the draft regulation, novel food produced using new technology would be subjected *to a simpler, clearer,*

Table 2.1 Regulatory Oversight on Cloning of Animals

Regulation	Example	Purpose of Cloning
Stringent laws	European Union—cloning is banned for food	Research, production of pharmaceuticals or medical devices and for conservation of rare breeds
	New Zealand, Japan—voluntary ban for food	Research
Moderate level of oversight	Australia	Restricted to breeding stock
Permissive regulation	United States—Guidance for Industry #179—food from clones safe to eat	Food—milk and meat from cattle and pigs
	Canada	Novel foods
	Argentina	Production of pharmaceuticals, not food

and more efficient authorization procedure so that safe and innovative food would reach the EU market faster (EUR-Lex-52013PC0893, 2013).

In the EU, marketing of food from clones is subject to premarket approval under the Novel Food Regulation (EUR-Lex-31997R0258, 1997). Scientific food safety assessment is done by the EFSA. Interestingly, the EFSA did conduct a scientific risk assessment of food from clones in 2008 and concluded that there is no indication of differences in terms of food safety in milk or meat of clones or their progeny when compared to that from conventionally bred livestock (European Food Safety Authority, 2008). However, **no European or foreign-food business operator has to date applied for authorization to market food from clones** (European Food Safety Authority, 2012).

Cloning is not banned for purposes where the technique can be justified such as for research, for conservation of endangered species, or for production of pharmaceuticals and medical devices. On September 8, 2015, The European Parliament in Strasbourg voted 529 to 120 for a ban on cloning of all farm animals, their descendants, and products derived from them, including imports (European Parliament News, 2015). The decision was based on consumer research findings indicating, *"A majority of EU citizens strongly oppose the consumption of food from animal clones or from their descendants and that a majority also disapprove of the use of cloning for farming purposes, on animal welfare and general ethical grounds."* The members moreover recommended that the Directive should be upgraded to a regulation for ease of enforcement. The scope of the Directive was extended to include all animals produced for farming purposes.

2.2.6.2 *United States of America*

In the United States, the regulation of animal cloning is under the jurisdiction of the FDA (http://www.fda.gov/cvm/cloning.htm or http://www.fda.gov/AnimalVeterinary/SafetyHealth/AnimalCloning/default.htm). In 2001, the CVM of the FDA on its website uploaded a request to all stakeholders to voluntarily refrain from introducing food from clones or their progeny, into the human or animal food supply, pending a risk-assessment process that it had initiated. The FDA also contracted with the National Academies of Sciences' National Research Council (NAS/NRC) to study the hazards in animal biotechnology including clones and their progeny, which was tabled in September 2002 (National Academy of Sciences, National Research Council, 2002). **Evaluation of all the available data with respect to the composition of milk and meat did not show any significant differences between clones and sexually reproduced cattle, swine, and goats,** prompting the FDA to conclude that **food from clones was safe to eat**. In January 2008, FDA released the final version of the risk assessment (Centre for Veterinary Medicine, U.S. Food and Drug Administration, 2008) as well as **Guidance for Industry #179** jointly with the Undersecretary of Marketing and Research Services and the United States Department of Agriculture (USDA) (FDA, 2008). However, in order to ensure a smooth transition in the market, and so that other governments could also develop their own regulatory systems, the USDA requested producers to keep food from clones out of the general food supply, although food from sexually reproduced progeny of clones has been freely entering the market. Meanwhile, the industry has developed a supply chain management system for meat that involves education, identification and traceability (all clones are identified in a registry), and financial incentives (carcasses are directed to food streams that will accept clones, and the owner can claim a refund exceeding the value of the issued carcass) (Rudenko, 2011). As there is no moratorium on using clones as breeding stock, sales of semen from bull clones have been happening internationally.

2.2.6.3 Canada

Similar to the United States, food and feed from clones and their progeny are defined as novel foods/feeds and requires premarket safety assessment in Canada (European Commission, 2013). Also, products and by-products produced by clones and their progeny are considered as new substances and manufacture and import of such substances require clearance from the Minister of the Environment as they are considered to be new substances under the Canadian Environment Protection Act. Preimport notification is required to import clones and their reproductive material. Registration of cloned animals is voluntary but animal clones can be identified by supplemental designations on registration documents. Animals used for research purposes are explicitly prohibited from entering the food chain or being released into the environment.

2.2.6.4 Argentina

In Argentina, **cloning is not considered to be different from other assisted reproduction techniques.** Cloning research for the production of calves was started by the Institute of Biology and Research Medicine around 1994 in association with the Roslin Institute of Edinburg, Scotland and later with a Japanese research group. In 2002, animal cloning was successfully done by a company called **Biosidus** which was able to make **transgenic cows expressing pharmaceuticals**. Commercial cloning services are available in Argentina for genetically engineering animals for the production of pharmaceuticals, but none have been approved for food. The government is in the process of determining its own policy on the technology (Global Agricultural Information Network Report, 2014a).

2.2.6.5 Brazil

Cloning research started in the late 1990s in Brazil, mostly focused on cattle. In order to provide a legal framework for the cloning activities in Brazil, the country is in the process of drawing up a draft proposal for regulating research, production, import and sale of cloned animals (European Commission, 2013). **Cloned animals are registered through breeding organizations**: since May 2009, the Ministry of Agriculture, Livestock, and Food Supply (MAPA) changed its regulation to allow the registration of cloned cattle under the Brazilian Zebu Cattle Association (ABCZ). The Brazilian Zabu represents about 90% of the cattle base in Brazil.

2.2.6.6 Australia

In Australia, cloning is carried out by public and private research institutes but is restricted to a small number of **elite breeding stocks of beef and dairy cattle, and sheep** in contained facilities (Global Agricultural Information Network Report, 2014b).

2.2.6.7 Others

In **New Zealand**, research in cloning is pursued by **AgResearch Limited** which has an explicit **voluntary moratorium ensuring that cloned animals or products do not enter the food chain**. In **Japan**, the Ministry of Agriculture, Forestry and Fisheries keeps records of cloned animals and has imposed a **voluntary ban on cloning of livestock** except for research purposes (European Commission, 2013).

KEY TAKEAWAYS

Laws and Public Policy on Reproductive Cloning in Animals:

US	Regulated by the CVM of the FDA, concluded in 2008 that **food from clones safe to eat**, released **Guidance for Industry#179** jointly with USDA and Undersecretary of Marketing and Research Services; no labeling requirement; no restriction on sales of clones as breeding stock
EU	2008—European Group on Ethics **find no ethical justification for cloning animals**, 2013 EC adopts two proposals—(1) there would be no cloning for farming purposes in the EU or import of clones, but research on clones, production of pharmaceuticals, and endangered species permitted (2) temporary ban on placing on the market live clones, embryos of clones or food from clones
	EFSA assessment of food safety in 2008 concluded that food from clones is no different from that of conventionally bred livestock
Canada	Food from clones defined as novel foods requiring premarket safety assessment. Animals for research purposes are explicitly prohibited from entering food chain
Argentina	Cloning not considered different from other assisted reproductive techniques

2.3 HUMAN CLONING

The possibility of extrapolation of nuclear transfer and cloning technology used to create livestock to other mammals, including humans, has attracted tremendous media attention and have brought to the fore public concerns on the moral and ethical implications of the technology. Many have cited potential benefits and purposes for cloning humans as listed below (The President's Council on Bioethics, 2002).

- *To produce biologically related children*—allows infertile couples, same sex couples, or individuals to have children of the same genetic make-up as themselves.
- *To avoid genetic disease*—in cases where both parents carried recessive mutations which would result in unviable or diseased offspring.
- *To get ideal tissues or organs for transplantation*—in order to avoid rejection, transplanted tissues/organs have to be matched in the donor and recipient. Cloning would ensure exact matches so as to avoid transplant rejection.
- *To replicate a loved one*—as a means of resurrecting a deceased or preserving the biological identity of a dying beloved.
- *To reproduce special individuals with talent, beauty, or intelligence*—as a means of preserving for posterity the unique characteristics of specific individuals.

Notwithstanding the merit of the stated purposes, there do appear to be two compelling reasons why human cloning should be done:

- *Treatment of diseases caused by mitochondrial disorders:*
  Ruth Deech, chairperson of the Human Fertilization and Embryology (HFE) Authority in 1998 cited before the Commons Science and Technology Committee, a possible application of full human cloning (The Human Genetics Advisory Commission & The Human Fertilization and Embryology Authority, 1998, p.22). A rare form of an **inherited disease, characterized by epilepsy and blindness, is caused by mutations in mitochondrial DNA**. By removing the

nucleus—minus the defective mitochondria—from an embryo created by in vitro fertilization in the normal way and placing it in a donated egg stripped of its own nucleus, a cloned baby could be created that would be the genetic offspring of its parents, but without the disorder (see Box 2.3).

- *Studying the mechanism and control of cellular differentiation:*
  Nuclear transfer experiments and the ability to derive embryonic stem cells from cloned human embryos and to control their differentiation into different cell types provide revolutionary new opportunities in biology and medicine. These methods make it possible to **study human genetic diseases** in entirely new ways. Cloned human cells could be used **to test new drugs** or could be used for cell therapies to treat diseases caused by defects in cell development, for example, motor neuron disease (MND), also called amyotrophic lateral sclerosis or Lou Gehrig disease.

Early reports of cloning in humans came from the Korean Seoul National University. Hwang et al. (2005) reported the creation of 31 cloned human embryos and the establishment of 11 stem cell lines from them (although this report was later declared as being fraudulent, see Chapter 4: Recombinant DNA Technology and Genetically Modified Organisms). The same year in Newcastle, UK, Alison Murdoch and her team also created a cloned embryo from human somatic cells (Stojkovic et al, 2005). The purpose of these experiments was not reproductive cloning but for deriving cell lines for therapy and research. In the UK, therapeutic cloning has been legal since 2001. Research is currently permitted on human embryos up to 14 days and has helped scientists **study early embryology**. Being able to clone human embryos would facilitate research into several other aspects such as **DNA damage and repair and genetic imprinting**. In 2005, Ian Wilmut was granted license to carry out human therapeutic cloning for studying and devising treatment options for MND. In 2002, a private US company based in Las Vegas, Nevada, called Cloneaid claimed to have created the first cloned human baby called "Eve" said to be genetically identical to her 31-year-old mother (Associated Press, 2002). However, as the company did not provide any test data, the claim remains unverified.

2.3.1 ETHICAL CONSIDERATIONS

Several ethical and moral concerns have arisen over the potential applications of nuclear transfer and cloning technology in human beings. Although it is evident that the technology holds ample potential in biology and medicine, **many feel that suitable legislation for banning its use, especially in human reproduction, is necessary**. Apprehensions regarding the technology include the following:

2.3.1.1 "Playing God"

Detractors of the technology are convinced that scientists are playing God, directing the creation of life itself. Arguments put forward by scientists in favor of the technique include (1) ever since plants and animals have been domesticated, man has used selective breeding to direct the course of evolution; (2) advances in medicine such as the use of antibiotics, vaccination and sophisticated surgical procedures, have all intervened with nature to extend life expectancy, and this technique is only an extension of scientific developments in the field; (3) the process of embryo splitting and

BOX 2.3 BABIES WITH THREE PARENTS

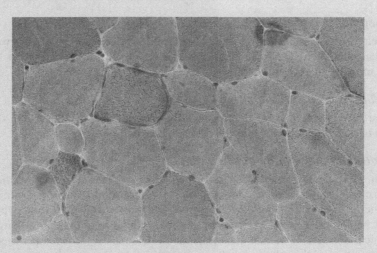

Very high magnification micrograph showing ragged red fibers, commonly abbreviated RRF, in a mitochondrial myopathy. Gomori trichrome stain.

(By Nephron (Own work) [CC BY-SA 3.0 (http://creativecommons.org/licenses/by-sa/3.0) *or GFDL* (http://www.gnu.org/copyleft/fdl.html)]*, via Wikimedia Commons)*

Mitochondria are organelles in the cell cytoplasm responsible for the generation of energy required to power metabolic reactions in the cell. Mitochondria contain DNA that encodes proteins which (along with some nucleus encoded proteins) directs the structure and function of the organelle. Mutations in the mitochondrial DNA result in defective mitochondria and have been implicated in over 50 human diseases, some of which cause disability and death. The list of such disorders includes stroke-like episodes, muscular dystrophy, fatal liver failure, mental retardation with intractable epilepsy, blindness, deafness, and diabetes. As mitochondria are passed down from mother to child (as only the sperm nucleus and not its cytoplasm enters the egg during fertilization), these diseases show maternal inheritance. If the cytoplasm of defective embryos were to be replaced with normal cytoplasm, it would be possible to ensure that mothers with such diseases do not pass on the defect to their offspring.

In 2008, a team of scientists led by Doug Turnbull of Newcastle University, UK, used SCNT to create embryos which technically had three parents. Using in vitro fertilization (IVF) technique, a fertilized egg was created by fusing the egg from the diseased mother and sperm of the biological father. Within hours of fertilization, the nucleus containing the DNA of the parents was removed and transplanted into an enucleated egg from a normal donor. Abnormal embryos discarded from fertility treatments were used by the Newcastle team to create 10 embryos in which the cytoplasm had been replaced. (The embryos were allowed to develop for 6 days before being destroyed.)

Although the achievement raised an ethical storm with several opponents of the technology fearing a slippery slope that would lead to the creation of "designer babies," the technique shows considerable promise to prevent the inheritance of genetic diseases. Consequently, UK considered an amendment to the Human Fertilization and Embryology Bill that would allow the treatment to be used simply by seeking the approval of the Human Fertilization and Embryology Authority (HFEA). The amendment was approved by the Parliament in February 2015 making UK the first country to approve laws that allow creation of babies from three persons (Gallagher, 2015).

Reference

Gallagher, J. (2015, February 24). (Health Editor) UK approves three-person babies. Retrieved from http://www.bbc.com/news/health-31594856.

cloning is **technically speaking not "unnatural"** as it happens all the time in the development of identical twins.

2.3.1.2 Commodification of life

Many fear that cloning of human beings may be done for the **sole purpose of providing organs and tissue** for transplantation. To create another person as a means to an end rather than as an end in themselves is **morally unacceptable**.

2.3.1.3 Legal/Social status of a clone

Apart from a fear of undermining the autonomy of the individual, issues regarding the legal rights and social status of a clone would have to be resolved.

2.3.1.4 Safety/Health and quality of life of the clone

Signs of aging in Dolly such as the reduction in telomeres compared to age-matched sheep have led to fears that clones may perhaps be **susceptible to age-related disorders** such as Parkinson's disease, Alzheimer's disease, diabetes, hypertension, heart disease, and several types of cancers. It is uncertain whether the technology can be applied without compromising human health and welfare. The ethical conduct of human experimentation demands respect for persons, beneficence, and justice. It is not clear if these basic tenets can be met in cloning experiments.

2.3.1.5 Eugenics

A recurrent topic in several science fiction movies and books (such as Aldous Huxley's *Brave New World*), many are afraid that cloning **could be used to change the nature and diversity of human populations**.

KEY TAKEAWAYS

Purpose of cloning humans:
- Treatment of diseases caused by mitochondrial disorders
- Studying the mechanism and control of cellular differentiation

Others:
- To produce biologically related children
- To avoid genetic disease
- To get ideal tissues or organs for transplantation
- To replicate a loved one
- To reproduce special individuals with talent, beauty, or intelligence

Ethical issues in cloning humans:
- "Playing God"
- Commodification of life
- Legal/Social status of a clone
- Safety/Health and quality of life of the clone
- Eugenics

2.3.2 LAWS AND PUBLIC POLICY ON REPRODUCTIVE CLONING IN HUMANS

In many countries, the policy toward cloning-to-produce-children (reproductive cloning) has been different from that concerning therapeutic cloning used to isolate stem cells. Although most find the former repugnant and many countries have called for a moratorium on such studies, the promise of radical, new therapies using stem cells has resulted in several countries advocating lighter regulations on therapeutic cloning. The United Nations held debates between 2001 and 2005 over a proposed international convention against cloning of human beings (Walters, 2004).

2.3.2.1 The United Nations Declaration on Human Cloning

The **proposal for an international ban on human cloning** was mooted by France and Germany in August 2001 to prevent those looking to clone humans from engaging in venue shopping, looking for nations that had not as yet banned reproductive cloning. The Vatican observer in the United Nation's (UN) Legal Committee wanted that the convention should include a ban on research cloning (therapeutic cloning) too, a stand that was supported by the United States and Costa Rica. Their argument was that early human embryos are persons too and as such have the right not to be harmed. Moreover, the research was unethical on various counts: **exploitation of women** as oocyte donors, possibly recruited from the poorest peoples, encouraging an international commercial market for oocytes and cloned embryos; the **research would divert funds** from research into other sources of stem cells and other perhaps more pressing problems. After several rounds of discussions in 2002 and 2003, a 2-year deferral of further debate was mooted by the members of the Organization of the Islamic Conference. Supporting this, 80 to 79 were the countries with the most liberal policies on stem cell research including Belgium, Brazil, China, France, Germany, Singapore, Switzerland, and United Kingdom (UK). Countries like UK, Belgium, China, Singapore, and Sweden already permitted/supported research cloning, hence did not want the practice to be banned. Following several months of lobbying among the different factions, in **March 2005**, the **General Assembly of the United Nations** voted for a **nonbinding declaration calling for a complete ban (including both reproductive and therapeutic) cloning in human beings**, with 46% for and 18% against the ban (United Nations, 2005). This **UN** *Declaration on Human Cloning* (see Box 2.4) under UN rules is now the official position of that body.

2.3.2.2 UNESCO Universal Declaration on the Human Genome and Human Rights

The United Nations Educational, Social and Cultural Organization (UNESCO) in 1993 established the Bioethics Programme led by the **International Bioethics Committee (IBC).** This committee has the Intergovernmental Bioethics Committee, consisting of representatives from 36 member states, as well as another 36 outside experts. One of the **nonbinding international agreements** that this program has sponsored is the *Universal Declaration on the Human Genome and Human Rights* adopted unanimously by the UNESCO General Conference in 1997 and ratified by the UN General Assembly in 1998. Guidelines for implementation of the Declaration were approved by UNESCO in 1999. This declaration calls for member states to undertake specific actions, such as the prohibition of *practices which are contrary to human dignity, such as reproductive cloning of human beings*. It also calls on the IBC to study *practices that could be contrary to human dignity, such as germline interventions* (UNESCO, 1997).

BOX 2.4 UNITED NATIONS DECLARATION ON HUMAN CLONING

The UN Declaration on Human Cloning is annexed to the Universal Declaration on the Human Genome and Human Rights. It sought to emphasize the Article 11 of the Universal Declaration on the Human Genome and Human Rights adopted by UNESCO on 9 December 1998 which states that *"practices that are contrary to human dignity such as reproductive cloning of human beings, shall not be permitted."* The declaration called for countries *"to prohibit all forms of human cloning inasmuch as they are incompatible with human dignity and the protection of human life."* It was adopted in the 59th General Assembly held in March 2005. The Assembly adopted the text by a vote of 84 in favor (including the United States and Germany) to 34 against (including United Kingdom, South Korea, and India), with 37 abstentions.

Reproduced below is the declaration:

1. *"Member States are called upon to adopt all measures necessary to protect adequately human life in the application of life sciences;*
2. *Member States are called upon to prohibit all forms of human cloning inasmuch as they are incompatible with human dignity and the protection of human life;*
3. *Member States are further called upon to adopt the measures necessary to prohibit the application of genetic engineering techniques that may be contrary to human dignity;*
4. *Member States are called upon to take measures to prevent the exploitation of women in the application of life sciences;*
5. *Member States are also called upon to adopt and implement without delay national legislation to bring into effect paragraphs (1) to (4);*
6. *Member States are further called upon, in their financing of medical research, including of life sciences, to take into account the pressing global issues such as HIV/AIDS, tuberculosis and malaria, which affect in particular the developing countries."*

The declaration is a powerful but nonbinding instrument that encouraged, but did not require, countries to pass laws conforming to its position. It has been criticized for being ambiguous enough to satisfy both the proponents and the opponents of human cloning. Consequently, the decision to adopt it remained highly controversial.

References

United Nations. (2005). General Assembly adopts United Nations Declaration on Human Cloning by vote of 84-34-37 [Press Release]. Retrieved from http://www.un.org/press/en/2005/ga10333.doc.htm.
United Nations General Assembly. (2005, March 23). Resolution adopted by the General Assembly: 59/280. United Nations Declaration on Human Cloning. Retrieved from http://www.nrlc.org/uploads/international/UN-GADeclarationHumanCloning.pdf.

2.3.2.3 Council of Europe Convention on Human Rights and Biomedicine, 1998

The Council of Europe that is an organization of 47 European and Central Asian nations has a **Bioethics Division** within its Legal Affairs field, guided by a Steering Committee on Bioethics (CDBI). The Council's *Convention on Human Rights and Biomedicine* (drafted in 1997 which came into effect in 1999) in *Article 18* specifically prohibits the creation of human embryos for research purposes (Council of Europe, 1997, CETS No. 164). It is the *first legally binding international text designed to preserve human dignity, rights, and freedoms through a series of principles and prohibitions against the misuse of biological and medical advances.* As an amendment to the Convention, the *Additional Protocol on the Prohibition of Cloning Human Beings with regard to the Application of Biology and Medicine, on the Prohibition of Cloning Human Beings* was introduced in 1998 and came into force on March 1, 2001. *"Any intervention seeking to create a human being genetically identical to another human being, whether living or dead"* is explicitly prohibited in *Article 1* of the Protocol (Council of Europe, 1998).

2.3.2.4 European Union Charter of Fundamental Rights, 2000

Signed on December 7, 2000, by the Presidents of the European Parliament, the Council, and the Commission, the *European Union Charter of Fundamental Rights* contains the fundamental rights and freedoms guaranteed by the European Convention for the Protection of Human Rights and Fundamental Freedoms and published by the Official Journal of the European communities December 18, 2000. The Charter is derived from the constitutional traditions common to the member states as general principles of Community Law. **The prohibition of the reproductive cloning of human beings is included in *Article 3* of the Charter** (European Union, 2000).

2.3.2.5 Legislation at the National Level

Only about **30 countries** have explicitly or implicitly **banned reproductive cloning of human beings** (UNESCO, 2004), although there is **no country that has *permitted* reproductive cloning.** There is therefore an *international consensus against reproductive cloning* in its current form, although countries differ in the regulation of cloning for therapeutic purposes. Three approaches are seen in the legislation prohibiting reproductive cloning:

1. Prohibit creation of a clone embryo.
2. Prohibit the implantation of a clone embryo into a womb.
3. Prohibit any attempt to artificially create a human being genetically identical to another human being, embryo, or fetus, dead, or alive.

Discussed here is the regulatory oversight in terms of national legislation, official guidelines and opinions in select countries, illustrative of the national treatment of human cloning (UNESCO, 2004).

2.3.2.5.1 United Kingdom—Human Reproductive Cloning Act 2001

In the UK, oversight for human biotechnologies is provided by the Human Fertilization and Embryology Authority (HFEA), established in 1991. Under the *HFE Act, 1990,* it monitors and licenses all research involving embryos including in vitro fertilization, storage of eggs, sperm, and embryos. The ***Human Reproductive Cloning Act 2001*** was passed to **explicitly prohibit reproductive cloning.** In discussions involving the Human Genetics Advisory Commission and public individuals and organizations subsequent to the birth of Dolly in 1997, Dame Ruth Deech, chairperson of the HFEA, had suggested allowable reasons (such as developing methods for treatment of mitochondrial diseases and developing methods of therapy for diseased or damaged tissues or organs) why embryo research may be warranted. In 2001, the HFE Act 1990 was **amended** to include the *HFE (Research purposes) Regulation 2001,* thus **allowing therapeutic cloning.** The first license under this regulation was granted to the researchers at the University of Newcastle in order to allow them to investigated treatments to Alzheimer's disease, Parkinson's disease, and diabetes. The 2001 Cloning Act has since been replaced by the *HFE Act 2008* which also allows experiments on hybrid human–animal embryos for transplantation research.

2.3.2.5.2 United States

There is currently **no federal law that bans human cloning completely**. Clinical research using cloning technology to clone a human being is subject to Food and Drug Authority (FDA) regulation under the *Public Health Service Act* and the *Federal Food, Drug, and Cosmetic Act* (Zoon, 2001).

The United States House of Representatives in 1998, 2001 (Human Cloning Prohibition Act of 2001-HR 2505), 2003 (Human Cloning Prohibition Act of 2003-HR 234), and 2007 (Human Cloning Prohibition Act of 2007-HR 2560) voted on the issue of a ban on human cloning but could not be passed due to division in the Senate.

Before starting clinical research using cloning technology, the sponsor must submit an investigational new drug application to FDA's Centre for Biologics Evaluation and Research. As per Title 21 of the Code of Federal Regulations, Part 312, the application must include the research plan, the authorization from a properly constituted institutional review board, a commitment from the investigators, and informed consent from all human subjects of the research.

The American Association for the Advancement of Science (AAAS) has engaged the public and various professional communities in discussions on various aspects of human cloning since 1997. The latest **statement on human cloning issued in 2013 endorses a legal ban on reproductive cloning due to the serious health risks associated with it in animals**, but **supports Stem Cell research including therapeutic cloning** given the immense potential that the technology holds in therapy (AAAS, 2013). It however **advocates federal oversight for both private and public sectors**. In 1997, a report addressing the scientific, religious, legal and ethical considerations on cloning human beings, and including several commissioned papers by various experts was prepared by the National Bioethics Advisory Committee at the behest of the President of the United States, Bill Clinton. This Committee was however dissolved in October 2000 and replaced by **The President's Council on Bioethics** by President George W. Bush. A comprehensive document on the ethical aspects of cloning *Human Cloning and Human Dignity: An Ethical Inquiry* was published by the Council in July 2002 (The President's Council on Bioethics, 2002).

The Bioethics Defense Fund found in a survey of 50 states (Nikas and Bordlee, 2011) that eight states (Arizona, Arkansas, Indiana, Michigan, North Dakota, Oklahoma, South Dakota, and Virginia) prohibit cloning for any purpose. Nine states (California, Connecticut, Illinois, Iowa, Maryland, Massachusetts, Missouri, Montana, and New Jersey) expressly allow "clone and kill" research, which means that the law prohibits implantation of a cloned embryo (reproductive cloning) but permits destruction of a cloned embryo for research. Arizona, Indiana, Louisiana, and Michigan expressly prohibit state funding of human cloning for any purpose, whereas in five others (California, Illinois, Missouri, Maryland, and New York) state fund may be used for funding embryonic stem cell research using surplus embryos from cloning and in vitro fertilization clinics.

2.3.2.5.3 Canada

In March 2004, the Act *Respecting Assisted Human Reproduction and Related Research (Assisted Human Reproduction Act)* was adopted. **Article 5 of the Act prohibits the creation of a human clone or transplant into human,** nonhuman life form or artificial device. The Act also prohibits creation of an in vitro embryo for purposes other than assisted reproduction procedures, and creation of human being from an embryo derived from a cell or part of a cell taken from an embryo or fetus. The legislation does however **permit embryonic stem cell research with supernumerary embryos with license**.

2.3.2.5.4 Australia

Cloning of human beings, as in *producing from one original*, duplicate(s) or descendant(s) that are *genetically identical to the original* is **prohibited by the *Gene Technology Act 2000***, which took effect in 2001. Regulations set out in this Act have also been incorporated into state laws in

Victoria, Western Australia, and South Australia. The *Prohibition of Human Cloning Act 2002* and the *Research Involving Human Embryos Act 2002* were passed in 2002 and applies to the creation of cloned embryos for reproduction, for implanting a human embryo clone in the body of a human or an animal, for export, or for import.

2.3.2.5.5 India

In India, there is **no specific law that bans cloning**. The Indian Council of Medical Research in 2000 issued a *Consultative Document on Ethical Guidelines for Biomedical Research on Human Subjects*. This document states that as the *safety, success, utility, and ethical acceptability* of cloning has not yet been established, research on cloning (through nuclear transplantation or embryo splitting), *with intent to produce an identical human being, as of today, is prohibited*. Cloning for research may still be considered on a case-by-case basis by the National Bioethics Committee.

KEY TAKEAWAYS

Laws and Public Policy on Reproductive Cloning in Humans:

United Nations (2005)	Declaration on Human Cloning (annexed to and elaborates on Article 11 of the UNESCO Universal Declaration on the Human Genome and Human Rights)—prohibits all forms of human cloning inasmuch as they are incompatible with human dignity and protection of human life
UNESCO (1997)	Universal Declaration on the Human Genome and Human Rights—prohibits practices contrary to human dignity such as reproductive cloning of human beings
Council of Europe (1998)	Convention of Human Rights and Biomedicine—Article 18 prohibits creation of human embryos for research purposes. Additional Protocol on the Prohibition of Cloning of Human Beings—Article1 prohibits creation of a human being genetically identical to another human being living or dead
European Union (2000)	Charter of Fundamental Rights—Article 3 prohibits reproductive cloning of human beings

National Laws:

UK 2001	Human Reproductive Cloning Act—prohibits reproductive cloning. In 2001, the Human Fertilization and Embryology Act 1990 modified to include Human Fertilization and Embryology (Research purposes) Regulation 2001, allowing therapeutic cloning. The HFEA 2008 allows hybrid human—animal embryos for transplantation
US	Currently no federal law that bans human cloning. Clinical research using human cloning regulated under the Public Health Service Act and the Federal Food, Drug and Cosmetic Act of the FDA
Canada 2004	Act Respecting Assisted Human Reproduction and Related Research (Assisted Human Reproduction Act)—Article 5 prohibits creation of human clone and implantation. Embryonic stem cell research permitted using supernumerary embryos
Australia 2000	Gene Technology Act prohibits human cloning. Also, Human Cloning Act 2002 and the Research Involving Human Embryos Act 2002
India	Currently no specific law bans human cloning. Ethical Guidelines for Biomedical Research on Human Subjects 2000 of ICMR cloning with intent to produce an identical human being is prohibited

2.4 SUMMARY

Cloning refers to the process of establishing a population of genetically identical individuals. The process is facilitated by new techniques developed through an understanding of genetics, cell biology, and in vitro culture of animal cells and tissues. A nucleus from a cell of a donor individual is extracted, transferred into an enucleated egg cell, and reprogramed to give rise to an embryo, which is then implanted in a surrogate mother to give an individual genetically identical to the donor. The technique has the potential to generate copies of elite animals to be used in livestock breeding. It can also be used to establish herds of transgenic animals for production of pharmaceuticals and specialty chemicals which can be isolated and purified from blood or milk. Cloning has been critiqued for the commodification of animals and health problems often observed in the clones. Although human cloning is technically possible, ethical considerations have prompted nations to prohibit the creation of cloned human embryos. An exception is the recent amendment of the Human Fertilization and Embryology Regulation in UK that allows the creation of babies with three parents in cases in which the mother has a disease caused by defective mitochondria.

REFERENCES

Technique

Altonn, H. (Reporter). (1998, July 22). Cloning breakthrough at UH—The scientists cloned more than 50 mice in three generations using the 'Honolulu technique'. *Honolulu Star-Bulletin*. Retrieved from http://archives.starbulletin.com/98/07/22/news/story1.html.

Briggs, R., & King, T. J. (1952). Transplantation of living nuclei from blástula cells into enucleated frogs' eggs. *Proceedings of the National Academy of Sciences of the United States of America, 38,* 455−463. Retrieved from http://www.pnas.org/content/38/5/455.long.

Campbell, K. H., McWhir, J., Ritchie, W. A., & Wilmut, I. (1994). Sheep cloned by nuclear transfer from a cultured cell line. *Nature, 380,* 64−66.

Schnieke, A. E., Kind, A. J., Ritchie, W. A., Mycock, K., Scott, A. R., Ritchie, M., ... Campbell, K. H. S. (1997). Human factor IX transgenic sheep produced by transfer of nuclei from transfected fetal fibroblasts. *Science, 278,* 2130−2133.

Willadsen, S. M. (1986). Nuclear transplantation in sheep embryos. *Nature, 320,* 63−65.

Wilmut, I., Schnieke, A. E., McWhir, J., Kind, A. J., & Campbell, K. H. S. (1997). Viable offspring derived from fetal and adult mammalian cells. *Nature, 385,* 810−813.

Applications of cloning animals

FDA. (2006, December 28). Use of edible products from animal clones or their progeny for human food or animal feed. Draft Guidance for Industry # 179. Retrieved from http://www.fda.gov/downloads/AnimalVeterinary/GuidanceComplianceEnforcement/GuidanceforIndustry/UCM052471.pdf.

Horse and Hound. (2003a, May 30). *First equine clone created.* Retrieved from http://www.horseandhound.co.uk/news/first-equine-clone-created-34061.

Horse and Hound. (2003b, August 7). *First cloned foal born in Italy.* Retrieved from http://www.horseandhound.co.uk/news/first-cloned-foal-born-in-italy-40987.

Minkel, J.R. (Reporter). (2007, August 2). Korean cloned human cells were product of "Virgin Birth"—fraudulent cloned cells were likely the first example of a human egg turned directly into stem cells [Scientific American]. Retrieved from http://www.scientificamerican.com/article/korean-cloned-human-cells/.

Rudenko, L., Matheson, J. C., & Sundlof, S. F. (2007). Animal cloning and the FDA—the risk assessment paradigm under public scrutiny. *Nature Biotechnology*, *25*, 39—43. http://dx.doi.org/10.1038/nbt0107-39 Retrieved from http://www.nature.com/nbt/journal/v25/n1/full/nbt0107-39.html.

Samdani, T. (2014). Xenotransplantation. *Medscape*. Retrieved from http://emedicine.medscape.com/article/432418-overview#a1.

Sample, I. (2012, October 1). (Science correspondent) GM cow designed to produce milk without an allergy causing protein [The Guardian]. Retrieved from http://www.psfk.com/2012/10/lactose-intolerant-drinkable-milk-cows.html.

Ethical issues in cloning animals

Brooks, K. R., & Lusk, J. L. (2011). U.S. consumers attitude toward farm animal cloning. *Appetite*, *57*, 483—492. Retrieved from http://www.westlaboratory.org/wp-content/uploads/2011/05/Brooks-2011-US-consumers-attitudes-toward-farm-animal-clioning.pdf.

Humane Society of the United States. (2003). *HSUS asks FDA to ban sale from cloned farm animals*. Retrieved from http://www.hsus.org/ace/15431.

Sullivan, B. (Technology correspondent). (2004). *Religions reveal little consensus on cloning*. MSNBC [Online article] Retrieved from http://www.nbcnews.com/id/3076930/#.VuZSmtDVDic.

Tucker-Foreman, C. (2002). Public interest perspective on animal cloning. In: Animal cloning and the production of food products: Perspectives from the food chain. Proceedings from a workshop sponsored by the Pew Initiative on Food and Biotechnology and the Center for Veterinary Medicine of the U.S. Food and Drug Administration.

Laws and public policy on reproductive cloning in animals

Center for Veterinary Medicine, U.S. Food and Drug Administration. (2008). *Animal Cloning: A risk assessment*. Retrieved from http://www.fda.gov/downloads/AnimalVeterinary/SafetyHealth/AnimalCloning/ucm124756.pdf.

EUR-Lex-52013PC0892. (2013). *Proposal for a directive of the European Parliament and of the Council on the cloning of animals of the bovine, porcine, ovine, caprine and equine species kept and reproduced for farming purposes /*COM/2013/0892 final-2013/0433(COD)*/*Retrieved from eur-lex.europa.eu/legal-content/EN/TXT/?uri=CELEX:52013PC0892.

EUR-Lex-52013PC0893. (2013). *Proposal for a Council Directive on the placing on the market of food from animal clones*. Retrieved from eur-lex.europa.eu/legal-content/EN/TXT/?uri=CELEX:52013PC0893.

EUR-Lex-31997R0258. (1997). Regulation (EC)No 258/97 of the European Parliament and the Council of 27 January 1997 concerning novel foods and novel food ingredients. Official Journal L 043, 14/02/1997 P.0001-0006. Retrieved from eur-lex.europa.eu/legal-content/EN/TXT/?uri=CELEX:31997R0258.

European Commission. (2013). *Commission staff working document—Impact Assessment*. Retrieved from http://ec.europa.eu/food/animals/docs/aw_other_aspects_cloning_impact_assessment_report_en.pdf.

European Commission Press Release Database. (2013, December 18). *Food: Commission tables proposals on animal cloning and novel food*. Retrieved from europa.eu/rapid/press-release_IP-13-1269_en.htm.

European Food Safety Authority. (2008). *Food safety, animal health and welfare and environmental impact of animals [1] derived* from cloning by Somatic Cell Nucleus Transfer (SCNT) and their offspring and products obtained from those animals [2]. doi: 10.2903/j.efsa.2008.767. Retrieved from http://www.efsa.europa.eu/en/efsajournal/pub/767.

European Food Safety Authority. (2012, July 5). *Animal cloning: EFSA reiterates safety of derived food products but underscores animal health & welfare issues*. Retrieved from http://www.efsa.europa.eu/en/press/news/120705.

European Parliament News. (2015, September 8). *EP wants animal cloning ban extended to offspring and imports*. Retrieved from http://www.europarl.europa.eu/news/en/news-room/20150903IPR91517/EP-wants-animal-cloning-ban-extended-to-offspring-and-imports.

FDA. (2008). Guidance for industry: Use of animal clones and clone progeny for human food and animal feed. Retrieved from http://www.fda.gov/downloads/animalveterinary/guidancecomplianceenforcement/guidanceforindustry/ucm052469.pdf.

Global Agricultural Information Network Report. (2014a). *Argentina Biotechnology Annual Report/ United States Department of Agriculture, Foreign Agriculture Service, August 2014, 27p. See Chapter 2 Pg 22–27*. Retrieved from gain.fas.usda.gov/Recent GAIN Publications/Agricultural Biotechnology Annual_Buenos Aires_Argentina_8-15-2014.pdf.

Global Agricultural Information Network Report. (2014b). *Australia Biotechnology Annual Report/ United States Department of Agriculture, Foreign Agriculture Service, September 2014, 27p. See Chapter 2 Pg 24–27*. Retrieved from gain.fas.usda.gov/Recent GAIN Publications/Agricultural Biotechnology Annual_Canberra_Australia_9-130-2014.pdf.

National Academy of Sciences, & National Research Council. (2002). *Animal biotechnology: Science-based concerns*. Washington D.C: National Academies Press.

Rudenko, L. (2011). *Animal biotechnology in the United States: the regulation of animal clones and genetically engineered animals. Challenges for agricultural research* (pp. 243–254). OECD Publishing. http://dx.doi.org/10.1787/9789264090101-21-en.

The European Group on Ethics in Science and New Technologies to the European Commission. (2008). *Ethical aspects of animal cloning for food supply—Opinion no 23*. Retrieved from http://stopogm.net/sites/stopogm.net/files/ethicalcloning.pdf.

Human cloning

Associated Press. (2002, December 30). Clonaid baby: A clone or a fake? [Online article] Retrieved from http://archive.wired.com/medtech/health/news/2002/12/57001.

Hwang, W. S., Roh, S. I., Lee, B. C., Kang, S. K., Kwon, D. K., Kim, S., et al. (2005). Patient-specific embryonic stem cells derived from human SCNT blastocysts. *Science, 308,* 1777–1783.

Stojkovic, M., Stojkovic, P., Leary, C., Hall, V. J., Armstrong, L., Herbert, M., ... Murdoch, A. (2005). Derivation of a human blastocyst after heterologous nuclear transfer to donated oocytes. *Reproductive BioMedicine Online, 11*(2), 226–231.

The President's Council on Bioethics. (2002). *Human cloning and human dignity: An ethical inquiry*. Retrieved from https://bioethicsarchive.georgetown.edu/pcbe/reports/cloningreport/index.html.

The Human Genetics Advisory Commission & The Human Fertilization and Embryology Authority. (1998). *Cloning issues in reproduction, science and medicine*. [Report] Retrieved from http://www.hfea.gov.uk/docs/Cloning_Issue_Report.pdf.

Laws and public policy on reproductive cloning in humans

AAAS. (2013). *American Association for the Advancement of Science Statement on Human Cloning*. Retrieved from http://www.aaas.org/page/american-association-advancement-science-statement-human-cloning.

Nikas, N.T. & Bordlee, D.C. (2011). Human cloning laws: 50 state survey. Bioethics Defense Fund Retrieved from http://bdfund.org/wordpress/wp-content/uploads/2012/07/CLONINGChart-BDF2011.docx.pdf.

Council of Europe. (1997). *Convention for the protection of Human Rights and Dignity of the Human Being with regard to the Application of Biology and Medicine: Convention on Human Rights and Biomedicine, CETS No 164*. Retrieved from http://www.coe.int/en/web/conventions/full-list/-/conventions/treaty/164.

Council of Europe. (1998). *Additional Protocol to the Convention for the Protection of Human Rights and Dignity of the Human Being with regard to the Application of Biology and Medicine, on the Prohibition of Cloning Human Beings. European Treaty Series No.168*. Retrieved from https://rm.coe.int/CoERMPublicCommonSearchServices/DisplayDCTMContent?documentId=090000168007f2ca.

European Union. (2000, December 18). *Charter of Fundamental Rights of the European Union* (2000/C 364/ 01). Retrieved from http://www.europarl.europa.eu/charter/pdf/text_en.pdf.

UNESCO. (1997). *Universal Declaration on the Human Genome and Human Rights.* Retrieved from http:// www.unesco.org/new/en/social-and-human-sciences/themes/bioethics/human-genome-and-human-rights/.

UNESCO. (2004). *National Legislation Concerning Human Reproductive and Therapeutic Cloning.* Retrieved from http://unesdoc.unesco.org/images/0013/001342/134277e.pdf.

United Nations. (2005). *General Assembly adopts United Nations Declaration on Human Cloning by vote of 84-34-37* [Press Release]. Retrieved from http://www.un.org/press/en/2005/ga10333.doc.htm.

Walters, L. (2004). The United Nations and human cloning: A debate on hold. *Hastings Centre Report,* 5–6. Retrieved from www.thehastingscenter.org/pdf/publications/hcr_jan_feb_2004_in_brief.pdf.

Zoon, K.C. (2001, March 28). *Letter to Associations* [Web article] Retrieved from http://www.fda.gov/ biologicsbloodvaccines/safetyavailability/ucm105853.htm.

FURTHER READING

Reviews and Books

Huxley, A. (1932). *Brave new world.* Harper & Row. A primer on Stem cells: http://stemcells.nih.gov/info/ basics/pages/basics10.aspx.

Human cloning and human dignity: An ethical inquiry. (2002). The President's Council on Bioethics. https:// bioethicsarchive.georgetown.edu/pcbe/reports/cloningreport/children.html.

Cloning issues in reproduction, science and medicine. (1998). A report from The Human Genetics Advisory Commission and The Human Fertilization and Embryology Authority. http://www.hfea.gov.uk/docs/ Cloning_Issue_Report.pdf.

The European Group on Ethics in Science and New Technologies to the European Commission (2008). Opinion No. 23 Ethical aspects of animal cloning for food supply. http://stopogm.net/sites/stopogm.net/ files/ethicalcloning.pdf.

Web Resources

De-extinction

National Geographic: http://www.nationalgeographic.com/deextinction/.

TED talks: http://tedxdeextinction.org/.

Human Cloning

Human Cloning: a need for a comprehensive ban: https://cbhd.org/cloning/position-statement

STEM CELL RESEARCH

3

Stem cell research is the key to developing cures for degenerative conditions like Parkinson's and motor neuron disease from which I and many others suffer. The fact that the cells may come from embryos is not an objection, because the embryos are going to die anyway.
-Stephen Hawking, English theoretical physicist, cosmologist, author

CHAPTER OUTLINE

3.1 Introduction 56
3.2 Sources of Stem Cells 57
    3.2.1 Embryonic Stem Cells... 58
    3.2.2 Nuclear Transfer—Embryonic Stem Cells.. 58
    3.2.3 Fetal Stem Cells.. 59
    3.2.4 Cord Blood Stem Cells or Neonatal Stem Cells.. 59
    3.2.5 Adult Stem Cells 59
    3.2.6 Induced Pluripotent Stem Cells 60
    3.2.7 Stimulus-Triggered Acquisition of Pluripotency (STAP) Cells 60
3.3 Benefit to Society 60
    3.3.1 Understanding Cellular Differentiation... 64
    3.3.2 Study Disease Progression... 65
    3.3.3 Regenerative Medicine.. 65
    3.3.4 Tissue Engineering... 66
    3.3.5 Grow Organs for Transplantation 66
    3.3.6 Alter Current Biomedical Practices for Treatment of Cancer......................... 66
    3.3.7 Identify Drug Targets and Test Potential Therapeutics 67
    3.3.8 Toxicity Testing.. 67
3.4 Ethical Issues in Stem Cell Research 68
    3.4.1 Moral Status of Embryos—Fetalistic Viewpoint.. 68
    3.4.2 Exploitation of Women—Feministic Viewpoint 70
3.5 Ethical Issues in Stem Cell Translation... 71
    3.5.1 Clinical Trials Using Stem Cells 71
    3.5.2 Guidelines for the Clinical Translation of Stem Cells................................... 71
3.6 Stem Cell Research Policy... 73
3.7 The Role of Politics and Public Opinion in Shaping Stem Cell Policy........................... 76
3.8 Summary 77
References 78
Further Reading 80

An Introduction to Ethical, Safety and Intellectual Property Rights Issues in Biotechnology.
DOI: http://dx.doi.org/10.1016/B978-0-12-809231-6.00003-X

Shinya Yamanaka, Director of the Center for iPS Cell Research and Application, Kyoto University, accepted a plaque of appreciation from **Francis Collins**, Director of the National Institutes of Health, and **Michael Gottesman**, Deputy Director for Intramural Research of the National Institutes of Health, at the National Institutes of Health in Montgomery County, Maryland State on January 14, 2010. (*By National Institutes of Health [Public domain], via Wikimedia Commons*)

3.1 INTRODUCTION

Stem cells are cells capable of mitotic cell division with the unique ability to differentiate into other cell types. They are found in virtually all multicellular organisms. A stem cell has **three characteristics**: (1) it can **divide indefinitely**; it is the only cell other than a cancer cell that is immortal, (2) it is **capable of self-renewal,** that means it can make identical copies of itself forever, and (3) it is **capable of changing or differentiating into all the different cell types** that make up the body.

Recent advances in stem cell research have led to a better understanding of the biochemical processes that drive cellular differentiation. In turn, this has led to an improvement in our understanding of a wide variety of diseases such as cancer, diabetes, and Alzheimer's and could potentially open up new avenues of disease control and therapy. However, stem cell research has been hounded by **ethical and moral objections**, mainly revolving around the **source of the cells**, so much so that in several countries, use of government funds for stem cell research [derivation of human embryonic stem (ES) cells] is banned. This chapter examines the potential of stem cell research to revolutionize medicine by way of radical new approaches to disease therapies, and attempts to put into perspective the religious, moral, and ethical objections to stem cell research.

3.2 SOURCES OF STEM CELLS

Based on origin, two types of mammalian stem cells exist, the ***embryonic stem cells*** derived from the blastocyst formed around 5–7 days after fertilization, and the ***adult stem cells*** that are found in adult tissues. Although the ES cells are more versatile as they are able to differentiate into several kinds of tissues (***pluripotent***) and consequently have more clinical applications, having to sacrifice preimplantation stage embryos for their derivation raises an ethical dilemma. Adult stem cells are found in several organs such as the brain, bone marrow, skeletal muscles and are responsible for replenishing specialized cells and maintaining normal turnover of cells in adult tissues. These cells have reduced differentiation potential and are capable of giving rise to only one type of cells (***unipotent***), or few related types of cells (***multipotent***) specific to the organ they are present in (see Fig. 3.1).

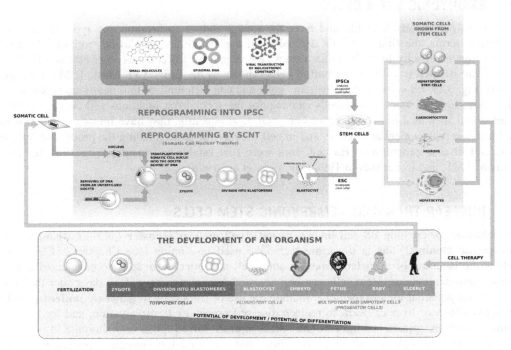

FIGURE 3.1 Differentiation potential of cells

(By Kaidor—Own work based on File: The development and the ways to rejuvenate cells—ru.svg, CC BY-SA 3.0, https://commons.wikimedia.org/w/index.php?curid=29004582.)

KEY TAKEAWAYS

Differentiation is the process of acquisition of the specialized functions of a cell.
Cell potency is the ability of a cell to differentiate into other cell types:

- Totipotency—ability of a single cell to divide and produce all the differentiated cells in an organism
- Pluripotency—ability of a cell to differentiate into cells of any of the three germ layers: endoderm (gastrointestinal tract and lungs), mesoderm (muscle, bone, blood, or urogenital tract), or ectoderm (epidermal tissues and nervous system)
- Multipotency—ability of the cells to differentiate into multiple, but limited cell types (e.g., a hematopoietic cell can differentiate into other blood cells such as lymphocytes, monocytes, or neutrophils)
- Unipotency—ability of a cell to divide and give rise to more cells of the same type

3.2.1 EMBRYONIC STEM CELLS

ES cells are particularly unstable and have the unique ability to become any type of adult cell. In technical terms, the first few cells derived from the zygote are said to be "totipotent" as they have the innate capacity to differentiate into a new organism. ES cells are derived from totipotent cells and are isolated from the inner ball of cells (known as the **Inner Cell Mass**, ICM) in a blastocyst (see Fig. 3.1). ES cells are "pluripotent" cells and can differentiate into nearly all types of cells in the body. In other words, these cells can acquire the specialized functions and characteristics of any adult cell such as a muscle, liver, or bone cell. Consequently, they have **immense use in medicine** for tissue replacement and repair, and in **cell biology** for understanding the process of differentiation and disease progression. ES cell lines were first established in mouse in 1981 (Evans & Kaufman, 1981) and in humans in 1998 (Thomson et al., 1998).

3.2.2 NUCLEAR TRANSFER—EMBRYONIC STEM CELLS

One technique for creating ES cell lines is **somatic cell nuclear transfer** (SCNT). Also known as "**therapeutic cloning**," this is the same technique used in cloning (see Chapter 2: Cloning). It involves the isolation of a nucleus from an adult cell and injecting it into an enucleated egg cell using a micromanipulator. Stem cells can be harvested from the embryo, 5−7 days after the procedure. Several potential uses in medicine include the use of stem cells to **generate patient-matched new organs** such as liver or heart, to replace diseased organs by transplantation, and for treatment of injuries such as spinal cord injuries and sports injuries. Given the severe shortage in organs for transplantation, the high rate of tissue rejection, and the need for immune suppressant drugs in transplant patients, patient-specific therapeutic cloning could **potentially simplify organ transplantation**. This is because the organ grown from stem cells derived from a particular patient would have cells **genetically identical** to the patient, thereby obviating the need for immune-suppressants, and **preventing tissue rejection**. The technique could possibly alleviate the suffering of millions of patients across the world.

3.2.3 **FETAL STEM CELLS**

Fetal stem cells were first isolated and cultured by John Gearhart and his team at the Johns Hopkins University School of Medicine in 1998 (Shamblott et al., 1998). These cells known as *primordial germ cells* are the precursors of eggs and sperms and were isolated from the gonadal ridges and mesenteries of 5–9-week fetuses obtained by therapeutic abortion. Embryonic germ (EG) cells isolated from them are found to be pluripotent. Problems associated with isolation of these stem cells are (1) they can be obtained **only from 8- to 9-week-old fetuses** and (2) EG cells have **limited proliferation** capacity.

3.2.4 **CORD BLOOD STEM CELLS OR NEONATAL STEM CELLS**

Hematopoietic stem cells necessary for the formation of red blood cells, white blood cells, and platelets, are present in the bone marrow. **Bone marrow transplantation** has been used in clinical practice for the treatment of several cancers such as leukemia, lymphomas, and certain blood and immune disorders such as thalassemia and anemia. Hematopoietic stem cells can also be **harvested and cultured from umbilical cord blood** (UCB) and used as an alternative for bone marrow transplantation. Unlike ES, UCB stem cells do not raise any ethical issues as they are harvested from umbilical cord discards after delivery of the baby. UCB cells can be stored in freezers for later use in transplantation. Since the establishment in 1993 of the first UCB bank in New York (New York Blood Center, NYBC, http://nybloodcenter.org/), large-scale **UCB-banking** has been established in several countries, example CryoSave (http://cryo-save.com/en/home). India has established a booming industry in stem cell banking—the first such private stem cell bank is LifeCell (http://www.lifecell.in/), which in association with CRYO-CELL International, USA, has a facility outside Chennai with offices all over India. Others include Jeevan Blood Bank and Research Centre, also in Chennai (http://www.jeevan.org/stem-cell-bank/), and Cryo-Save India Private Limited, Bangalore.

3.2.5 **ADULT STEM CELLS**

Adult stem cells or somatic stem cells are present in several tissues and organs such as the bone marrow, brain, spinal cord, liver, skin, epithelial lining of the digestive tract, cornea, retina, gums, and teeth. The **primary role of these cells is to maintain the tissue/organ** by replenishing dead or damaged cells. Isolating these cells has proved challenging because their numbers are small (as low as **1 in 10,000**), they are **difficult to identify**, and their **locations are difficult to predict**. Adult stem cells have been **used to treat blood disorders and skin repair**, but their use in therapy has been limited due to the cells being only unipotent.

In January 2002, adult stem cells capable **transdifferentiation** (giving rise to tissues other than their tissue of origin) was isolated from human bone marrow by Catherine Verfaillie of the University of Minnesota (Jiang et al., 2002). These **multipotent adult progenitor cells** were capable of prolonged self-renewal and could give rise to muscle, cartilage, bone, liver, and different types of neurons and brain cells. Studies in mice also indicated the plasticity of adult stem cells. In mice, neural stem cells were found to differentiate into skeletal muscle, heart, lung, blood, and skin after transplantation. These reports raised hopes of isolating pluripotent stem cells from adults and circumventing the ethical dilemma associated with the only known reliable source of pluripotent cells, the ES cells.

3.2.6 INDUCED PLURIPOTENT STEM CELLS

These stem cells are **adult cells that have been reprogramed** to behave like ES cells. In addition to being able to divide and self-propagate, induced pluripotent stem cells (iPSCs) are pluripotent as they can be induced to form cells of other types. The creation of iPSCs was first reported in mouse in 2006 (Takahashi & Yamanaka, 2006), and a year later in humans (Takahashi et al., 2007; Yu et al., 2007) (for details of the technique, see Box 3.1). These reports generated a lot of excitement in the scientific world since it opened up the possibility of getting **autologous (patient specific) stem cells**. These could be used in tissue/organ repair, with none of the drawbacks of tissue rejection or the need for immune suppression currently associated with transplantation from heterologous (related) donors. Also, since it did not involve sacrificing embryos, the technology had **fewer ethical issues**.

3.2.7 STIMULUS-TRIGGERED ACQUISITION OF PLURIPOTENCY (STAP) CELLS

A technique for generating pluripotent stem cells from adult cells that was remarkable for its simplicity and did not involve transfer of the "Yamanaka factors" was published in two papers in *Nature* in January 2014 by Haruko Obokata and coworkers of the RIKEN Centre for Developmental Biology, Kobe, Japan, in association with Charles Vacanti of Harvard Medical School, Boston, United States. The technique involved stressing cells in mild acidic conditions (pH 5.7) at 37°C for 25 minutes. The stem cells induced by this method, known as stimulus-triggered acquisition of pluripotency cells, were superior to ESCs since they could produce not just embryonic tissues, but also placental tissues. However, the results could not be replicated in other laboratories and the papers were retracted in June 2014 on the basis of scientific fraud by the lead author (see Box 3.2).

3.3 BENEFIT TO SOCIETY

Stem cells offer radical new therapies and new ways of studying human diseases. By manipulation of the hormonal and cultural conditions, stem cells can be coaxed to develop into a variety of cells, tissues, and even organs. Culturing them under in vitro conditions helps scientists answer basic questions on the genetics, biochemistry, and physiology of disease causation and progression, and allows them to test new drugs. Transplanting stem cells to patients could help repair or replace diseased tissues and organs, alleviating human suffering and improving the quality of life. Summarized in Table 3.1 are some examples of tissue and organ regeneration from stem cells. Many of the applications are currently in the clinical trials stage (see Table 3.2). Also discussed below are some of the applications of stem cells.

BOX 3.1 INDUCED PLURIPOTENT STEM CELLS (iPSC)

The possibility to reprogram mature adult cells into pluripotent stem cells was first demonstrated in 2006 by Shinya Yamanaka's laboratory in Kyoto University, Japan (Takahashi & Yamanaka, 2006). The group identified 24 genes as being important for the functioning of embryonic stem cells (ESCs) which when transferred to mouse fibroblast cells could induce them to behave like embryonic cells. By a process of elimination, they found that there were **four genes that encode transcription factors** which were critical for reprograming the fibroblast cells: *Oct4, Sox2, cMyc*, and *Klf4* (often referred to as the **"Yamanaka factors"**). When transferred to mouse fibroblast cells using retroviral vectors, these factors could reactivate an ES cell specific gene *Fbx15*. Validation of the technique came in 2007 from his own lab as well as from two independent reports from Harvard, MIT, and UCLA in which the factors were used to reprogram mouse fibroblast cells to ESCs as evidenced by the activation of another functionally important marker gene *Nanog*. More significantly, the four factors could be used to transform human fibroblasts into pluripotent stem cells. Although Yamanaka and coworkers at Kyoto University used a retroviral system to deliver the four factors into human fibroblast cells (reported in *Cell* in November 2007, Takahashi et al., 2007), a group led by James Thomson at University of Wisconsin-Madison, used lentiviruses to transform human fibroblasts with *Oct4, Sox2, Nanog*, and another gene *Lin28* (reported in *Science* in November 2007, Yu et al, 2007). Subsequently, there have been several reports from researchers in Hong Kong and China on the induction of iPSCs from epithelial cells, including renal epithelial cells from urinary cell cultures. In 2009, generation of iPSCs without transfer of genes, but by repeated treatment of cells with specific proteins (referred to as "protein-induced pluripotent stem cells" or "piPSCs") was reported by Sheng Ding and coworkers (Zhou et al., 2009).

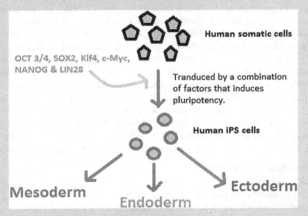

Induction of pluripotent stem cells by gene transfer: (By Humanips - Own work, CC BY-SA 3.0, https://commons.wikimedia.org/w/index.php?curid=22765759)

Problems associated with the technology include:

- *Low success rate*—Less than 1% of transformed cells are stem cells
- *Time taken for generation of stem cells*—The programing takes a few weeks in mouse and even longer in human
- *Safety*—Two of the four genes used for the induction of pluripotency, *cMyc* and *Klf-4*, are **oncogenic**. In Yamanaka's experiments, 20% of the chimeric mice developed cancer. Alternatives to the retroviral vectors used in the process, including adenoviruses, plasmids or direct DNA/protein delivery systems have been investigated to avoid problems of tumorigenicity associated with retroviral vectors. Issues regarding safety currently being addressed by researchers include the potential for genetic and epigenetic abnormalities, tumorigenicity, and immunogenicity of transplanted cells.

For his pioneering work in stem cell research, Shinya Yamanaka was awarded the 2012 Nobel Prize along with Sir John Gordon.

(Continued)

BOX 3.1 (CONTINUED)

References

Takahashi, K., & Yamanaka, S. (2006). Induction of pluripotent stem cells from mouse embryonic and adult fibroblast cultures by defined factors. *Cell*, *126*(4), 663–676. http://dx.doi.org/10.1016/j.cell.2006.07.024. *PMID* 16904174.

Takahashi, K., et al. (2007). Induction of pluripotent stem cells from adult human fibroblasts by defined factors. *Cell*, *131*(5), 861–872. http://dx.doi.org/10.1016/j.cell.2007.11.019. *PMID* 18035408.

Yu, J., Vodyanik, M. A., et al. (2007). Induced pluripotent stem cell lines derived from human somatic cells. *Science*, *318*(5858), 1917–1920. http://dx.doi.org/10.1126/science.1151526. *PMID* 18029452.

Zhou, H., Wu, S., Joo, J. Y., et al. (2009). Generation of induced pluripotent stem cells using recombinant proteins. *Cell Stem Cell*, *4*(5), 381–384. http://dx.doi.org/10.1016/j.stem.2009.04.005. *PMID* 19398399.

BOX 3.2 ETHICAL AND SCIENTIFIC MISCONDUCT IN STEM CELL RESEARCH

Hwang Woo Suk, a scientist at the Seoul National University, South Korea, shot to international fame with the publication of a paper in *Science* in 2004, which reported the successful cloning of 30 human embryos through SCNT and the derivation of a stem cell line (Hwang et al., 2004). A second landmark publication was made the next year (also in *Science*) on the creation of 11 patient-specific ES cell lines (Hwang et al., 2005). The papers raised high hopes in South Korea fueling the country's aspiration to become the world's leading stem cell research center. Hwang was feted as the "Pride of Korea" and he was appointed head of a new World Stem Cell Hub set up in 2005. In June 2005, the Ministry of Science and Technology conferred the title of "Supreme Scientist" with a prize of US$15 million. However, Hwang's claim to fame proved to be short lived as reports of ethical and scientific violations created a media storm. Although the problems started with reports of ethical violations in the procurement of eggs for the study, investigations revealed that much of the data in the two papers was fabricated. Seoul National University where Hwang worked, found that all 11 stem cell lines that Hwang reported were fabricated. Both papers were retracted by the journal on January 11, 2006 (Kennedy, 2006), and he was expelled from the university in March 2006. In May 2006, Hwang was charged with embezzlement of research funds to the tune of $3 million and of having fabricated data for applying for research funds. He was handed a 2-year suspended sentence on October 26, 2009 for the offence. In spite of the legal battles, Hwang continued his research and was the first to report the successful cloning of a dog (Snuppy) using SCNT in 2005.

Exploitation of Women in Stem Cell Research

Within months of the publication of the Hwang's first paper, reports emerged of ethical violations with respect to the manner in which the eggs for research were procured (Baylis, 2009). Hwang reportedly used 242 eggs from 16 women in the 2004 *Science* paper and 185 eggs from 18 women in the 2005 *Science* paper. A fertility expert, Sung-Il Roh, at the MizMedi Hospital, however reported to *Nature* in November 2005 that he alone had purchased and supplied to Hwang 313 mature eggs from 21 donors for the 2004 paper, and 900 eggs from 62 donors for the 2005 paper. In an ensuing enquiry by the Ministry of Health and Welfare and the Seoul Central District Prosecutor's Office, it emerged that Hwang had acquired over 2200 eggs from around 120 women, and additionally used retrieved eggs (537 immature eggs) from excised ovaries (57 whole and 56 partial ovaries) for his experiments. The Seoul National University Investigational Committee, the National Bioethics Committee, and the Ministry of Health and Welfare found a number of ethical issues with the procurement of eggs in Hwang's work:

1. *Failure to disclose adequately the purpose for which the eggs were to be used*: Many of the donors believed that the eggs were going to be used in fertility research and did not know about stem cell research. In some cases, eggs from donors undergoing fertility treatment were graded and the better quality eggs were used for research while the lesser quality ones were used for the fertility treatment, possibly reducing the chances of a successful pregnancy. This was not revealed to the patients. Some of the patients were not told that ovaries would be removed, and others were not aware that the excised ovaries would be used in research.

(Continued)

BOX 3.2 (CONTINUED)

2. *Failure to explain potential harm involved in the procedure to research participants*: Several short-term side effects ranging from nausea to fluid in abdomen and ovarian hyperstimulation syndrome, as well as long-term effects such as infertility, result from the hormone treatments prior to egg donation. This was not adequately communicated to most of the donors.

3. *Monetary compensation for eggs*: Egg providers were paid either in cash or in kind-benefits in the form of discounted fertility treatments. While receiving financial compensation for research participation is not illegal in most countries, it is discouraged to prevent commodification of life.

4. *Procurement of eggs through coercion*: The investigation revealed that two female researchers of Hwang's team provided a total of 31 eggs for his research. Although documented as "voluntary," the voluntariness of the donation has been suspect. Hwang allegedly circulated consent forms in 2003 to female researchers in his laboratory in order to procure eggs. While altruism and national pride may have prompted some of them to sign, it appears that the possibility of career advancement was a significant motivator.

The investigation revealed clear violations of the existing Bioethics and Biosafety Act in South Korea and resulted in Hwang Woo Suk losing his license to derive stem cells from cloned human embryos in March 2006. A moratorium on human ES cell research was enforced by the National Bioethics Committee in South Korea, which ended only in 2009.

Scientific Fraud and Stimulus Triggered Acquisition of Pluripotent Stem Cells

Two papers describing a revolutionary new technique to generate stem cells from adult cells, published in *Nature* in January 2014 with lead author Haruko Obokata of RIKEN Centre for Developmental Biology, Kobe, Japan, ran into problems within a month of publication due to alleged use of duplicated images in the papers, and numerous failed attempts to replicate the results (see The Rise and Fall of STAP). There were allegations of images from an earlier paper (published in 2011) and the Ph.D. thesis of Obokata having been reused, and of the images in the two 2014 papers having been edited. Although initially dismissed by coauthor (and corresponding author of one of the studies), Charles Vacanti of Harvard Medical School, as a "mix up of some panels," another coauthor, Teruhiko Wakayama, wanted the papers to be withdrawn, prompting both *Nature* and RIKEN to look into the allegations in March 2014. RIKEN announced on April 1, 2014, that it held its employee Haruko Obokata guilty of *research misconduct* and the Director Ryoji Noyori held her coauthors *gravely responsible* for negligence in failing to verify their findings. The incident was a set-back to the efforts of the Japanese Government and Prime Minister Shinzo Abe to promote gender equality in the work-force, as also the efforts to promote Japan's cutting edge research and development in biosciences to further economic interests. A paper in *Nature* can only be retracted if all the authors recommend it. Vacanti and Obokata disputed the allegations, with Obokata confident that she would be able to reproduce her results. However, with tests done by one of the coauthors indicating a genetic mismatch between the stem cell lines created using the technique, and the adult mouse from which they were purportedly derived, the two papers were finally retracted by Obokata on June 4, 2014. Although Obokata was given time till November 2014 by RIKEN to reproduce her results, she was unable to do so and she resigned from RIKEN in December 2014. Tragically, another coauthor, Yoshiki Sasai, committed suicide on August 5, 2014; he blamed the media "bashing" for his suicide in a note left for his family.

References

Baylis, F. (2009). For love or money? The saga of Korean women who provided eggs for embryonic stem cell research. *Theoretical Medicine and Bioethics, 30*(5), 385–396. Retrieved from https://www.researchgate.net/publication/26854342_For_love_or_money_The_saga_of_Korean_women_who_provided_eggs_for_embryonic_stem_cell_research).

Hwang, W. S., et al. (2004). Evidence of a pluripotent human embryonic stem cell line derived from a cloned blastocyst. *Science, 303*(5664), 1669–1674. http://dx.doi.org/10.1126/science.1094515. *PMID* 14963337. (**Retracted, see** PMID 16410485).

Hwang, W. S., et al. (2005). Patient-specific embryonic stem cells derived from human SCNT blastocysts. *Science, 308*(5729), 1777–1783. http://dx.doi.org/10.1126/science.1112286. *PMID* 15905366. (**Retracted, see** PMID 16410485).

Kennedy, D. (2006). Editorial Retraction. *Science, 311*(5759). 335b–335b. http://dx.doi.org/10.1126/science.1124926. *PMID* 16410485.

The Rise and Fall of STAP. *Nature* [Specials and supplements archive]. Retrieved from http://www.nature.com/news/stap-1.15332.

Table 3.1 Examples of Regeneration of Tissues and Organs From Stem Cells

Year	Tissue/ Organ	Source of Stem Cells	Purpose	Laboratory
2011	Kidney	iPSC from skin cells	Transplantation	Jamie Davies, Edinburgh University, Scotland
2011	Rat organs in mice	(Blastocyst complementation)	Transplantation	Hiromitsu Nakauchi, University of Tokyo, Japan
2012	Optic cup	Human ESCs in 3D cultures	Retinal transplants	Yoshiki Sasai, RIKEN Centre for Developmental Biology, Japan
2013	Liver	Three types of stem cells	Organ transplantation and testing drugs	Takanori Takebe, Yokohama City University School of Medicine
2013	Blood vessels	Human IPSCs from skin cells	Restoring blood flow in diabetic foot	Massachusetts General Hospital, Boston and Harvard Stem Institute, Harvard University
2013	Bio engineered kidney in rats	Endothelial and kidney cells seeded on a decellularized scaffold of rat kidney	Transplantation	Harald Ott, Massachusetts General Hospital Centre for Regenerative Medicine, Boston, USA
2013	Muscles	Myosatellite stem cells	In vitro meat	Mark Post, Maastricht University, UK
2014	Stomach	Human stem cells	Studying *Helicobacter pylori* infections	James Wells, Pluripotent Stem Cell Facility, Cincinnati Children's Hospital Medical Centre, Ohio, USA
2014	Precursors of sperm and eggs	Human ESC and iPS cells	Curing infertility	Azim Surani, University of Cambridge, UK; Jacob Hanna, Weizmann Institute of Science, Rehovat, Israel

3.3.1 UNDERSTANDING CELLULAR DIFFERENTIATION

Cells in the embryo have in their nucleus the gene expression program which directs the developmental fate of all subsequent cell types. Scientists today are trying to decipher the set of **gene regulators** that determine cellular differentiation in order to understand the underlying circuitry which results in the evolution of new cell types from existing cells. Efforts revolve around determining the key genes or the key components which determine a particular cell type that distinguishes it from all others. The rationale for this effort is that once the components that give identity to a cell type are known, it would be possible to **switch developmental pathways**, thereby changing the identity of the resulting cell type. Thus an understanding of the regulatory pathways of that founding gene expression program could provide a model for the **understanding of all cell types** and, in turn, provide us with the background for potentially understanding all diseases. It would also aid in **preventing and treating birth defects**.

3.3.2 STUDY DISEASE PROGRESSION

For several diseases such as Parkinson's and Alzheimer's disease, diagnosis of the disease in the initial stages is difficult and **symptoms of the disease appear only after the affected cells have undergone irrevocable changes**. One of the anticipated outcomes of stem cell research is that it may be possible to follow the changes that occur in gene expression and trace the biochemical and physiological changes occurring in diseased cells from start to finish if cells from a diseased patient could be cloned (by inserting the DNA from a diseased individual into the nucleus of an egg cell). This could potentially lead to early detection, improved diagnosis, and therapeutic strategies for the prevention and control of several genetic diseases.

Studying the progression of genetic diseases by tissue sampling is often impossible as it further compromises the health of patients especially for diseases that affect the brain or the heart. Ian Wilmut and his team want to create ES cell lines by cloning cells from people with amyolotropic lateral sclerosis, a currently incurable neurodegenerative condition. Cloning would make it possible to create cultures of motor neurons from these patients and facilitate an investigation on the cause of the disease and allow testing of new therapies.

In 2012, an international 5-year project, **StemBANCC** (http://stembancc.org/), was initiated at the University of Oxford that collaborated 35 academic—industry partners interested in using stem cells for drug discovery. The project aims to generate a **library of 1500 iPS cell-lines from 500 people** characterized with respect to their genetic, protein and metabolic profiles, which would be available to researchers to *study diseases, develop new treatments and test the efficacy and safety of new drugs*.

3.3.3 REGENERATIVE MEDICINE

Stem cells can be used to replace damaged or dead cells in organs—a process known as **therapeutic regeneration**. The first demonstration of a clinical application of this technology came in 2012, when researchers at the Cedars-Sinai Medical Center and Johns Hopkins University reported extraction of stem cells from a patient following a heart attack, growing them in a petri-dish and returning them to the patient's heart (Makkar et al., 2012). The treatment reduced scarring and led to **regrowth of the heart tissue**. Another example is an FDA-approved clinical trial for **macular degeneration** (a condition that leads to blindness): it was found that in over half of 18 patients, the treatment with ES cells improved eye sight and the therapy was safe in the long term (Schwartz et al., 2015).

Regenerating tissues and organs from human stem cells has applications in **studying human diseases for which animal models do not exist**. An example is that of **human gastric diseases** such as ulcers and cancers caused by chronic infections of *Helicobacter pylori*. As the bacterium has little effect on animals, studying the disease has been difficult. In 2014, scientists led by James Wells at the Pluripotent Stem Cell Facility, Cincinnati Children's Hospital Medical Centre, were able to grow **pea sized mini stomachs** in vitro, which could be infected by injecting *H. pylori*, and behaved similar to ordinary human stomachs. The in vitro grown stomachs were hollow with an interior lining folded into glands and pits similar to a real stomach (Zastrow, 2014).

3.3.4 TISSUE ENGINEERING

Growing **body parts such as ears, noses, wind-pipes, and tear ducts for surgical replacement** of damaged or defunct organs are possible by growing stem cells on **biocompatible scaffolds to form 3-D structures**. Tissue engineering has become a reality due to advances in the field of **biomaterials**. A number of different materials from **proteins to plastic polymers** have been used to make scaffolds which are then seeded with cells and treated with growth factors to assemble tissues and organs. Alternatively, donor organs are stripped of cells and the remaining **collagen scaffold** is used to grow a new organ. This latter technique has the advantage of being able to **repurpose surgical discards** to generate custom built organs by stripping the original cells and **seeding with patient specific cells** to avoid organ rejection. The world's first tissue-engineered organ transplant was in 2008, when Claudia Castillo received a **trachea** (wind pipe) made by growing cells from her body over a scaffold of the wind pipe taken from a recently deceased donor (Owens, 2011). Other organs grown in this manner include **bladders and urethras**. Considerable research has gone into the making of **hearts for transplantation** (Maher, 2013).

3.3.5 GROW ORGANS FOR TRANSPLANTATION

Difficulties in organ transplantation including the shortage of organs and risk of rejection could be circumvented by **creating replacement organs from patients own cells**. In 2011, Jamie Davies and coworkers at Edinburgh University, Scotland, were able to grow **kidneys** in the laboratory (using amniotic fluid and animal fetal cells) that were **half a centimeter long** (about the size of the kidneys in an unborn baby) which they hoped would grow to full-size if transplanted into a human (The Telegraph, 2011).

In another approach called "**blastocyst complementation**," researchers at the Center for Stem Cell Biology and Regenerative Medicine at the University of Tokyo, Japan, headed by Hiromitsu Nakauchi, found that stem cells from rats could be introduced into mice to give **chimeras** (animals with bodies made up of genetically different cells) from which rat organs could be harvested for transplantations. Pending approval by the regulatory authorities, the researchers plan to **grow human organs, for instance the pancreas, in pigs, for harvest and transplant** to diabetic patients (Ryall, 2013).

Growing meat in vitro, using myosatellite cells (adult stem cells which repair muscle tissue) attached to Velcro and stretching them (to mimic the way cells grow) in a serum containing medium, was demonstrated by Mark Post of Maastricht University, the Netherlands, in 2013. The meat strips (around 20,000 tiny strips) could be pressed into a patty to make a burger (dubbed "frankenburger" and "Googleburger") at an estimated cost of £250,000 (Woollaston, Reilly, & McDermott, 2013). This is nevertheless an area that is rapidly developing and giving rise to new ventures.

3.3.6 ALTER CURRENT BIOMEDICAL PRACTICES FOR TREATMENT OF CANCER

The **cancer stem cell theory** holds that there are stem cells that are necessary for the continuance of the tumor. If that theory is correct, the development of anticancer therapeutics would need to take a completely new approach. It would mean that instead of the current use of therapeutics against cancer cells, therapies would target the cancer stem cells that are feeding the tumor (Blanpain, 2015). This could have a radical and very positive effect on cancer treatment in the next decade.

3.3.7 IDENTIFY DRUG TARGETS AND TEST POTENTIAL THERAPEUTICS

The **safety and efficacy of potentially useful new chemotherapeutic agents** could be tested on cells and tissues derived by cloning without the need for extensive preliminary testing in animals or exposure of humans to highly experimental drugs. It would also be possible to study the interactions between genes and drugs. Other applications include the possibility to investigate the manner in which **pathogens interact with specific cell types** and use of viruses for cellular transformation and gene therapy.

In 2013, a team of scientists led by Takanori Takebe from the Yokohama City University School of Medicine reported the **formation of liver buds in vitro** from a mixture of three different kinds of stem cells: hepatocytes (iPSCs differentiated into liver cells), endothelial stem cells (from UCB, to form blood vessels), and mesenchymal stem cells (to form connective tissues) (Takebe et al., 2013). The liver buds **approximately 4 mm across** were formed by the stem cells self-organizing into a functional organ and when transplanted into mice connected to the host blood vessels and continued to grow (Baker, 2013). The technique could possibly help in regenerative medicine, although issues such as whether the organ would continue to be functional and/or whether it would transform into tumors need to be addressed before this potential application can be realized. As the **liver buds are functional, they could possibly be used for tests such as metabolizing drugs and for toxicity analysis**, reducing the need for animal experimentation in preclinical trials.

Testing drugs for neurological condition such as Alzheimer's disease, Parkinson's disease and Huntington's disease in animal models have been ineffective partly due to the animals not having the same disease characteristics as humans. Hartung et al. at the Johns Hopkins University were able to create "**mini-brains**" by stimulating iPSCs to grow into brain cells (Dockrill, 2016). The technique could possibly replace large numbers of animals in testing drugs.

3.3.8 TOXICITY TESTING

Animal models of many diseases **do not accurately reflect** the disease in humans and due to differences in metabolism, toxicological studies in animals often do not help in predicting toxicity in humans. The use of stem cells would reduce the need for animal and human experimentation as human cells and tissues could be used in toxicity testing.

KEY TAKEAWAYS

Sources of Stem Cells	Applications of Stem Cells
• Embryonic stem cells • Nuclear transfer—embryonic stem cells (therapeutic cloning) • Fetal stem cells • Cord blood stem cells or neonatal stem cells • Adult stem cells • Induced pluripotent stem cells (iPSC)	• Understand cellular differentiation • Study disease progression • Regenerative medicine • Tissue engineering • Grow organs for transplantation • Alter current biomedical practices for treatment of cancer • Identify drug targets and test potential therapeutics • Toxicity testing

3.4 ETHICAL ISSUES IN STEM CELL RESEARCH

For the potential that stem cells offers in revolutionizing medical practice to become a reality, **considerable research needs to be done**. Stem cells are **unstable and fragile**, so researchers need many stem cell lines. The main ethical challenge associated with stem cell research is with the **source of the cells**. Although some advances have been made in the use of adult stem cells, almost all current procedures involve the use of ES cells isolated from 5- to 7-day-old embryos, and EG cells derived from immature aborted fetuses. At present, there are **four main sources of human embryonic stem cells** (hESCs) used in research:

1. surplus embryos from in vitro fertilization (IVF) clinics
2. embryos created in the laboratory from donated eggs and sperm
3. embryos created through SCNT (therapeutic cloning/research embryos)
4. aborted fetuses (EG cells).

Several questions regarding the bioethics of human ES cells involve **religious and cultural viewpoints** and the manner in which life itself is defined. For many people who advocate the "**utilitarian**" **ethical framework** (see Chapter 5: Relevance of Ethics in Biotechnology), use of embryos in stem cell research is **not an insurmountable problem** as they believe that embryos, under certain conditions, may be sacrificed for beneficial purposes. There are significant **reasons of beneficence** for pursuing stem cell research. For others however, having to "kill" embryos constitute the main reason to radically oppose human stem cell research. In their view, **embryos should never be used as a mere means to the ends of others, however valuable these may be**. A far less controversial approach that has emerged in several countries, including the United States, capitalizes on the "**discarded—created distinction**." While it is considered unethical to create an embryo merely for deriving hESCs, surplus embryos from IVF, which in any case were not going to be implanted in a woman's womb, could now be put to good use in furthering scientific research (an argument on the "**nothing is lost**" principle). One **drawback** of this specious solution is the **inability to create patient-specific-hESCs necessary for transplantation and regenerative medicine**. Central to the discussion of the ethics in stem-cell research are **two important viewpoints**: (1) that which **pertains to the embryo** (*fetalistic*), and (2) that which **pertains to women** (*feministic*).

3.4.1 MORAL STATUS OF EMBRYOS—FETALISTIC VIEWPOINT

The first question posed by bioethicists is that of the moral status of the embryos used for deriving the ES cells. According to the **Jewish religion, embryos not in the body, such as the in vitro embryos, have no status at all**, while **Muslims consider an embryo as having a moral status only after 40 days**, when the bones are knit. According to the Christian faith, **life begins at the time of conception** so embryos are also persons with souls. So for a practicing Catholic, the killing of embryos (which have the "inherent potential" to become a human being) for the purpose of isolating stem cells is akin to murder. It is morally unacceptable to take a life, even if it is to save another. The issue is further complicated by extrapolation of religious and ethical sentiments regarding abortion, euthanasia, and right to life/death.

Scientists argue that **most religions confuse "fertilization" and "conception."** By definition, "conception" is the formation of a zygote capable of survival and maturation under normal conditions. Embryos arising from fertilization in a Petri dish are only a **ball of cells** and have **no**

possibility of ever giving rise to a human being unless implanted in a womb. Moreover, even if these cells are genetically human, they **do not show any of the characteristics of persons**—they are neither **conscious** nor **self-aware** and, consequently, do not have an independent moral status. Others have argued that as the derived cells would continue to grow and multiply in the recipient of a transplantation, the embryo is not technically "killed."

An intermediate position in the human ES cell debate is one that makes a **"use-derivation"** distinction. Using hESCs is considered ethically acceptable, but their derivation is not—because it involves killing human embryos. In 2001, US president George Bush vetoed the use of federal funds for derivation of new stem cell lines for research and other purposes. In his words,

> "As a result of private research, more than 60 genetically diverse stem cell lines already exist" I have concluded that we should allow federal funds to be used for research on these existing stem cell lines "where the life and death decision has already been made. This allows us to explore the promise and potential of stem cell research" without crossing a fundamental moral line by providing taxpayer funding that would sanction or encourage further destruction of human embryos that have at least the potential for life".
>
> **(Bush, 2001)**

Possible **objections to the use of embryos** (surplus IVF embryos and research embryos) have been discussed by Guido deWert and Christine Mummery (deWert & Mummery, 2003) under the *principle of proportionality* (in relation to an important goal), the *slippery slope argument* (where acceptance or moral justification of a practice X will lead inevitably to an acceptance of an "undesirable" practice Y), and the *principle of subsidiarity* (that no other suitable alternatives exist).

- *Principle of proportionality*: The argument here is that since the sacrificing of the embryos **serves a higher goal of new therapies** which could benefit a large number of patients, research into hESCs **cannot be considered disproportional**.
- *Slippery slope argument*: Two variants of the slippery slope argument have been considered—the "empirical" version, which supposes that an acceptance of a practice "X," would inevitably lead to the acceptance of an undesirable practice "Y," so to prevent "Y," "X" should be banned; and the "logical" version, in which a justification of a practice "X," automatically implies an acceptance of an undesirable practice "Y." In the context of ES cells, the "empirical" version **anticipates that acceptance of the isolation of hESCs for replacement therapy would lead to an acceptance of cosmetic and frivolous applications**, while the "logical" version is afraid that a **moral justification of creating embryos for the derivation of hESC or therapeutic cloning could automatically result in an acceptance of embryos with germ line modification and/or reproductive cloning**.
- *Subsidiarity argument*: This argument presumes that **other sources of stem cells would be intrinsically superior to hESCs** and would be preferable to the use of hESCs. While the need to avoid ethical issues does provide sufficient motivation to develop methods to generate stem cells from sources other than embryos, many scientists are of the opinion that research in hESCs would have to continue for the present. This is because hESCs are easier to generate than any other type of stem cells and large number of stem cell lines would be required for validation of potential applications.

3.4.2 EXPLOITATION OF WOMEN—FEMINISTIC VIEWPOINT

The second major objection to stem cell research pertains to the **requirement for eggs or oocytes**. To be effective, current success rates being low, a **large number of eggs are required**. The egg donation procedure is uncomfortable and potentially painful and carries some **medical risk**. Women must undergo hormone treatments to stimulate ovulation, counseling sessions to understand the risks involved, and a medical procedure in which a needle is inserted into the vagina to remove eggs from the ovary. A small percentage of donors develop ovarian-hyperstimulation syndrome, which in rare cases can cause kidney failure. It also raises fears of exploitation of women, a justifiable fear going by media reports of stem cell research in Korea by Hwang Woo Suk and coworkers (see Box 3.2). In a speech of the Holy See to the UN in 2003, Archbishop Migliore (2003) stated, "*the process of obtaining these eggs, which is not without risk, would use women's bodies as mere reservoirs of oocytes, instrumentalizing women and undermining their dignity.*" The UN Declaration on Human Cloning (see Box 2.7 in Chapter 2: Cloning) also stresses this point and calls upon Member States to take **measures to prevent the exploitation of women** in the application of life sciences.

Scientists insist that the dependence on women for donation of eggs is **likely to be a short-term problem**. Alternative sources being developed include creating **artificial eggs** by reprogramming adult cells, differentiation of ES cells into germ cells which produce oocytes (ovary stem cells), and **using nonhuman oocytes** such as rabbit or frog eggs. **Improvement in the iPSC technology** that reprograms adult cells into stem cells and does not therefore require oocytes could probably circumvent this ethical issue.

Banning stem cell research on the basis of ethical issues has also been criticized as being immoral. Ethicists Katrien Devolder and Julian Savulescu argue that **scientists have a moral obligation to do research that could potentially save lives**, and that the United Nations Declaration on Human Cloning is unethical and should be withdrawn in order to allow nations to pursue therapeutic cloning. To quote:

"To fail (omit) to do beneficial research is as wrong as doing harmful research. To fail to release a drug which will save 100,000 lives is morally equivalent to killing 100,000 people. To fail to develop a drug which will save 100,000 lives is morally equivalent to failing to release it. We may not be able to point to those people whose lives would have been saved but their lives are no less valuable because they are in the future or they are anonymous. Cloning research could result in treatments for common diseases like heart disease, stroke and cancer. It has a considerable potential to save hundreds of thousands if not millions of lives. Through a failure of moral imagination, we may continue to hold back cloning research and be responsible for the deaths of many people who perished while we delayed the development of treatments. This research is of enormous potential benefit to humanity. This provides a strong prima facie case in favour not just of allowing it, but positively supporting it through permissive legislation and generous public funding. The laws which prevent such life-saving research may be, in a moral sense, lethal".

(Savulescu, 2007)

3.5 ETHICAL ISSUES IN STEM CELL TRANSLATION

Although stem cells have the potential to cure many human ailments, it is **important to note that other than hematopoietic cells obtained from bone marrow transplants (that have been routinely used for over 50 years to treat several blood and immune related disorders such as thalassemia, leukemia, and lymphoma), no other type of stem cells can currently be used as a therapy.** Stem cell therapies are currently in the **clinical trials stage of development for several conditions** such as age-related macular degeneration (AMD) and spinal cord injuries. Despite this, media attention focused on the potential application of the technology has spawned a number of clinics and medical centers offering stem cell therapies. This has led to **another ethical problem** associated with stem cell research: that of **"unproven stem cell interventions being marketed directly to patients as stem cell therapies."**

3.5.1 CLINICAL TRIALS USING STEM CELLS

Clinical trials on human patients establishing the therapeutic use of stem cells began in 2013. The first such trial approved by the government was conducted in Japan: researchers at the RIKEN Centre for Developmental Biology, and the Institute of Biomedical Research and Innovation, Kobe, treated a 70-year-old woman with AMD by surgically implanting a sheet of retinal cells grown from iPSCs generated from her skin cells (Gallagher, 2013). Other clinical trials with human stem cells currently being conducted are summarized in Table 3.2.

3.5.2 GUIDELINES FOR THE CLINICAL TRANSLATION OF STEM CELLS

An important concern emerging in the field of stem cell clinical translation is that of **"stem cell tourism"** that has resulted in patients often **seeking treatment in countries with more liberal policies which permit untested procedures in the guise of medical innovations**.

Specific guidelines for the ethical use of stem cell research in clinical applications has been formulated by the International Society for Stem Cell Research (ISSCR, www.isscr.org) Task Force, a multidisciplinary group of stem cell researchers, clinicians, ethicists and regulatory officials from 13 countries. The *Guidelines for the Clinical Translation of Stem Cells* was published in 2008 in *recognition of the urgent need to address the problem of unproven stem cell interventions being marketed directly to patients.* The problem as identified in the guidelines is reproduced below:

"Numerous clinics around the world are exploiting patients' hopes by purporting to offer new and effective stem cell therapies for seriously ill patients, typically for large sums of money and without credible scientific rationale, transparency, oversight, or patient protections. The ISSCR is deeply concerned about the potential physical, psychological, and financial harm to patients who pursue unproven stem cell-based "therapies" and the general lack of scientific transparency and professional accountability of those engaged in these activities.

The marketing of unproven stem cell interventions is especially worrisome in cases where patients with severe diseases or injuries travel across borders to seek treatments purported to be stem cell-based "therapies" or "cures" that fall outside the realm of standard medical practice. Patients seeking medical services abroad may be especially vulnerable because of insufficient

Table 3.2 hESC and iPSC-Based Products in Clinical Trials

Cell Type (Product Name)	Company or Group	Trial Location	Disease	Stage of Trial	Cell Delivery	Status of Trial
hESC-derived RPE (MA09-hRPE)	Ocata Therapeutics	United States	Dry AMD	Phase I/II	Cell suspension	Active, not recruiting
hESC-derived RPE (MA09-hRPE)	Ocata Therapeutics	United States	Stargardt	Phase I/II	Cell suspension	Active, not recruiting
hESC-derived RPE (MA09-hRPE)	Ocata Therapeutics	United Kingdom	Stargardt	Phase I/II	Cell suspension	Recruiting
hESC-derived RPE (MA09-hRPE)	CHABiotech (licensed from Ocata)	Korea	Dry AMD	Phase I/II	Cell suspension	Recruiting
hESC-derived RPE (MA09-hRPE)	CHABiotech (licensed from Ocata)	Korea	Stargardt	Phase I	Cell suspension	Active, not recruiting
hESC-derived RPE (MA09-hRPE)	University of California, Los Angeles (with Ocata's cells)	United States	MMD	Phase I/II	Cell suspension	Not yet recruiting
iPSC-derived RPE (autologous)	Rikagaku Kenkyūsho (RIKEN) Institute	Japan	Wet AMD	Phase I	Monolayer sheet (no membrane)	On hold
hESC-derived RPE (PF-05206388)	Pfizer	United Kingdom	Wet AMD	Phase I	Membrane-immobilized monolayer sheet	Recruiting
hESC-derived RPE (Opregen)	Cell Cure Neuroscience	Israel	Dry AMD	Phase I/II	Cell suspension	Recruiting
hESC-derived $CD15^+ISL\text{-}1^+$ cardiac progenitors	Assistance publique, Hôpitaux de Paris	France	Severe heart failure	Phase I	Cells embedded in fibrin patch	Recruiting
hESC-derived pancreatic endoderm (VC-01)	Viacyte	United States	Type I diabetes	Phase I/II	PEC-01 cells encapsulated in a medical device	Recruiting
hESC-derived oligodendrocyte progenitors (AST-OPC1)	Asterias Biotherapeutics	United States	Spinal cord injury	Phase I	Cell suspension	Completed (took over from Geron)
hESC-derived oligodendrocyte progenitors (AST-OPC1)	Asterias Biotherapeutics	United States	Spinal cord injury	Phase I/II	Cell suspension	Recruiting

AMD, *age-related macular degeneration;* hESC, *human embryonic stem cell;* iPSC, *induced pluripotent stem cell;* MMD, *myopic macular degeneration;* RPE, *retinal pigment epithelium.*
Reprinted by permission from Macmillan Publishers Ltd: [NATURE REVIEWS DRUG DISCOVERY] (Kimbrel EA and Lanza E. Current status of pluripotent stem cells: moving the first therapies to the clinic. Nature Reviews Drug Discovery **14**, *681–692 doi: 10.1038/nrd4738), 1 (2015).*

local regulation and oversight of host clinics. Some locales may further lack a system for medical negligence claims, and there may be less accountability for the continued care of foreign patients".

(International Society for Stem Cell Research, 2008)

The guidelines comprise of 40 recommendations that ensure that the basic tenets of biomedical ethics are upheld. This includes written informed consent from donors of cells for allogenic use (recommendation 3) and from patients involved in clinical trials (recommendation 28), considerations of social justice (recommendation 35−38), and adequate testing through preclinical trials (recommendation 11 and 12) to establish efficacy (recommendation 13−15) and safety (recommendations 16−19). The guidelines also emphasize the importance of reporting of the intervention (recommendation 20) and peer-review (recommendation 23), as well as a risk−benefit analysis that compares stem cell therapies being developed with the best available therapies (recommendation 24−26), so as to ensure beneficence (recommendation 27). The guidelines recommend regulatory oversight (recommendation 21 and 22) by a body independent of the investigators.

These guidelines recognize the fact that many medical innovations which have benefitted patients have been introduced into medical practice without formal clinical trial process, and in very specific cases, stem cell interventions may be warranted in a small number of seriously ill patients. Recommendation 34 specifically addresses such cases where clinician−scientists may provide unproven stem cell interventions outside of the context of a formal clinical trial. It specifies the conditions of strict documentation of the justification and approval of the procedure, voluntary informed consent of the patient, and adequate measures to ensure safety and control of adverse events that may occur.

A new and updated draft *Guidelines for Stem Cell Research and Clinical Translation* dated June 26, 2015, proposed by the ISSCR Task Force, currently under discussion (Kimmelman & Daley, 2015), also incorporates elements of the ISSCR *Guidelines for the Conduct of Human Embryonic Stem Cell Research* which had been proposed in 2006 (International Society for Stem Cell Research, 2006).

3.6 STEM CELL RESEARCH POLICY

There is no international consensus on hESC research. The 2005 **UN Declaration on Human Cloning** (see Chapter 2: Cloning) was limited to a ban on all forms of human cloning that are contrary to *human dignity*, and left the interpretation of the declaration to member countries. The **United Nations Educational, Social and Cultural Organization** (UNESCO) has observed three different positions (UNESCO, 2004) with regard to national legislation and regulations concerning ES cell research:

1. *Generally prohibit research on embryos (with some specific exceptions) and/or creation of embryos for research purposes* (Example: Poland and Italy)
2. *Permit research on supernumerary embryos produced by fertility treatment, but prohibit creation of embryos for research purposes* (Example: France and Germany)
3. *Permit creation of embryos for research purposes with strict conditions* (Example: Japan and Sweden)

Positions 1 and 2 are generally regarded as prohibiting therapeutic cloning although position 1 is ambiguous as the "specific exception" could be interpreted as "developing new therapies by research on embryos." Position 3 is interpreted as permitting therapeutic cloning.

In Europe, Article 18 of the **European Convention of Human Rights and Biomedicine** forbids the creation of embryos for all research purposes (see Chapter 2: Cloning). In most countries, existing laws on reproduction and creation of embryos for IVF are being modified to reflect the regulation of SCNT. For instance, in 2001, **United Kingdom** (UK) amended the Human Fertilization and Embryology Act of 1990 (since amended as the Human Fertilization and Embryology Act 2008) to permit the creation of embryos for research purposes in order to better study the progression of certain genetic disorder with a view of developing therapies. Under this act, the Human Fertilization and Embryology Authority (HFEA) would monitor and license clinics that carry out IVF or other assisted reproductive technology procedures, regulate human embryo research and storage of reproductive materials, and license creation of embryos for research for programs outlined in the UK law (Small & Doherty, 2011).

In **Australia**, the Research Involving Human Embryos Act, 2002, and the Human Cloning Act 2002, prohibit SCNT for reproductive or therapeutic purposes (Australian Stem Cell Centre, 2010). Australia also forbids the importation of cloned, parthenogenetic (an embryo not derived through fertilization), androgenic, or chimeric embryos (embryos having nonhuman cells). Trading in human eggs, sperm, or embryos is a crime attracting prison terms of up to 15 years depending on the offense. Australia's laws do however allow research on surplus embryos from IVF clinics with consent from donors, subject to oversight by the National Health and Medical Research Council Licensing Committee.

In countries that lack specific legislation, regulatory oversight may be provided by specific provisions in relevant guidelines. For example, in **Canada**, legislation to regulate assisted reproductive technologies and embryo research is still being debated, but guidelines have been issued by the Canadian Institute for Health Research (Canadian Institute for Health Research, 2013). As per these guidelines, research involving derivation, in vitro study and clinical trials of hES cell lines require review and approval by the central Stem Cell Oversight Committee, by local Research Ethics Boards and, where appropriate, animal ethics committees. The Canadian guidelines require a medical rationale for the research, informed consent of the donors, protection of donor's privacy and prohibition of payment to donors. The Canadian guidelines also prohibit use of public funds for making embryos solely for research or for research combining hESCs and nonhuman embryos.

In **India**, National Guidelines for Stem Cell Research were issued in 2007 by the Indian Council of Medical Research in association with the Department of Biotechnology and revised and updated in December 2013 (Indian Council of Medical Research, 2013). The current guidelines recognize three categories of experiments (Permitted, Restricted, and Prohibited) and ensure regulatory oversight by the Institutional Committee for Stem Cell Research (IC-SCR) and the National Apex Committee for Stem Cell Research and Therapy (NAC-SCRT). The Guidelines however cover only stem cell research (basic and translational) and not stem cell therapy. The only stem cell therapy currently allowed by law in India is that of hematopoietic bone marrow transplants; all others are considered as experimental therapies and requires clearance by regulatory bodies that oversee clinical trials in the country.

In the **United States**, federal funding for research on embryos is banned through the Dickey-Wicker Amendment of 1996, named after Representatives Jay Dickey and Roger Wicker. The amendment bans use of federal funds in research that exposes human embryos to risk or injury greater than that allowed for research on in utero fetuses. The amendment thus prohibits use of federal funds for the creation of human embryos for research purposes or research in which human embryos are destroyed or discarded. The Dickey-Wicker Amendment has appeared in every Labor—HHS appropriation bill since 1996. Interpreting this amendment, the National Institute of Health (NIH) in August 2000 released guidelines for research using human pluripotent stem cells that stipulated that hESCs can only be developed with private funding from surplus frozen embryos created for fertility treatments, with the consent of the donor. In 2001, President George W. Bush prohibited the use of federal funds for creation of new hES cell lines (National Institute of Health, 2009a). The funds could be used to conduct research on adult stem cells, or on the approximately 60 cell lines that had been established before 2001. In June 2005, the Stem Cell Research Enhancement Act (H.R. 810), which would allow federal funds for derivation of ESC from surplus IVF embryos, was passed by both the House and the Senate, but was vetoed by President Bush. A second attempt at passing the bill was made in 2007, but was thwarted by Presidential veto. In both instances, the bill did not have the two-third majority necessary to override the veto. The 2001 Executive order was reversed by President Barack Obama issuing an executive order titled *Removing Barriers to Responsible Scientific Research Involving Human Stem Cells* in 2009. Implementing this executive order 13505, in July 2009, the NIH published the ***Stem Cell Research Guidelines*** (National Institute of Health, 2009b), which specified the kind of research in hESC using NIH funds (National Institute of Health, 2015). Eligible for funds would be hES cell lines obtained from spare embryos from IVF clinics with donor consent. It also mandated the setting up of a *"new Registry listing hESCs eligible for use in NIH funded research. All hESCs that have been reviewed and deemed eligible by the NIH in accordance with these Guidelines will be posted on the new NIH Registry"*. The guidelines prohibited the use of NIH funds for derivation of hESC from human embryos under the Dickey amendment, and the use of the funds for hESCs derived from other sources including SCNT, parthenogenesis, and/or IVF embryos created for research. Thus, Obama's revocation of Bush's policy broadened the hES cell resources for researchers, but they are still unable to create their own lines (only lines from donated spare IVF embryos permitted) using NIH funds. This seriously limits research on patient specific hES lines which are necessary for therapeutic applications without immunosuppressant drug requirements.

In addition to the NIH, guidelines for stem cell research in the United States have also been proposed by the National Academies. In its 2005 ***Guidelines for Human Embryonic Stem Cell Research*** (National Academies Press, 2005) in order to manage the ethical and legal concerns with respect to the technology, the National Academies of Science advocated the establishment of **Embryonic Stem Cell Research Oversight (ESCRO) Committees**. In its 2007 *Amendments to the National Academies' Guidelines for Human Embryonic Stem Cell Research*, the composition and responsibilities of the ESCRO Committees was further elaborated.

KEY TAKEAWAYS

International and Regional Stem Cell Research Policy:

UN 2005 Declaration on Human Cloning (annexed to and elaborates on Article 11 of the UNESCO Universal Declaration on the Human Genome and Human Rights)—prohibits all forms of human cloning inasmuch as they are incompatible with human dignity and protection of human life

Council of Europe 1998 Convention of Human Rights and Biomedicine—Article 18 prohibits creation of human embryos for research purposes.

National Stem Cell Research Policy:

UK 2001—modified the Human Fertilization and Embryology Act 1990 to include Human Fertilization and Embryology (Research purposes) Regulation 2001, allowing therapeutic cloning. Since amended as the HFEA 2008, licenses creation of embryos for research purposes

Australia 2002—Human Cloning Act, and the Research Involving Human Embryos Act 2002 prohibits SCNT for reproductive and therapeutic purposes, research on surplus embryos from IVF clinics with consent of donors, permitted

Canada 2004—Act Respecting Assisted Human Reproduction and Related Research (Assisted Human Reproduction Act)—Embryonic stem cell research permitted using supernumerary embryos from IVF clinics subject to review and approval of Stem Cell Oversight Committees

India—National Guidelines for Stem Cell 2007 of ICMR updated in 2013. Regulatory oversight provided by the Institutional Committee for Stem Cell Research (IC-SRC) and the National Apex Committee for Stem Cell Research and Therapy (NAC-SCRT)

United States—Federal funding for research on embryos is banned through the Dickey-Wicker Amendment of 1996. 2001 President Bush prohibits use of federal funds for creation of new hES cell lines. Reversed by President Obama's 2009 order, allows federal funds to be used for hES cell lines obtained from spare embryos from IVF clinics with donor consent, subject to the 2009 NIH Stem Cell Research Guidelines.

3.7 THE ROLE OF POLITICS AND PUBLIC OPINION IN SHAPING STEM CELL POLICY

The course of stem cell research has been heavily influenced by religious and political leanings of the public and policy makers. Discussions on human ES cells have **polarized opinions** into those who are **opposed to stem cell research** as sacrificing embryos is ethically unjustifiable and those who **support stem cell research** on the grounds that denying treatment to the needy is morally wrong. The obvious solution to the ethical dilemma would be to **develop stem cell therapies that did not require the stem cells to be derived from embryos**. This has been the objective of politicians and policy makers in the United States. In banning the use of federal funds for the creation of new ES cell lines and permitting the use of federal funds for research only on the 64 cell lines already in existence, the Bush administration in 2001 aimed to increase spending on research that did not involve killing embryos. In the United States, opinions on the ethics of stem cell research are largely an extension of the "pro-life/pro-choice" debate on abortion. Although in 1973, the US Supreme Court legalized abortion, opponents of abortion mobilized opposition to stem cell research on the same grounds of destruction of life (Pew Research Center, 2008a). Surveys conducted by the Pew Forum on Religion & Public Life and the Pew Research Center for the People & the Press between 2002 and 2007 (Pew Research Center, 2008b), found that the issue of stem cell research clearly **divided Americans along political lines** with the majority of Democrats (60%) and Independents (55%) considering it more important to conduct stem cell research that could potentially lead to new cures than it is to avoid destroying embryos, whereas only 37% of Republicans held that opinion. The survey also found **significant differences in opinion based on religion.** Among white evangelical Protestants, 57% were opposed to embryogenic stem cell research, whereas only 31% favored it. The majority of religiously unaffiliated (68%), white mainline Protestants (58%), and non-Hispanic Catholics (59%) supported stem cell research. Overall, the 2007 poll concluded that 51% of Americans say that it is more important to conduct stem cell research, whereas 35% feel it is more important to not destroy embryos. Between 2004 and 2008, attempts to allow federal funds to be used for research using surplus embryos from IVF clinics were consistently vetoed by President Bush (Republican), despite reports that many of the existing hES cell lines are dead or contaminated with mouse cells thereby making it impossible to use them for therapeutic applications. It was only in 2009 with the election of a Democrat, President Obama, that (by virtue of an executive order) **federal funds could be used for developing human embryogenic stem cell lines from surplus IVF embryos.**

In countries where public controversies regarding the use of embryos in stem cell research are absent, policies have been supportive of use of embryos both in therapeutic cloning and stem cell translational research. **China**, for example, has one of the most permissive hESC policies in the world. In **South Korea** and **Japan**, major advancements in stem cell research have resulted from **very flexible policies regarding research.** Scientists in these countries have pioneered stem cell technologies and enjoyed support from their governments both in recognition and funding (although not without controversies, see Box 3.2). In **Japan, conditional regulatory approval for clinical translation** can be obtained by **proving safety in a small number of patients**, obviating the time and expense associated with clinical trials, and paving the way for the country to be a leader in the use of iPS cells for the treatment of retinal diseases. South Korea has been attempting to restore its image following the rise and fall of Hwang Woo Suk (see Box 3.2) and continues to promote therapeutic cloning for stem cell research with government funds supporting two main areas—spinal cord damage and chronic conditions such as arthritis. These aim to help Korean companies capture a potential market for treatments. Singapore has made concerted efforts to attract scientists from around the globe and has over 40 groups working on stem cell research earning it the reputation of "Asia's stem cell center."

In Europe, Germany and Italy have among the most restrictive policies on stem cell research banning the creation of embryos for stem cell research although research on imported embryos is permitted. Therapeutic cloning is legal in the UK, Sweden, Belgium, and Spain— the **UK is regarded as Europe's leader in stem cell research**, therapeutic cloning by SCNT having been legalized in 2004, and experiments in animal–human hybrid embryos being permitted since 2008.

3.8 SUMMARY

Stem cells are special cells found in bodies of animals, which have the unique capacity to self-perpetuate and to give rise to other types of cells in the body. The importance of stem cells in medicine lies in the ability of the cells when transplanted to a patient to repair tissue damage and reverse tissue degeneration occurring in several genetic diseases such as Alzheimer's, Parkinson's, muscular dystrophy, diabetes, arthritis, blindness, and several blood and immune disorders. Although stem cells can be extracted from several different types of tissue, the most reliable source is the ICM of 5–7-day-old embryos. The need to sacrifice embryos for generating ES cell lines has raised several ethical issues pertaining to the moral status of embryos and whether they can be used as a means to an end. Techniques to induce stem cell characteristics in adult cells are being developed and could provide a solution to the ethical dilemma facing ES cell research. Clinical application, in the form of stem cell therapies, are currently in the stage of clinical trials in several countries and are expected to treat several diseases for which there is no known effective treatment. Other than hematopoietic cells obtained from bone marrow transplants no other type of stem cells can be at present used as a therapy.

REFERENCES

Evans, M. J., & Kaufman, M. H. (1981). Establishment in culture of pluripotential cells from mouse embryos. *Nature, 292,* 154–156.

Jiang, Y., Jahagirdar, B. N., Reinhardt, R. L., Schwartz, R. E., Keene, C. D., Ortiz-Gonzalez, X. R., ... Verfaille, C. M. (2002). Pluripotency of mesenchymal stem cells derived from adult marrow. *Nature, 418,* 41–49.

Shamblott, M. J., Axelman, J., Wang, S., Bugg, E. M., Littlefield, J. W., Donovan, P. J., Blumenthal, P. D., Huggins, G. R., & Gearhart, J. D. (1998). Derivation of pluripotent stem cells from cultured human primordial germ cells. *Proceedings of the National Academy of Sciences of the United States of America. 1998, 95,* 13726–13731, *Nov (23).*

Takahashi, K., Tanabe, K., Ohnuki, M., Narita, M., Ichisaka, T., Tomoda, K., & Yamanaka, S. (2007). Induction of pluripotent stem cells from adult fibroblast by defined factors. *Cell, 131*(5), 861–872.

Takahashi, K., & Yamanaka, S. (2006). Induction of pluripotent stem cells from mouse embryonic and adult fibroblast cultures by defined factors. *Cell, 126*(4), 663–676.

Thomson, J. A., Itskovitz-Eldor, J., Shapiro, S. S., Waknitz, M. A., Swiergiel, J. J., Marshall, V. S., & Jones, J. M. (1998). Embryonic stem cell lines derived from human blastocysts. *Science, 282,* 1145–1147.

Yu, J., Vodyanik, M. A., Smuga-Otto, K., Antosiewicz-Bourget, J., Frane, J. L., Tian, S., ... Thomson, J. A. (2007). Induced pluripotent stem cell lines derived from human somatic cells. *Science, 318*(5858), 1917–1920.

Benefit to Society

Baker, M. (2013, July 3) (Reporter) *Miniature human liver grown in mice.* Retrieved from http://www.nature.com/news/miniature-human-liver-grown-in-mice-1.13324.

Blanpain, C (2015, April 2) *Cancer: a disease of stem cells?* Retrieved from http://www.eurostemcell.org/factsheet/cancer-disease-stem-cells.

Dockrill, P. (2016, February 16) (Reporter) *These lab-grown 'mini-brains' could help replace animal testing this year.* Retrieved from http://www.sciencealert.com/these-lab-created-mini-brains-could-help-replace-animal-testing-as-soon-as-this-year?perpetual=yes&limitstart=1.

Maher, B. (2013, July 3) (Reporter) *Tissue engineering: How to build a heart.* Retrieved from http://www.nature.com/news/tissue-engineering-how-to-build-a-heart-1.13327.

Makkar, R. R., Smith, R. R., Cheng, K., Malliaras, K., Thomson, L. E., Berman, D., . . . Marbán, E. (2012). Intracoronary cardiosphere-derived cells for heart regeneration after myocardial infarction (CADUCEUS): A prospective, randomised phase 1 trial. *Lancet.*, *379*(9819), 895−904 . http://dx.doi.org/10.1016/S0140-6736(12)60195-014.

Owens, B (2011, July 8) (Reporter) *What's new about new synthetic organs?* [Newsblog] Retrieved from http://blogs.nature.com/news/2011/07/whats_new_about_new_synthetic.html.

Ryall, J. (2013, June 20) (Reporter) *Human organs 'could be grown in animals within a year'*. Retrieved from http://www.telegraph.co.uk/news/science/science-news/10132347/Human-organs-could-be-grown-in-animals-within-a-year.html.

Schwartz, S. D., Regillo, C. D., Lam, B. L., Eliott, D., Rosenfeld, P. J., Gregori, N. Z., . . . Lanza, P. (2015). Human embryonic stem cell-derived retinal pigment epithelium in patients with age-related macular degeneration and Stargardt's macular dystrophy: Follow-up of two open-label phase 1/2 studies. *The Lancet, 385* (9967), 509−516. http://dx.doi.org/10.1016/S0140-6736(14)61376-3.

Takebe, T., Sekine, K., Enomura, M., Koike, H., Kimura, M., Ogaeri, T., ... Zheng, Y.-W. (2013). Vascularized and functional human liver from an iPSC-derived organ bud transplant. *Nature, 499*, 481−484. http://dx.doi.org/10.1038/nature12271.

The Telegraph (2011, April 11) *Scientists create human kidneys from stem cells*. Retrieved from http://www. telegraph.co.uk/news/health/news/8443740/Scientists-create-human-kidneys-from-stem-cells.html.

Woollaston, V., Reilly, R., & McDermott, N. (2013, August 5) (Reporters) 'At least it tastes of meat!': World's first test-tube artificial beef 'Googleburger' gets GOOD review as it's eaten for the first time. Retrieved from http://www.dailymail.co.uk/sciencetech/article-2384715/At-tastes-meat--Worlds-test-tube-artificial-beef-Googleburger-gets-GOOD-review-eaten-time.html.

Zastrow, M. (2014, October 29) (Reporter) *Tiny human stomachs grown in the lab*. Retrieved from http://www.nature.com/news/tiny-human-stomachs-grown-in-the-lab-1.16229.

Ethical issues in stem cell research

Archbishop Migliore (2003, September 30) *Holy See's call for a ban on all human cloning*. Retrieved from http://www.catholic.org/featured/headline.php?ID=385.

deWert, G., & Mummery, C. (2003). Human embryonic stem cells: Research, ethics and policy. *Human Reproduction, 18*(4), 672−682 . Retrieved from http://humrep.oxfordjournals.org/content/18/4/672.full.

George W. Bush: "Fact Sheet: Embryonic Stem Cell Research," August 9, 2001. Online by Gerhard Peters and John T. Woolley, *The American Presidency Project*. Retrieved from http://www.presidency.ucsb.edu/ws/?pid=79025.

Savulescu, J. (2007). *Ethics of stem cell and cloning research*. Retrieved from http://www.bep.ox.ac.uk/__data/assets/pdf_file/0017/9008/Ethics_of_Stem_Cell_and_Cloning_Research.pdf.

Ethical issues in stem cell translation

Gallagher, J (2013, July 19) (Health and Science Reporter) *Pioneering adult stem cell trial approved by Japan*. Retrieved from http://www.bbc.com/news/health-23374622.

International Society for Stem Cell Research. (2006, December 21) *Guidelines for the Conduct of Human Embryonic Stem Cell Research*. Retrieved from http://www.isscr.org/docs/default-source/hesc-guidelines/isscrhescguidelines2006.pdf.

International Society for Stem Cell Research (2008, December 3) *Guidelines for the Clinical Translation of Stem Cells*. Retrieved from http://www.isscr.org/docs/guidelines/isscrglclinicaltrans.pdf.

Kimmelman, J. & Daley, G.Q. (2015, June 26) Guidelines for stem cell science and clinical translation (draft) *International Society for Stem Cell Research*. Retrieved from http://www.isscr.org/docs/default-source/hesc-guidelines/isscr-draft-guidelines-for-stem-cell-science-and-clinical-translation.pdf?sfvrsn=2.

Stem cell research policy

Australian Stem Cell Centre (2010) *Ethics & law of stem cell research: Fact sheet 6*. Retrieved from http://www.stemcellfoundation.net.au/docs/fact-sheets/fact-sheet-6---law-and-ethics-of-stem-cell-research.pdf?sfvrsn=13.

Canadian Institute for Health Research (2013) Updated Guidelines for Human Pluripotent Stem Cell Research into the second edition Tri-Council Policy Statement: Ethical Conduct for Research Involving Humans (TCPS 2). Retrieved from http://www.pre.ethics.gc.ca/pdf/eng/consultation/Stem_Cell_Integration_Final_EN.pdf.

Indian Council of Medical Research (2013) National Guidelines for Stem Cell Research. Indian Council of Medical Research Department of Health Research and Department of Biotechnology. Retrieved from https://www.ncbs.res.in/sites/default/files/policies/NGSCR%202013.pdf.

National Academies Press (2005) Guidelines for Human Embryonic Stem Cell Research. Retrieved from http://www.nap.edu/catalog/11278/guidelines-for-human-embryonic-stem-cell-research.

National Institute of Health (2009a) Human Embryonic Stem Cell Policy Under Former President Bush (Aug. 9, 2001–Mar. 9, 2009). In Stem Cell Information [World Wide Web site]. Bethesda, MD: National Institutes of Health, U.S. Department of Health and Human Services, 2009. Retrieved from http://stemcells.nih.gov/policy/pages/2001policy.aspx.

National Institute of Health (2009b) 2009 Guidelines on Human Stem Cell Research. In Stem Cell Information [World Wide Web site]. Bethesda, MD: National Institutes of Health, U.S. Department of Health and Human Services, 2011. Retrieved from http://stemcells.nih.gov/policy/pages/2009guidelines.aspx.

National Institute of Health (2015) Policy. In Stem Cell Information [World Wide Web site]. Bethesda, MD: National Institutes of Health, U.S. Department of Health and Human Services, 2015. Retrieved from http://stemcells.nih.gov/policy/Pages/Default.aspx.

Small, S. and Doherty, K. (2011, December 14) *Regulation of stem cell research in the United Kingdom*. Retrieved from http://www.eurostemcell.org/regulations/regulation-stem-cell-research-united-kingdom.

UNESCO (2004) *National legislation concerning human reproductive and therapeutic cloning*. Retrieved from http://unesdoc.unesco.org/images/0013/001342/134277e.pdf.

The role of politics and public opinion in shaping stem cell policy

Pew Research Center (2008a) *Stem cell research at the crossroads of religion and politics*. Retrieved from http://www.pewforum.org/2008/07/17/stem-cell-research-at-the-crossroads-of-religion-and-politics/.

Pew Research Center (2008b) *Declining majority of Americans favor embryonic stem cell research*. Retrieved from http://www.pewforum.org/2008/07/17/declining-majority-of-americans-favor-embryonic-stem-cell-research/.

FURTHER READING

Books and reviews

Katrien Devolder (2015). *The ethics of embryonic stem cell research* (p. 176). New York: Oxford University Press.

Kimbrel, E. A., & Lanza, E. (2015). Current status of pluripotent stem cells: Moving the first therapies to the clinic. *Nature Reviews Drug Discovery, 14*, 681–692. doi: 10.1038/nrd4738 http://www.nature.com/nrd/journal/v14/n10/full/nrd4738.html.

Okano, H., Nakamura, M., Yoshida, K., Okada, Y., Tsuji, O., Nori, S., Ikeda, E., Yamanaka, S., & Miura, K. (2013). Steps toward safe cell therapy using induced pluripotent stem cells. *Circulation Research, 112*, 523–533. *doi: 10.1161/CIRCRESAHA.111.256149* http://circres.ahajournals.org/content/112/3/523.

Web resources

Stem Cell Policy: World Stem Cell Map-MBBNet: http://www.mbbnet.umn.edu/scmap.html.

Tissue engineering and regenerative medicine: http://www.nibib.nih.gov/science-education/science-topics/tissue-engineering-and-regenerative-medicine.

The heart makers: https://www.youtube.com/watch?v=pd3TFB0wOI0.

Bioengineered kidney makes urine: https://www.newscientist.com/article/dn23382-kidney-breakthrough-complete-lab-grown-organ-works-in-rats/.

Concerns about stem cell tourism: https://www.youtube.com/watch?v=SgHBi1y1KG8.

RECOMBINANT DNA TECHNOLOGY AND GENETICALLY MODIFIED ORGANISMS

All the flowers of all the tomorrows are in the seeds of today
-**Indian Proverb**

CHAPTER OUTLINE

4.1 Introduction 84
4.2 Gene Therapy 86
    4.2.1 Technique 88
    4.2.2 Application of Gene Therapy 88
    4.2.3 Severe Combined Immune Deficiency (SCID) 89
    4.2.4 Challenges in Gene Therapy 89
4.3 Gene Editing Therapy 90
4.4 GMOs and the New Biotech Industry 91
    4.4.1 Recombinant Proteins 91
    4.4.2 Recombinant Antibodies 91
    4.4.3 Pharming 97
4.5 Genetically Modified Crops (GM Crops) 100
    4.5.1 Health and Safety Concerns 103
    4.5.2 Environmental Concerns 104
    4.5.3 Ethical and Socioeconomic Concerns 105
4.6 Transgenic Animals 105
    4.6.1 Disease Models 106
    4.6.2 Increasing Food Production 106
    4.6.3 Producing Proteins for Industry 108
    4.6.4 Vector Control 109
    4.6.5 Pets 109
    4.6.6 Hypoallergenic Pets 111
4.7 Synthetic Organisms 111
4.8 Challenges in Applications of rDNA Technology and GMO Release 112
    4.8.1 Effect on Environment 112
    4.8.2 Effect on Biodiversity 112
    4.8.3 Effect on Sociocultural Norms 113
    4.8.4 Effect on Socioeconomic Status of Farmers 113

An Introduction to Ethical, Safety and Intellectual Property Rights Issues in Biotechnology.
DOI: http://dx.doi.org/10.1016/B978-0-12-809231-6.00004-1

4.9 Objections to Genetic Engineering and GMOs 114

    4.9.1 Proportionality 114

    4.9.2 Slippery Slope 114

    4.9.3 Subsidiarity 114

4.10 GMOs and "Biopolitics" 117

4.11 Summary 121

References 122

Further Reading 126

The first gene therapy patients Cynthia Kisik and Ashanti DeSilva in 1992 with the pioneer physicians of gene therapy: (from left) French Anderson, MD; R. Michael Blaese, MD; and Kenneth Culver, MD.

(Photo used with permission from the **Immune Deficiency Foundation**).

4.1 INTRODUCTION

Born with a rare genetic disorder known as Severe Combined Immune Deficiency (SCID) caused by a mutation in a gene that encodes an enzyme adenosine deaminase (ADA), Ashanti DeSilva could not fight infections and had a bleak chance of survival. ADA is required to prevent toxic build up in white blood cells which give the body immunity to diseases. In 1990, when Ashanti was 4 years old, doctors William French Anderson, Michael Blaese, and Kenneth Culver inserted a synthetic DNA containing a normal ADA gene into her bone marrow cells, which allowed her to mount a normal immune response to infections. Gene therapy which has helped Ashanti and several others overcome genetic defects is a powerful example of recombinant DNA technology.

The genesis of recombinant DNA technology lay in the discovery of a unique class of enzymes known as the "**restriction endonucleases.**" Discovered in 1970 by Daniel Nathans, Werner Arber and Hamilton Smith at the Johns Hopkins University, Baltimore, these enzymes had the **ability to cleave DNA at specific sites** (a discovery that earned them the Nobel Prize in Medicine in 1978) (Nobel Media, 2014). Restriction endonucleases could be isolated from several different strains of bacteria and were each distinct in that they specifically recognized and cleaved DNA at different "target" sites. Three years later, using these enzymes, Stanley Cohen of Stanford University in collaboration with Herbert Boyer of University of California, San Francisco, cut circular extra nuclear DNA molecules known as "**plasmids**" that are present in multiple copies and can be easily isolated from common bacteria such as *Escherichia coli* (*E. coli*). By **inserting a "foreign" DNA fragment** into a **cut plasmid** and **returning it into a bacterial host**, they were able to obtain several copies of the DNA insert as the bacteria multiplied in a culture broth (Cohen, Chang, Boyer, & Helling, 1973). The process referred to as "**DNA cloning**" (see Fig. 4.1) has grown to be the key technique powering biotechnological advancements.

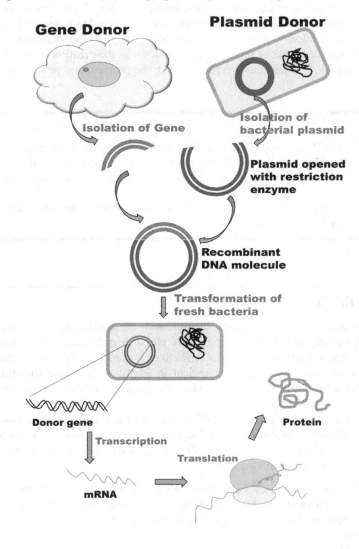

FIGURE 4.1

Gene cloning.

As DNA is ubiquitous, found in living organisms across taxonomic classes, genes from any organism (of animal, plant, bacterial, or viral origin) could be **cloned in bacteria**. With the discovery of more cloning vectors, such as plasmids that could integrate into plant genomes, and several viruses, it was soon possible to **clone genes in plant and animal hosts** also. Thus, potentially, "designer" chimeric organisms containing specific genes conferring characteristics of choice could be created in the laboratory. This technology is referred to as *recombinant DNA (rDNA) technology* or *genetic engineering (GE)*. On the one hand, rDNA technology provides scientists the tools to devise gene therapy to treat diseases (as in the case of Ashanthi), precisely breed livestock and crops with superior agronomic characteristics, and create "synthetic" life forms for meeting environmental challenges such as oil spills and managing plastic waste. On the other hand, it can possibly open a proverbial Pandora's Box creating monsters defying the definition of "life." This chapter aims to examine the scope of this rDNA technology in the creation of genetically modified organisms (GMOs) and aims to put into perspective both the promise and public perceptions associated with it.

KEY TAKEAWAYS
- Recombinant DNA Technology is defined by the Encyclopedia Britannica as "the joining together of DNA molecules from different organisms and inserting it into a host organism to produce new genetic combinations that are of value to science, medicine, agriculture and industry."
- New organisms created using rDNA technology (also known as Genetic Engineering) are called GMOs or transgenic organisms.

4.2 GENE THERAPY

Many diseases arise due to changes in genes (**mutations**); these diseases can be due to alterations in single gene (**monogenic disorders**) which include blood disorders such as hemophilia and thalassemia, or multiple genes (**polygenic**) such as Alzheimer's disease. Cancer is another group of disorders that arise due to gene mutations. It is logical to attempt and **repair the mutated genes** to treat the disease; however, this has been found to be surprisingly difficult despite efforts since 1972.

Gene therapy is the use of segments of DNA (genes) to prevent or treat a disease caused by gene mutation. (The definition excludes use of drugs to treat diseases caused by specific gene mutations as in some cancers; these are referred to in Oncology as targeted therapy where the gene product is the target.) Gene therapy is of two types:

1. *Somatic gene therapy*: Transfer of a section of DNA to a tissue other than sperm or egg. The effect will be seen **only in the body of the recipient** and **will not be inherited** by progeny.
2. *Germ-line gene therapy*: Transfer of **gene into sperm or egg**; the effects will be **inherited** by future generations.

For ethical reasons, gene therapy in humans is restricted to somatic gene therapy. Somatic gene therapy is of two types depending on where the manipulation (insertion of the gene into relevant cells) is done:

1. *Ex vivo*: This means the exterior; relevant cells are **modified outside the body** and infused back. For instance, stem cells can be removed, infected with gene carrying viruses that enter the cells and insert the gene into the nucleus, and the cells can then be infused back.
2. *In vivo*: Here the vector carrying the gene is **injected into the body** where it goes to the target tissues; this is a "blind" procedure.

Gene therapy can be used for two broad groups of diseases:

1. **Replace a defective gene** which is not working, i.e., not producing the relevant protein (*gene augmentation therapy*). Examples include immune deficiency disorders, hematological disorders, and cystic fibrosis (see Fig. 4.2A).
2. **Inhibit a defective gene** which is over-expressed and causing disease such as cancer (*gene inhibition therapy*). The inhibition can be directly of the gene, or the new gene can produce a protein which blocks the product of the mutated gene (see Fig. 4.2B).

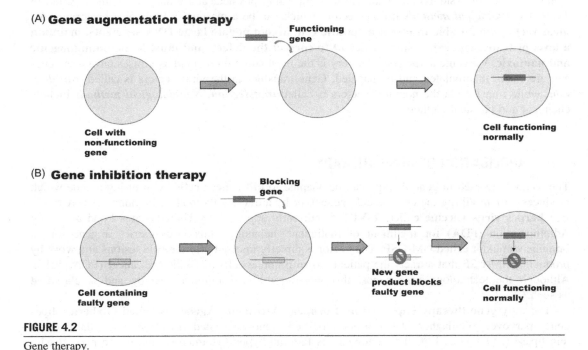

(A) **Gene augmentation therapy**

Functioning gene

Cell with non-functioning gene

Cell functioning normally

(B) **Gene inhibition therapy**

Blocking gene

Cell containing faulty gene

New gene product blocks faulty gene

Cell functioning normally

FIGURE 4.2

Gene therapy.

Table 4.1 Gene Delivery Systems

Biological	Chemical Transfection	Physical Transfection
Viral Vectors	Lipid complexes	Electroporation (electric current making pores)
Adenoviruses	Calcium phosphate	Microinjection
Adeno-associated viruses (AAV)	DEAE-dextran	Ballistic particles ("gene guns")
Alphaviruses		
Herpesviruses		
Lentiviruses		
Retroviruses		
Vaccinia viruses		
NonViral Vectors		
Naked DNA		
Oligonucleotides		

4.2.1 TECHNIQUE

Introduction of viable functioning genes into living cells has proved to be difficult but several methods (delivery systems) are available now including physical and biological methods listed in Table 4.1. *Biological methods* use a "vector" which can be a plasmid, a virus or a bacterium. **An ideal vector should able to target a specific cell, accommodate large DNA segments, maintain a level of transgenic expression sufficient to correct the defect, and must be nonimmunogenic and nontoxic.** Virus mediated gene delivery is the most common method as viruses have less genes and are easily manipulated and propagated. Gene transfer via the viral vectors is called *transduction*, while transfer via the nonviral vectors is called *transfection*. *Nonbiological methods* include chemical and physical methods.

4.2.2 APPLICATION OF GENE THERAPY

The technology used in gene therapy can be adapted to **kill cancer cells** by inserting a gene which produces a toxin killing the cell (suicide genes), or by attracting the body's immune cells. A modified **herpes virus vaccine called T-VEC** is currently approved by United States Food and Drug Administration (FDA) for treatment of malignant melanoma; this virus carries a gene for an immune stimulant called GM-CSF. The vector is directly injected into the skin lesions and work by producing GM-CSF that attracts the patient's own immune cells that kills the cancer (FDA, 2015). Although the technology uses genes, this works more like immune therapy and is classified as such.

The **only gene therapy approved by European Medicines Agency is called Glybera** (alipogene tiparvovec) (Gallagher, 2012). It is a modified adeno-associated virus that carries the lipoprotein lipase (LPL) gene. LPL deficiency causes familial hyperchylomicronemia, which present with recurrent pancreatitis, leading to early onset diabetes and cardiovascular complications. The vector is injected directly into muscle where the gene gets inserted into the nucleus of muscle cells and produces the enzyme.

Other medical conditions where gene therapy has been used with limited success include:

- *Immune deficiency diseases*: SCID, X-linked immune deficiency (X-SCID) (gene: IL2RG), ADA deficiency (ADA-SCID), Wiskott–Aldrich syndrome
- *Hematological diseases*: Hemophilia, Beta-thalassemia
- *Ophthalmological*: Leber congenital amaurosis (gene involved RPE65), Choroideremia
- *Metabolism*: LPL deficiency
- *Cancer*: Leukemia/lymphoma

4.2.3 SEVERE COMBINED IMMUNE DEFICIENCY (SCID)

No story of gene therapy would be complete without mention of the "Bubble boy." David Vetter was born in 1971 in Texas, USA, with SCID; his immune system was so weak that even a minor infection could have killed him. Since his brother had died of infection, doctors were aware that he would also suffer from this condition, and right from his birth, he lived in a specially constructed sterile plastic bubble till he died at the age of 12. NASA even designed a special suit for him so that he could walk outside. Gene therapy was in its infancy but it was thought he had a chance of a bone marrow transplant and underwent this procedure using stem cells donated by his sister. However, four months later, he died of lymphoma caused by Epstein Barr virus, thought to be introduced through the stem cells donated by his sister.

It was only in 1990 that another victim of SCID, this time caused by ADA deficiency, 4-year-old Ashanthi DeSilva, underwent gene therapy. This was only partially successful but techniques have since then been refined. A recent report showed that 14 of the 16 children who received this experimental therapy 9 years previously were living normal lives (Geddes, 2013).

4.2.4 CHALLENGES IN GENE THERAPY

Problems associated with gene therapy include the following:

- the **genetic cause of the disease must be known**, the **gene cloned and ready for use**
- a **vector must be found** to carry the gene safely into the target cells and to insert the gene in the correct location in the nucleus
- the **gene must be inserted in sufficient amounts to produce enough protein** for the whole body and for the life time of the patient.

Although these issues lead to failure of therapy, one problem that can lead to death of a patient due to therapy is **induction of cancer**. If the gene is incorrectly inserted near a cancer causing gene (oncogene), it could stimulate the oncogene with serious outcome. In one such case, a patient treated with retroviral vector for X-SCID developed leukemia; it was found that the virus had inserted near LMO-2 gene on chromosome 11. This gene is responsible for development of both the lymphoid and myeloid series and is the site of a translocation that occurs in leukemia. Many such cases have been reported and continue to be a vexing issue. **Gene therapy trials were stopped pending review after the death of 18-year-old Jesse Gelsinger**, suffering from ornithine transcarbamoylase deficiency. He was infused with an adenovirus containing the corrective genes; this resulted in an overwhelming immune reaction resulting in multiorgan failure and death. However, despite these setbacks, trials on humans are on and the website https://clinicaltrials.gov/ currently lists more than **4500 ongoing trials**.

4.3 GENE EDITING THERAPY

Gene editing therapy involves **altering the sequence of the in situ human genome** to achieve a therapeutic effect (unlike gene therapy which involves addition of new genes to human cells). This includes the correction of mutations, the addition of therapeutic genes to specific sites in the genome, and the removal of mutated genes or genome sequences. Powerful new molecular tools that have been developed that can "edit" genes include ZFN (zinc finger nucleases), TALEN (transcription activator like effector nucleases), meganucleases, and the recently refined CRISPR/Cas9 system. Essentially, these use naturally occurring enzyme systems to cut and manipulate the genome at predefined locations to correct errors. Applications include antiviral strategies, immunotherapies, and the treatment of monogenic hereditary disorders. A detailed description of the method is beyond the scope of this book and the interested reader is referred to a recent review (Maeder and Gersbach, 2016).

The first success of this approach in cancer was reported in November 2015. A 1-year-old girl, Layla, who suffered from refractory leukemia was infused with gene-edited donated immune T cells called UCART19. Using TALEN, specific genes were edited in order to make these T-cells behave in two specific ways. First, the cells became invisible to a powerful anti-leukemia drug that would usually kill them, and second, they were reprogrammed to only target and fight against leukemia cells. After one infusion and months of isolation, doctors at Great Ormond Street Hospital, London, have declared Layla leukemia free (LePage, 2015).

The first study in humans of this technology was in HIV infection. A biotech company, Sangamo BioSciences in Richmond, California, used ZFN gene-edited cells to treat 12 people with HIV. Blood of the patients was extracted, ZFNs used to cut out a gene for a protein on T-cells targeted by HIV and the blood infused back. The results were positive—50% of patients could stop taking HAART (Reardon, 2014). At present, more than 70 patients have been recruited for this trial.

Despite more than four decades of work, classical gene therapy has had limited success. It is hoped that with the advent of powerful new gene editing tools, the promise of gene therapy will be fulfilled.

KEY TAKEAWAYS
- Gene therapy is the use of segments of DNA (i.e., genes) to prevent or treat a disease caused by gene mutation.
- Targeted therapy against gene mutations as in cancer treatment use drugs (and not DNA sequences) and cannot be called gene therapy.
- Despite initial promises, gene therapy remains largely experimental restricted to few medical centers.

- It has been successful in few monogenic disorders especially in nonmalignant disorders.
- Difficulties faced include stable transfection of the target cells with an active gene, and the occasional occurrence of leukemia and multiorgan failure.
- It is hoped that new genome-editing technologies will dramatically improve our abilities to alter gene function and push this field forward.

4.4 GMOs and the New Biotech Industry

Shortly after the publication of Stanley Cohen's and Herbert Boyer's cloning experiments, a venture capitalist Robert A. Swanson approached Boyer with a proposal for commercial exploitation of the technology. Swanson's enthusiasm and faith in the commercial viability of the idea was so contagious that the **first biotech industry "Genentech"** (http://www.gene.com/) was created on April 7, **1976**. Since then, the biotech industry has grown: as of 2015, about 670 public companies and over 200 thousand employees generate around 133 billion US dollars of biotech revenue (Statista, 2016). The key regions for the global biotech industry are the United States and Europe.

4.4.1 RECOMBINANT PROTEINS

Genentech produced the **first human protein, somatostatin, in *E. coli* cells in 1977**. The next year the company cloned the human insulin gene and in 1982, the first recombinant DNA drug to be marketed, **"Humulin,"** was licensed to Eli Lilly and Company. **Human growth hormone** cloned by Genentech in 1979, received US Food and Drug Authority (FDA) approval in 1985 to be marketed as Protropin (Somatrem for injection), a growth hormone for children with growth hormone deficiency. It became the first recombinant pharmaceutical product to be manufactured and marketed by a biotechnology company. Subsequently the company received FDA approval for **interferon alpha-2a for treatment of hairy cell leukemia** (licensed to Hoffmann-La Roche Inc. as Referon-A) in 1986, and in 1987 to market Activase, a **tissue−plasmogen activator** to dissolve blood clots in patients with acute myocardial infarction (Genentech, n.d.). Many more pharmaceuticals produced by Genentech and other companies representative of the pharmaceuticals produced using rDNA technology approved for marketing in the United States and European Union (EU) are listed in Table 4.2.

Pharmaceutical compounds produced in living systems are known as **"biopharmaceuticals"** or **"biologics."** This term includes recombinant proteins produced in different host organisms, as well as medicines and diagnostics produced from cells and cell cultures, such as monoclonal (and recombinant) antibodies and vaccines. Globally, around 170 billion US dollars were spent on biopharmaceuticals in 2013 and is expected to exceed 220 billion US dollars by 2017 (Statista, 2016).

4.4.2 RECOMBINANT ANTIBODIES

Antibodies (also known as immunoglobulins, "Ig") are large Y-shaped glycoproteins used by the immune system and are produced by the B-lymphocytes. Typically, an antibody is made up of four

Table 4.2 Recombinant DNA Technology in Production of Recombinant Proteins for Pharmaceutical Applications

A: Proteins Expressed in Microbial Hosts

Product	Company	Therapeutic Indication	Date Approved
Recombinant Hormones			
Insulin			
Humulin (rh insulin produced in *E. coli*)	Eli Lilly	Diabetes mellitus	1982 (US)
Lantus (insulin glargine, long-acting rh insulin analog produced in *E. coli*)	Aventis	Diabetes mellitus	2000 (EU and US)
Exubera (rh insulin produced in *E. coli*)	Pfizer (New York)/Aventis (Kent, UK)	Diabetes mellitus	2006 (EU and US)
Human Growth Hormone			
Protropin (r hGH differing from hGH only in containing an additional N-terminal methionine residue; produced in *E. coli*)	Genentech	hGH deficiency in children	1985 (US)
Humatrope (r hGH produced in *E. coli*)	Eli Lilly	hGH deficiency in children	1987 (US)
Omnitrop (somatropin, rh GH produced in *E. coli*)	Sandoz (Basel)	Certain forms of growth disturbance in children and adults	2006 (EU and US)
Other Hormones			
Glucagen (rh glucagon produced in *S. cerevisiae*)	Novo Nordisk	Hypoglycemia	1998 (US)
Fortical (r salmon calcitonin produced in *E. coli*)	Upsher-Smith Laboratories (Minneapolis, MN, USA)/ Unigene (Fairfield, NJ, USA)	Postmenopausal osteoporosis	2005 (US), 2003 (EU)
Recombinant Growth Factors			
Granulocyte-Macrophage Colony-Stimulating Factor			
Neupogen (filgrastim, r GM-CSF differing from human protein by containing an additional N-terminal methionine; produced in *E. coli*)	Amgen	Chemotherapy-induced neutropenia	1991 (US)
Leukine (r GM-CSF, differing from the native human protein by one amino acid, Leu23; produced in *E. coli*)	Immunex (now Amgen)	Autologous bone marrow transplantation	1991 (US)

Table 4.2 Recombinant DNA Technology in Production of Recombinant Proteins for Pharmaceutical Applications *Continued*

A: Proteins Expressed in Microbial Hosts

Product	Company	Therapeutic Indication	Date Approved
Recombinant Interferons and Interleukins			
Interferon-α			
Alfatronol (rhIFN-α-2b produced in *E. coli*)	Schering-Plough	Hepatitis B, C and various cancers	2000 (EU)
Interferon-β			
Betaferon (r IFN-β1b differing from human protein by C17S substitution; produced in *E. coli*)	Schering AG	Multiple sclerosis	1995 (EU)
Others			
Proleukin (r IL-2, differing from human molecule in absence of an N-terminal alanine and a C125S substitution produced in *E. coli*)	Chiron	Renal-cell carcinoma	1992 (US)
Recombinant Vaccines			
Hepatitis B			
Recombivax (r HBsAg produced in *S. cerevisiae*)	Merck	Hepatitis B prevention	1986 (US)
Tritanrix-HB (combination vaccine containing r HBsAg produced in *S. cerevisiae* as one component)	GlaxoSmith Kline (GSK, Brentford, UK)	Vaccination against hepatitis B, diphtheria, tetanus, and pertussis	1996 (EU)
Hexavac (combination vaccine containing rHBsAG produced *S. cerevisiae* as one component)	Aventis Pasteur (Lyon, France)	Immunization against diphtheria, tetanus, pertussis, hepatitis B, polio, and *H. influenzae* type b	2000 (EU)
Ambirix (combination vaccine containing rHBsAg produced in *S. cerevisiae* as one component)	GlaxoSmith Kline (GSK, Brentford, UK)	Immunization against hepatitis A and B	2002 (EU)
Others			
Gardasil (quadrivalent human papillomavirus (HPV) recombinant vaccine; contains major caspid proteins from four HPV types, produced in *S. cerevisiae*)	Merck	Vaccination against diseases caused by HPV	2006 (US)

(Continued)

Table 4.2 Recombinant DNA Technology in Production of Recombinant Proteins for Pharmaceutical Applications *Continued*

B: Proteins Expressed in Animal Cell Lines

Recombinant Blood Factors

Factor VIII

Recombinate (rh Factor VIII produced in an animal cell line)	Baxter Healthcare (Deerfield, IL, USA)/ Genetics Institute	Hemophilia A	1992 (US)
Bioclate (rh Factor VIII produced in CHO cells)	Aventis Behring (King of Prussia, PA, USA)	Hemophilia A	1993 (US)
Helixate NexGen (octocog α; rh Factor VIII produced in BHK cells)	Bayer	Hemophilia A	2000 (EU)
Advate (octocog-α rh Factor VIII produced in CHO cell line grown in serum-free medium free from animal products)	Baxter (Leverkusen, Germany)	Hemophilia A	2004 (EU), 2003 (US)

Other Blood Factors

Benefix (rh Factor IX produced in CHO cells)	Genetics Institute	Hemophilia B	1997 (US and EU)

Recombinant Thrombolytics and Anticoagulants

Tissue Plasmogen Activator

Activase (alteplase, rh tPA produced in CHO cells)	Genentech	Acute myocardial infarction	1987 (US)
Tenecteplase (also marketed as Metalyse) (TNK-tPA, modified r tPA produced in CHO cells)	Boehringer Ingelheim (Ridgefield, CT, USA)	Myocardial infarction	2001 (EU)

Others

Xigris [drotrecogin-α; rh activated protein C produced in a mammalian (human) cell line]	Eli Lilly (Indianapolis, IN)	Severe sepsis	2001 (US), 2002 (EU)

Table 4.2 Recombinant DNA Technology in Production of Recombinant Proteins for Pharmaceutical Applications *Continued*

B: Proteins Expressed in Animal Cell Lines

Recombinant Growth Factors

Erythropoietin

Epogen (rh EPO produced in a mammalian cell line)	Amgen	Anemia	1989 (US)

Other Recombinant Products

Bone Morphogenic Proteins

Infuse bone graft (contains dibotermin α, a rh BMP-2 produced in CHO cells placed on an absorbable collagen sponge	Wyeth	Treatment of acute open tibial shaft fracture	2004 (US)

Recombinant Enzymes

Pulmozyme (dornase-α, r DNase produced in CHO cells)	Genentech	Cystic fibrosis	1993 (US)

Nucleic Acid-Based Products

Vitravene (fomivirsen, an antisense oligonucleotide)	Isis Pharmaceuticals (Carlsbad, CA, USA)	Treatment of cytomegalovirus retinitis in AIDS patients	1998 (US), 1999 (EU)
Macugen (pegaptanib sodium injection, a synthetic pegylated oligonucleotide that specifically binds vascular endothelial growth factor)	Eyetech (New York)/Pfizer	Neovascular, age-related macular degeneration	2006 (EU), 2004 (US)

(For a complete list of biopharmaceuticals approved in the US and EU prior to 2006 see http://www.nature.com/nbt/journal/v24n7/fig_tab/nbt0706-769_T1.html) Adapted by permission from Macmillan Publishers Ltd: [Nature Biotechnology] (Walsh, G. (2006) Biopharmaceutical benchmarks 2006. Nature Biotechnology **24**, 769–776. doi: 10.1038/nbt0706-769), © (2006).

linked peptide chains, two of which are long with 450 to 550 amino acid residues (known as the *heavy* chains), whereas two are short with 210–220 amino acid residues (known as the *light* chains). The assembly of the protein presents two distinct domains—the *variable* domain present at the "tips" of the Y-shaped molecule, which is the antigen binding site; and the *constant* domain at the "base" of the molecule that specifies the class of the antibody and is integral to antibody action such as neutralization, agglutination, precipitation, or complement activation. By means of somatic recombination occurring in B lymphocytes between gene segments specifying various regions of the peptide chains, a large number of different types of Ig molecules are generated which by virtue of different variable domains, can recognized different antigens. Each B-lymphocyte can produce only one type of Ig molecule, but in the event of antigen–antibody recognition (as happens during an infection), clonal propagation of the lymphocyte results in a large amount of the specific antibody to be produced in order to mount a defense. Immunization against diseases can be done either through *active immunization* or *vaccination* (in which antigenic substances, such as inactivated viruses or coat proteins, are injected into the body so as to train the immune system to produce specific antibodies against the antigen in the event of an infection), or *passive immunization* in which the antibodies are directly injected into the patient. The latter strategy is currently finding favor in clinical practice in the treatment of several cancers, inflammatory diseases such as arthritis, psoriasis, and auto-immune disorders. Monoclonal antibodies (mAbs) are also crucial to many diagnostic tests using ELISA (Enzyme Linked ImmunoSorbent Assay), RIA (RadioImmunoAssay), and as imaging and immunosensor agents.

mAb production at a commercial scale could be achieved using the procedure developed by Kohler and Milstein (1975). Referred to as the "**hybridoma technology**," the technique involves immunizing a mouse with an antigen of interest and then **fusing the B-lymphocytes from the spleen with immortal cancer (myeloma) cells**. The hybridomas created by the fusion secrete antibodies into the surrounding fluid. For production of the mAbs, the hybridomas are either **injected into the abdomen of a second mouse** (the ascites method), or **cultured in bioreactors** (the in vitro method). **Ethical issues, specifically animal welfare concerns, are associated with hybridoma technology** since animals need to be euthanized for extraction of the hybridomas from the spleen, and the ascites method causes considerable distress and pain to the animals. **Problems for therapeutic applications in humans include adverse reactions** (human antimurine antibody or **HAMA response**) that limit the duration of effective antibody treatment. Creation of human mAbs is hampered by the absence of a suitable human myeloma fusion partner, and mostly relies on human mouse fusions, which are unstable.

An **alternative to the use of animals** has been the production of *recombinant antibodies (rAbs)* which are **created using rDNA technology**. The first step in this direction came from the demonstration by McCafferty, Griffiths, Winter, and Chiswell (1990) that antibody fragments could be displayed on bacteriophages. **Antibody phage display libraries** could be created by **introducing DNA sequences specifying antibody fragments into phage genomes by means of suitable vectors**. For instance, the antibody variable region genes can be fused to the *pIII* phage coat gene. Vectors carrying the antibody variable region are introduced into bacteria which are then infected with modified phages. The **daughter phages produced on lysis will display the functional antibody fragments linked to the phage coat proteins**. Antibody display has since been possible in other microorganisms including yeast (Boder & Wittrup, 1997). In yeast display platforms, the antibody variable region genes are **fused to the yeast *Aga2p* gene**. The protein product from this gene forms a linkage with a cell-wall component Aga1p. The **antibody-Aga2p fusion protein is displayed** on the surface anchored by the association of the Aga2p with Aga1p. Full length antibodies too can now be displayed using platforms such as the *E. coli*-based E-clonal systems (Mazor, Van Blarcom, Iverson, & Georgiou, 2008).

Antibodies that bind to a specific antigen target can be isolated from an antibody display library by techniques such as fluorescence-activated cell sorting, ELISA, or paramagnetic beads. rAbs that remain bound to the antigen are **removed and characterized**. The **genes for the antibody are cloned in expression vectors** in suitable expression systems. Full length antibody expression is mostly achieved in yeast or mammalian cell lines, whereas antibody fragments are mostly expressed in bacterial systems (Echko & Dozier, 2010). Table 4.3 lists some of the mAbs used in cancer therapy and as vaccines.

General production methods for the manufacture of nonanimal rAbs can be broken down into five general steps:

(1) creation of an antibody gene library;
(2) display of the library on phage or cell surfaces;
(3) isolation of antibodies against the antigen of interest;
(4) modification of the isolated antibodies; and
(5) scaled up production of selected antibodies in a cell culture expression system
(Johns Hopkins Bloomberg School of Public Health, 2016).

Advantages of rAbs include the following:

- the ability to isolate antibodies against very specific antigens;
- to be able to switch between isoforms of the antibody once selected for antigen specificity; and
- the advantage of being an entirely nonanimal technology.

Production of rAbs can also be done more rapidly (around 8 weeks) than conventional methods (4−6 months) once an antibody display library has been created.

4.4.3 PHARMING

The initial reports of production of biopharmaceuticals involved expressing proteins in bacterial cells. However, the technology soon evolved to produce these compounds in transgenic animals and plants (also known as "**molecular farming**" or "**pharming**"). Basic steps involve **growing the transgenic plant or animal, harvesting, and downstream processing for extraction and purification of the protein**. There are a number of advantages in pharming over microbial systems:

- Many proteins require the addition of sugar moieties (glycosylation) to be active, which does not happen very efficiently in microbes.
- Many proteins need to be folded into tertiary and quaternary structures which microbes may not be able to do.
- The economics for mass production too are better for pharming, since scale up only requires a larger area under the crop, or increasing the number of animals, and avoids many of the issues of bioprocess engineering in large fermenters.

Table 4.3 Examples of Therapeutic Monoclonal Antibodies

Product	Company	Therapeutic Indication	Date Approved
Rituxan, MabThera (rituximab)	Genentech and Biogen-Idec	Diffuse large B-cell, CD20-positive, non-Hodgkin's lymphoma	2006 (FDA), 1998 (EMA)
Gardasil (**Quadrivalent Human Papillomavirus (Types 6, 11, 16, 18) Recombinant Vaccine**)	Merck & Co., Inc.	Vaccination of females 9 to 26 years of age for prevention of diseases caused by Human Papillomavirus (HPV)	2006 (FDA), 2006 (EMA)
Avastin (bevacizumab)	Genentech, Inc.	Locally advanced, metastatic or recurrent nonsmall cell lung cancer	2006 (FDA), 2005 (EMA)
Herceptin (trastuzumab)	Genentech, Inc	HER2 overexpressing metastatic gastric or gastroesophageal	2010 (FDA), 2000 (EMA)
Erbitux (cetuximab)	ImClone LLC	Recurrent locoregional disease and/or metastatic squamous cell carcinoma of the head and neck (SCCHN)	2011 (FDA), 2004 (EMA)
Xgeva (denosumab)	Amgen Inc.	Giant cell tumor of bone	2013 (FDA), 2011 (EMA)
Gazyva, Gazyvaro (obinutuzumab)	Genentech, Inc	Chronic lymphocytic leukemia (CLL)	2013 (FDA), 2014 (EMA)
Arzerra (ofatumumab)	GlaxoSmithKline	Chronic lymphocytic leukemia (CLL)	2014 (FDA), 2010 (EMA)
Cyramza (ramucirumab)	Eli Lilly and Company	Advanced gastric or GEJ adenocarcinoma	2014 (FDA), 2014 (EMA)
Unituxin (dinutuximab)	United Therapeutics Corporation	Pediatric patients with high-risk neuroblastoma	2015 (FDA), 2015 (EMA)
Keytruda (pembrolizumab)	Merck Sharp and Dohme Corporation	Metastatic nonsmall cell lung cancer (NSCLC)	2015 (FDA), 2015 (EMA)
Daralex, Darzalex (daratumumab)	Janssen Biotech, Inc.	Multiple myeloma	2015 (FDA), 2016 (EMA)
Opdivo (nivolumab)	Bristol-Myers Squibb Company	Advanced renal cell carcinoma	2015 (FDA), 2015 (EMA)
Portrazza (necitumumab)	Eli Lilly and Company	metastatic squamous nonsmall cell lung cancer (NSCLC)	2015 (FDA), 2016 (EMA)
Empliciti (elotuzumab)	Bristol-Myers Squibb Company	Multiple myeloma	2015 (FDA), 2016 (EMA)
Tecentriq (atezolizumab)	Genentech, Inc.	Locally advanced or metastatic urothelial carcinoma and metastatic nonsmall cell lung cancer (NSCLC)	2016 (FDA)
Lartruvo (olaratumab)	Eli Lilly and Company	Soft tissue sarcoma (STS)	2016 (FDA)

Table 4.4 Comparison of Production Systems for Recombinant Human Pharmaceutical Proteins

System	Overall Cost	Production Timescale	Scale-up Capacity	Product Quality	Glycosylation	Contamination Risks	Storage Costs
Bacteria	Low	Short	High	Low	None	Endotoxins	Moderate
Yeast	Medium	Medium	High	Medium	Incorrect	Low risk	Moderate
Mammalian cell cultures	High	Long	Very low	Very high	Correct	Viruses, prions and oncogenic DNA	Expensive
Transgenic animals	High	Very long	Low	Very high	Correct	Viruses, prions and oncogenic DNA	Expensive
Plant cell cultures	Medium	Medium	Medium	High	Minor differences	Low risk	Moderate
Transgenic plants	Very low	Long	Very high	High	Minor differences	Low risk	Inexpensive

Reprinted by permission from Macmillan Publishers Ltd: [Nature Reviews Genetics] (Ma, J. K-C., Drake P.M.W. & Christou P. (2003) Genetic modification: The production of recombinant pharmaceutical proteins in plants. Nature Reviews Genetics **4**, *794–805. doi:10.1038/nrg1177),* © *(2003).*

- Many proteins can be expressed in specific parts of the plant, or milk/body fluids (blood, urine, semen) from livestock; purification and downstream processing are relatively simple. Plants have an added advantage over animals in that many animal viral contaminants can be avoided (see Table 4.4).

Transgenic plants for plant molecular farming (PMF) can be created by transfer of gene of interest into cultured plant cells **by direct methods** such as biolistics (using a gene gun), electroporation (making holes in the cell wall using electric pulses, so that the cells are able to take up DNA), **or biological methods** (such as using *Agrobacterium* mediated gene transfer). The first recombinant plant derived pharmaceutical protein was the **human growth hormone produced in callus cells of tobacco and sunflower** (Barta et al., 1986). Soon after in 1989, **rAbs** were produced by crossing two individual transgenic plants each expressing single immunoglobulin long and short chains (Hiatt, Cafferkey, & Bowdish, 1989). The first recombinant protein produced for commercial purpose was **avidin (egg protein)** expressed in transgenic maize (Hood et al., 1997). In 2012 the US FDA approved the first drug produced in plant cells for use in humans: **Elelyso** (taliglucerase alfa) to treat Gaucher disease was produced by Protalix Biotherapeutics in carrot cells (Maxmen, 2012). The list of products produced by PMF include: **human pharmaceutical proteins** (e.g., human interferon-alpha produced in turnip, human serum albumin produced in tobacco); **antibodies** [e.g., against hepatitis B, and *Streptococcus mutans* produced in tobacco (approved in EU as CaroRx)]; **enzymes** (e.g., beta-glucuronidase and Trypsin produced in maize); **vaccine antigens** (e.g., Hepatitis B antigen and Norwalk virus in tobacco and potato); and **vaccine antigens for veterinary use** (e.g., Newcastle disease virus HN proteins in tobacco suspension cell culture) (reviewed by Liew & Hair-Bejo, 2015).

Problems associated with PMF include (Obembe, Popoola, Leelavathi, & Reddy, 2011):

- Low yield (solutions include optimizing the transcript expression by use of suitable promoters and codon usage; optimizing protein's stability)
- Incorrect glycosylation
- Choice of suitable host plants
- High cost of downstream processing

Transgenic animals such as sheep, goats, pigs, and cows have been successfully used for molecular farming of proteins used in treatment of several human diseases. For example, sheep have been used to produce **alpha 1 antitrypsin** used in the treatment of emphysema, **tissue plasminogen activator** and **fibrinogen**. **Clotting factors factor VIII and IX** used to treat hemophilia have been produced from sheep, pig and cows (reviewed in Maksimenko, Deykin, Khodarovich, & Georgiev, 2013). The **first FDA approval for a drug isolated from the milk of transgenic goats was issued in 2009 to GTC Biotherapeutics for the production of human antithrombin (ATryn)** used to treat thrombosis (Vedantam, 2009). Today, several thousand companies distributed around the globe produce pharmaceuticals using rDNA technology (for a list see: http://biopharmguy.com/links/company-by-location-biologics.php).

4.5 GENETICALLY MODIFIED CROPS (GM CROPS)

The contribution to advancements due to rDNA technology in the agriculture sector has been as remarkable as in the health sector. "Transgenic" or "Genetically modified" crops (crop plants into which genes from related/unrelated plants, microbes, or even animals have been transferred) have **improved cultivation practices, reduced pesticide and herbicide usage, and improved food quality**. Transgenic plants were first created in 1983 independently by four groups:

- Mary-Dell Chilton's group at the Washington University, St. Louis, Missouri [kanamycin (antibiotic)-resistant *Nicotiana plumbaginifolia*] (Bevan, Flavell, & Chilton, 1983);
- Jeff Schell and Marc VanMontagu at Rijksuniversiteit in Ghent, Belgium (kanamycin- and methotrexate-resistant tobacco) (Herrera-Estrella, Depicker, van Montagu, & Schell, 1983);
- Robert Fraley, Stephen Rogers, and Robert Horsch at Monsanto Company, St. Louis, Missouri (kanamycin-resistant petunia plants) (Fraley et al., 1983); and
- John Kemp and Timothy Hall at University of Wisconsin (bean gene in sunflower) (Murai et al., 1983).

Commercial cultivation of GM crops began in 1986 after to the introduction of important agronomic traits, such as tolerance to herbicides which helped farmers control competing weeds in cropping areas; the US Environmental Protection Agency (EPA) approved the release of **"Round-up Ready" tobacco able to tolerate the herbicide "Round-up" (glyphosate)** which kills broad leaved weeds. Since then, in major food/feed crops (**corn, soybean, and canola**), and in a fiber crop (**cotton**), GM varieties carrying genes for **herbicide tolerance and/or insect resistance have been commercially cultivated** in many countries. According to statistics released in 2014 by the International Service for the Acquisition of Agribiotech Applications (ISAAA, http://www.isaaa.org),

GM crops are grown in an estimated area of 181.5 million hectares by farmers in 28 countries (Fig. 4.3 summarizes the data for 2015), including not just the developed countries, but also the developing countries like Brazil, Argentina, and Burkino Faso (James, 2015). Information on transgenic crops which have been approved for commercial release in at least one country is available in the **OECD BioTrack Product Database** (www.oecd.org/biotech). The database includes description of each transgenic product by its Unique Identifier and includes details of its approvals in countries. GM crops under development in various laboratories slated for release in the coming years include nutritionally enhanced rice (the "golden rice" containing genes for the synthesis of provitamin A), cereal crops with enhanced water use efficiency, and disease and insect resistant fruits and vegetables.

Although the adoption of GM crops for food and feed has been rather rapid in the United States, adoption in the European Union (EU) has been difficult. In the 1990s, Europe had been beset by a number of food crises (such as the **"mad-cow" epidemic in 1999 and 2000** that affected the beef market). European consumers have been rather more apprehensive of the safety of GM

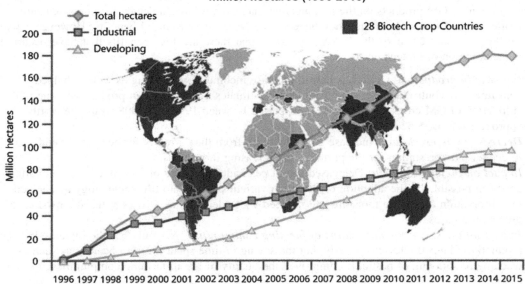

Up to ~18 million farmers, in 28 countries planted 179.7 million hectares (444 million acres) in 2015, a marginal decrease of 1% or 1.8 million hectares (4.4 million acres) from 2014.

Source: Clive James, 2015.

FIGURE 4.3

Global area of biotech crops.

Reproduced with permission from www.isaaa.org.

foods and feeds perceiving them to be "un-natural." Although in **1994 EU approved** commercial cultivation of tobacco engineered to be resistant to the herbicide bromoxynil (making it the first commercial GM crop to be marketed in Europe), **in 1997 Austria banned Syngenta's Bt176 maize** due to possible effects on nontarget insects such as butterflies as well as transfer of antibiotic resistance genes to humans and animals. Following this, **several other EU nations also banned the cultivation of GMOs**. During the **period 1999 to 2004**, spearheaded by France and Greece, **the EU declared a "de facto" moratorium on the approval of new GM crops. In 2003, the United States, Canada, and Argentina approached the World Trade Organization (WTO) claiming that the EU ban on GMOs was an unfair trade barrier. In 2006, the WTO ruled that EU restrictions on genetically engineered crops violated international trade rules.** Subsequently, the EU laws on labeling and traceability were modified and GMO approvals resumed. **In 2010, the only GM crops approved for cultivation in Europe were MON810, a Bt expressing maize conferring resistance to the European corn borer, and a potato called Amflora**, approved only for industrial applications. According to the 2014 report of ISAAA, five EU countries—Spain, Portugal, Romani, Slovakia, and Czech grow GM crops. Spain grew 92% of all the Bt maize in EU in approximately 131,000 ha (ISAAA, 2014). More than 30 GMOs or derived food and animal feed products have been approved for marketing in the EU since 1994, but consumer acceptance has been a problem—most European supermarkets choose not to stock products containing GM products on the grounds that many clients would decide to shop elsewhere.

Critics of GM crops claim that they have not delivered the promised benefits. However, Clive James, Founder and Chair of the ISAAA, believes that there are four benefits that have occurred since the adoption of biotech crops.

- *Increase in productivity:* This has helped increase income from farming as well as helped consumers by controlling the price of food. He estimates that increase in productivity due to the cultivation of GM crops during the 12-year period between 1996 and 2008 would be worth approximately US$ 52 billion.
- *Decrease in deforestation:* Increase in productivity from the currently farmed area would serve to decrease deforestation for the purposes of increasing farm lands.
- *Impact on the environment*: Decreased use of pesticides to a tune of 50% in cotton cultivation has been possible by the adoption of Bt cotton varieties. Also, "no-till" technology has helped in conservation of surface moisture and reduced water demand, as well as reduced emissions of greenhouse gases.
- *Impact on the socioeconomic status of farming communities:* Not only has the increase in productivity helped alleviate poverty, but the saving in time spent on farm operations such as weeding or spraying of pesticides, improved the quality of life in farming communities.

Studies on the trends in GM approvals in the period 1992–2014 indicate that there is in increase in the number of countries granting approvals for commercial cultivation for food and feed. This is perhaps an *indication of understanding and acceptance of countries to enhance regulatory capability to benefit from GM crop commercialization* (Aldemita, Reano, Solis, & Hautea, 2015).

Faced with the challenge of feeding an anticipated world population of 9 billion people by 2050, many believe that new varieties of crop plants with higher yield/quality, responsive to intensive cropping practices, and able to withstand the problems of climate change and

deteriorating agricultural lands, can be created only by GE. This is because the success of conventional breeding depends on useful genes being present in sexually compatible wild and weedy varieties of crop plants. In the event of these genes not being present, rDNA technology offers the possibility of transferring useful genes from unrelated plant, bacterial, or animal sources into crop plants. Nevertheless, there are several concerns regarding the adoption of GM crops raised by several organizations such as the Greenpeace International (http://www.greenpeace.org) and Organic Consumers Association (https://www.organicconsumers.org/), which are summarized below.

4.5.1 HEALTH AND SAFETY CONCERNS

Plants are known to contain various metabolites. GE could result in unforeseen changes in metabolism that could potentially alter such constituents or produce newer toxicants. This is of special concern for crops developed for pest resistance and herbicide tolerance.

4.5.1.1 Toxicity potential

In 1998, in a televised interview a scientist at the Rowett Institute, Aberdeen, Scotland, Arpad Pusztai, disclosed unpublished results of experimental rats fed with **GM potato engineered to contain the *Galanthus nivalis* agglutinin (GNA) lectin gene for insecticidal properties**. Apparently, the rats showed stunted growth, changes in the intestinal lining and damage to the immune system. The effect was attributed to the GNA since feeding rats unmodified potatoes with GNA showed similar effects. However, increases in crypt length in jejunum and decreases in cecal mucosa thickness were suggested to result not from the presence of GNA, but because of the genetic transformation of the potato (Ewen and Pusztai, 1999). Subsequently, it has been suggested that the viral origin of the CaMV promoter used to drive expression of the gene could perhaps be responsible for the changes. As the information was not immediately published in peer reviewed literature, there was initially considerable confusion as to his claims. In the ensuing publicity storm, Pusztai was removed from his job, and his unpublished work at Rowett Institute was reviewed and found to be flawed by the Royal Society of United Kingdom (Smith, 2010). The work was finally published as a letter in *The Lancet* in 1999. However, the controversy did nothing to allay public concerns about food safety and the reliability of scientific advice coming as it did shortly after the "mad-cow" BSE crisis in Europe.

4.5.1.2 Allergenicity

The allergenicity potential of new proteins expressed by the transgene inserted into the plant is a major food safety concern, especially in crops engineered to be resistant to insects. For instance, **the StarLink variety of GM maize was found to have allergenic properties in United States, EU, and Japan** and its use restricted to feed purposes by regulatory authorities.

4.5.1.3 Contamination of food

Crop plants such as maize, canola, and soybean are being engineered to produce pharmaceuticals, vaccines, and substances for industrial applications such as plastics, industrial lubricants and nonfood polymers. In the absence of adequate labeling and segregation mechanisms, there is the danger of contamination of food and feed with these products. Ensuring regulation of these crops is a challenge as exemplified by the case of **StarLink corn developed by AgroEvo (Aventis).** The variety was resistant to corn borer as it had been modified with a gene that encodes the Bt protein Cry9c. The US Environmental Protection Agency had approved cultivation of the variety in 1998, but due to concerns of allergenicity, had restricted its use to animal feed. However, **traces of StarLink corn were identified in taco shells manufactured by Kraft Foods in 2000, resulting in food recalls** and significant disruption of the food supply (Pollack, 2000).

4.5.1.4 Nutritional composition

Genetic modification of plants may result in alteration in metabolism due to the introduction of new genes. This in turn could affect the nutritional status of food. For instance, accumulation of xanthophylls and prolamines was observed in GM rice and could possibly cause nutritional imbalances in consumers.

4.5.1.5 Antibiotic resistance—potential for gene transfer

Antibiotic genes are used as selectable markers to create transgenic plants. This raises concerns as to whether these genes could be transferred to gut microbes from GM foods and potentially create antibiotic resistant pathogens.

4.5.2 ENVIRONMENTAL CONCERNS

Genes may be transferred from the GM crop to wild and weedy relatives by drifting pollen grains. This could lead to effects such as:

4.5.2.1 Genetic pollution

Since a sizable proportion of GM varieties in several crops contain genes conferring tolerance to herbicides, this could result in **"superweeds."** An often quoted example of genetic pollution is the case of Percy Schmeiser, a Canadian farmer whose fields were contaminated with "Round-up Ready" canola by pollen from a nearby GMO farm. The incident caught media attention because Monsanto successfully argued in a lawsuit that Schmeiser had violated their patent rights (see Chapter 13: Relevance of Intellectual Property Rights in Biotechnology, Box 13.5).

4.5.2.2 Loss of biodiversity

Introducing GM varieties in new habitats could result in the transfer of new genes to the traditional varieties and wild relatives of the crop. It could also **change ecosystems by affecting nontarget organisms**. In 1999, the journal *Nature* published the findings of scientists at Cornell University that the **larvae of monarch butterfly fed on milkweed leaves covered in pollen from Bt maize did not grow as well as those on control** (Losey, Rayor, & Carter, 1999). The subsequent public outcry resulted in major research collaboration between six groups in United States and Canada to examine this. Their findings published in *PNAS (USA)* in 2001 indicated that toxicity in pollen was dependent on the particular *Bt* gene used: expression of Bt in pollen from event Bt176 was higher than other events containing the *cry1Ab* gene (Sears et al., 2001). As Bt176 is no longer available in US maize varieties, overall risk to Monarch butterfly populations is estimated to be low.

4.5.2.3 Emergence of resistance

Toxin genes from a bacterium *Bacillus thuringiensis* (Bt) have been engineered into several crops in order to make them resistant to insect attacks. In the absence of adequate measures, **insects could develop resistance to the toxin**, thereby rendering futile the entire plant-incorporated-pesticide technology.

4.5.3 ETHICAL AND SOCIOECONOMIC CONCERNS

Many people consider GE as a violation of the natural order of life. Also of concern to several **vegetarians and religious groups** is the transfer of animal genes to food plants. Acceptance of GM technology for food and agriculture has been different across continents and countries. Although it is **well accepted in United States, Australia, and Argentina**, perhaps because of the importance of agricultural exports to these countries, **Europe and United Kingdom have been less enthusiastic about GM food** mainly due to doubts about its safety.

Modern agricultural biotechnology is generally developed by **multinational private companies like Monsanto and Syngenta**, and the varieties themselves are subject to IPR protection. Unlike conventional varieties, **seeds cannot be farm-saved and reused**. In efforts to reinforce IPR, technologies such as genetic use restriction technologies (GURT) [dubbed as "terminator technologies" by organizations such as RAFI (now ETC Group), http://www.etcgroup.org] have been proposed by plant biotech companies. Developing countries including India are worried that such **monopolies would influence the cost of seeds and possible exploitation of small farmers**. It would also undercut local production.

KEY TAKEAWAYS
- Transgenic plants were first created in 1983, independently reported by four groups.
- Commercial cultivation of GM crops began in 1986—US EPA approved glyphosate tolerant "Round-up Ready" tobacco.
- GM crops grown in 28 countries in around 180 million hectares.
- Major crops include corn, soybean, canola, and cotton.
- Major traits include herbicide tolerance and insect resistance.
- Others include nutritional quality (e.g., Golden Rice), viral resistance (e.g., papaya ring spot resistance), and water use efficiency.

4.6 TRANSGENIC ANIMALS

Animals with foreign genes integrated in their genomes have been created either by nuclear transfer technique or by microinjection of embryos. In the nuclear transfer technique, DNA with the gene of interest is transferred to cultured cells using viral vectors, transformed cell lines are selected, the nucleic from transformed cells are isolated and transferred into enucleated eggs. The embryos are then implanted in surrogate mothers (see Fig. 4.4, upper panel). Alternatively, transgenic animals

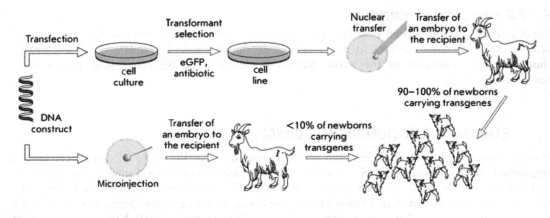

FIGURE 4.4

Creation of transgenic animals: Scheme for producing transgenic animals using the methods of **nuclear transfer (upper panel)** and **intranuclear microinjection of DNA (lower panel)**.

Maksimenko, O. G., Deykin, A. V., Khodarovich, Y. M., & Georgiev, P. G. (2013). Use of Transgenic Animals in Biotechnology: Prospects and Problems. Acta Naturae, 5(1), 33–46.

can be created by direct injection of DNA with the gene of interest into zygotes and transferring the embryo into the recipient (Fig. 4.4, lower panel). Transgenic animals may be created for various purposes as discussed in the following sections.

4.6.1 DISEASE MODELS

One early application of transgenic animals has been to use them as disease models, for example: the **Alzheimer's Mouse** that expresses mutant amyloid plaque protein and shows the pathology of Alzheimer's disease (Games et al., 1995; Elder, GamaSosa, & DeGasperi, 2010). The model was used to make the first vaccine that cleared plaques from mouse brain and restored function (Schenk et al., 1999). In the mid-1980s, scientists at Harvard in partnership with DuPont designed the **OncoMouse** specifically engineered to develop cancer (Stewart, Pattengale, & Leder, 1984). Another transgenic animal model is the **Parkinson's Fly** which has mutant forms of α-synuclein and exhibits the essential features of the human disorder (Feany and Bender, 2000). **One deficiency of these models is that many therapies effective in the models do not translate into cures in human systems.**

4.6.2 INCREASING FOOD PRODUCTION

A second application of genetically engineering of animals has been for increasing food yield. **Growth hormone genes** have been transferred to animals in order to **increase size**; for instance, the **"Superpig"** engineered to produce more meat by transfer of ovine growth hormone (Miller et al., 1989; Marshall, 2006), and **GM Salmon** reported to grow significantly faster than unmodified salmon (see Box 4.1).

BOX 4.1 AQUADVANTAGE SALMON

Size comparison of an AquAdvantage® Salmon (background) vs a nontransgenic Atlantic salmon sibling (foreground) of the same age. Both fish reach the same size at maturity but the nontransgenic salmon will take twice as long to grow to the mature size.

(Photo courtesy: AquaBounty Technologies.)

AquAdvantage® Salmon is a breed of fast growing Atlantic salmon created by introducing into its genome a growth hormone gene from Pacific Chinook salmon with a promoter of the ocean pout antifreeze protein. The genetically modified fish was developed by AquaBounty Technologies, Maynard, Massachusetts, United States (https://aquabounty.com/). AquAdvantage salmon grow to market size (4−5 kg) in 16−18 months whereas the unmodified salmon takes 3 years. In the United States, the Food and Drug Administration (FDA) regulates genetically engineered animals under the new animal drug provisions of the Federal Food, Drug and Cosmetic Act (FFDCA) and the FDA's regulations for new animal drugs. The company, AquaBounty Technologies, applied for market approval for AquAdvantage salmon in 1995. The FDA completed its food safety assessment in 2010 (FDA, 2015a), and released its environment impact statement by the end of 2012 (FDA, 2012), but the company received approval from the FDA only on 19 November 2015 (Ledford, 2015). Although deliberated upon for 20 years, the decision nevertheless made headlines as it represented approval for the first genetically modified animal for food purposes. The approval permits AquAdvantage salmon to be raised *"only in land-based contained hatchery tanks in two specific facilities in Canada and Panama. The approval does not allow AquAdvantage Salmon to be bred or raised in the United States. In fact, under this approval, no other facilities or locations, in the United States or elsewhere, are authorized for breeding or raising AquAdvantage Salmon that are intended for marketing as food to US consumers"* (FDA, 2015b). Canada approved AquaBounty's hatchery in Bay Fortune, Prince Edward Island, for egg production for commercial purposes in November 2013 (AquaBounty Technologies, 2013).

Aquaculture of Atlantic salmon conventionally uses net pens, and in North America, cultivation is mostly in coastal waters off Washington, British Columbia, Maine, and Atlantic Canada (New Brunswick, Nova Scotia, and Newfoundland). AquaBounty Technologies have established a 100-ton/year facility in the highlands of Panama, and another in Prince Edward Island, Canada, where the breeding stock are kept. Containment in both facilities is ensured by multiple physical barriers such as metal screens on tank bottoms, tank covers, nets, jump fences, screened tank overflows, and closed septic systems, designed to prevent escape of fish or eggs. The facility in Canada is indoor, whereas the one in Panama has tanks with net covers to prevent birds from predating on the fish. The entire facility is enclosed in chain-link fencing with restricted access, video surveillance cameras and motion detectors. A staff member lives on site. A geographical containment feature is provided by

(Continued)

BOX 4.1 (CONTINUED)

the high river and ocean temperatures not supporting the survival of escaped transgenic salmon. Both facilities have been vetted by FDA, and all steps in the product cycle will be subjected to regulatory oversight by the FDA. In addition to the physical barriers, biological barriers have also been incorporated to ensure that natural populations of Salmon will not be affected. The salmon grown in the Panama facility will be sterile triploid females created by treating the eggs so that 98.9% of them are triploid. AquaBounty is conducting research to develop a method ensuring 100% sterile production fish.

Public acceptance of GM salmon is yet to be revealed. Several retailers in the United States, for instance, Costco Wholesale (http://www.costco.com/) have made public statements that they will not be stocking the item (Costco, 2015). Many others are of the opinion that the GM salmon should be labeled, even though the FDA approval only requires voluntary labeling by producer. With global output for farmed Atlantic salmon being over 2 million tons, others are of the opinion that the 100 ton/year AquaBounty facility is not in any case going to make an impact on salmon sales. Moreover, in the two decades, it has taken to obtain market approval for AquAdvantage salmon, fast growing varieties of salmon with comparable growth performance have been bred by conventional methods. Such claims have however not been scientifically proven, given that no side-by-side growth trials have been performed anywhere in the world. The AquAdvantage salmon technology is innovative and the transgene can be inserted into the genome of other fast-growing salmon to boost their performance.

For global salmon production figures see http://www.marineharvest.com/globalassets/investors/handbook/2015salmon-industry-handbook.pdf

References

AquaBounty Technologies (2013, November 25). *AquaBounty cleared to produce salmon eggs in Canada for commercial purposes.* Retrieved from https://web.archive.org/web/20140412025311/http://www.aquabounty.com/documents/press/2013/20131125.pdf.

Costco (2015, November 20) Costco statement on GM salmon. Retrieved from http://webiva-downton.s3.amazonaws.com/877/4d/7/6857/Costco_GMO_salmon_statement.pdf.

FDA (2012, May 4). *Preliminary Finding Of No Significant Impact—AquAdvantage® Salmon: For Public Comment.* [Archived document] Retrieved from http://www.fda.gov/downloads/AnimalVeterinary/DevelopmentApprovalProcess/GeneticEngineering/GeneticallyEngineeredAnimals/UCM333105.pdf.

FDA (2015a, November 19). FDA has determined that the AquAdvantage Salmon is as safe to eat as non-GE salmon. Retrieved from http://www.fda.gov/ForConsumers/ConsumerUpdates/ucm472487.htm.

FDA (2015b, November 19). FDA takes several actions involving genetically engineered plants and animals for food. Retrieved from http://www.fda.gov/NewsEvents/Newsroom/PressAnnouncements/ucm473249.htm.

Ledford, H. (2015, November 23). (Reporter). *Salmon approval heralds rethink of transgenic animals.* Retrieved from http://www.nature.com/news/salmon-approval-heralds-rethink-of-transgenic-animals-1.18867.

4.6.3 PRODUCING PROTEINS FOR INDUSTRY

In addition to medicinal products, recombinant proteins for use in several industries can be produced in transgenic farm animals. For example, **spider silk**, known to be the **strongest material by weight, has been expressed in goat milk**. In 2001, scientists at Nexia Biotechnologies, Canada, inserted genes for spider silk proteins into goat embryos and were able to express the genes in the mammary glands of females. The protein could be purified from the milk and spun into fine thread. Trademarked as **"BioSteel,"** the polymer could be electrospun into nanofibers and nanomeshes. Although Nexia Biotechnologies went bankrupt in 2009, research in the area has continued in the laboratory of Randy Lewis at University of Wyoming and now at Utah State University (Hirsch, 2013).

4.6.4 **VECTOR CONTROL**

Several classes of insects are pests that cause damage to crops and livestock, or vectors that transmit disease-causing viruses in plants and animals including human beings. Methods such as the **Sterile Insect Technique (SIT)** have been developed for population suppression. Unlike insect control by use of pesticides, SIT is species specific and environmentally benign. It is based on mass rearing of millions of insects followed by radiation mediated sterilization and release of male insects which produce nonviable eggs on mating with females in the receiving environment. The target pest population declines with multiple releases of sterile insects over a period of time. The technique was used effectively in the 1950s to control screwworms that parasitized and decimated livestock herds in America affecting red meat and dairy supplies in Mexico, Central, and South America. It has since been used to control several insect pests such as the Mexican and West Indian Fruit fly in Northern Mexico, Pink boll worm in California, and the Mediterranean fruit fly, among others.

Alphey and Andreasen (2002) at Oxford University pioneered a method of genetically inducing sterility in mosquitoes, *Aedes aegypti*, which is the vector responsible for the spread of disease caused by viruses such as dengue, chikungunya, and Zika virus. The technique known as **Release of Insects with Dominant Lethal (RIDL) genes** uses GE to introduce a repressible dominant lethal gene into mosquitoes (see Box 4.2). **Field testing of RIDL has been conducted since 2009 by the company Oxitec, United Kingdom (Oxford Insect Technologies**, http://www.oxitec.com, founded by Alphey and partly owned by Oxford University). Tests have been conducted in **multiple locations** including the Cayman Islands, Malaysia, Panama, and Brazil. Pending regulatory approval trials are slated to be conducted in Florida Keys and India. Advantages of the technique include:

1. *"Self-limiting" technique*—the genetic modifications are not perpetuated in the wild population. Neither the genetically engineered mosquito nor its progeny survive.
2. *Species specific*—the released sterile mosquitoes do not mate with other species, so only the target insects are eliminated. Unlike chemical control, beneficial nontarget species such as bees and butterflies are not affected.
3. *Nontoxic and nonallergenic*—since no new gene for a toxin is introduced into the genetically engineered mosquitos, they will not affect other organisms such as pets, birds, or spiders which may ingest the transgenic insects.

4.6.5 **PETS**

A unique type of fish was developed by Zhiyuan Gong at the National University of Singapore in 1999 by integrating the gene for **green fluorescent protein from jellyfish into zebrafish** (Gong et al., 2003). The fish is a **brightly fluorescent green** under both natural as well as ultraviolet light. Other colors could be created by expressing the **red fluorescent protein from sea anemone**. The fish was introduced into the market trademarked as **GloFish** (https://www.glofish.com/) by Yorktown Technologies, L.P., Austin, Texas, as a tropical aquarium curiosity. Since it was not intended as food, the US FDA did not deem it necessary for regulatory clearance, and the GloFish became the first transgenic animal to be marketed in the United States.

BOX 4.2 OXITEC RIDL MOSQUITOES

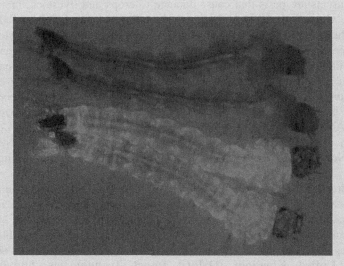

Oxitec mosquito larvae with color marker and wild mosquito with no marker.

(Photo courtesy: Oxitec Limited)

Mosquitoes are known to be carriers of agents that cause diseases such as malaria, Yellow fever, West Nile fever, dengue, chikungunya, and Zika viral fever, among many others, resulting in morbidity and mortality in the human population. Insect vector control has been an important part of methods to control the incidence and spread of these diseases. Through GE, a population suppression strategy known as RIDL has been developed by Oxitec Limited, United Kingdom (http://www.oxitec.com).

Aedes aegypti mosquitoes were transformed using transposon vectors which had a *tTAV* (tetracycline-repressible transcriptional activator) gene with a *hsp70* promoter, controlled by the *tetO* operator. In the absence of tetracycline, the tTAV protein binds to the operator and causes the expression of more tTAV protein in a positive feedback loop (Phuc et al., 2007). High levels of expression of tTAV protein are toxic to the cell possibly due to its similarity to other key transcription factors. In the presence of tetracycline, the tTAV protein complexes with the tetracycline and is no longer able to bind to the *tetO* operator. Consequently, only basal levels of tTAV protein are produced which is nontoxic to the cell. A single copy of the construct in the cell is sufficient for expression, so the gene is said to be "dominant." The transposon vector also has a marker gene, *DsRed2* which allow transformed cells/larvae to be identified by red fluorescence. The construct conforms to a dominant conditional (tetracyclin repressible) lethal system. Studies have shown that in the presence of 30 μg/mL of tetracycline, there is no significant difference in the survival of the transgenic and wild type mosquitoes, but in the absence of tetracycline, only 3%−4% of transgenics survive.

Trials indicated that the transgenic mosquitos (strain OX513A) could be used in mosquito control. The transgenic insects could be bred in the presence of tetracycline in net facilities, and when released would mate with wild type mosquitoes in the receiving environment. Since tetracycline is absent in the wild, neither the transgenics, nor their progeny would be able to survive. The strategy has the added advantage of being species specific (as the mosquitos will not mate with insects of other species), and self-limiting (as the transgenics as well as their progeny are killed). Although this system is bisexual, causing lethality in both male and female insects, for added safety, the larvae are mechanically separated on the basis of gender and only transgenic male mosquitos (which do not bite humans and feed only on plant sap) is released into the environment. [By addition of a female specific promoter (such as the vitellogenin gene) to a dominant lethal gene, female specific lethality can be achieved so that only transgenic male insects are released.]

(Continued)

BOX 4.2 (CONTINUED)

Testing of RIDL mosquitoes has already been carried out in laboratory and semi field conditions in Brazil, Malaysia, Mexico, and Cayman Islands. Typically, in a trial, the RIDL strain is mass reared with >2 million eggs a week with the release generation reared in 240 trays a week. The male pupae (which are smaller than the female pupae) are manually separated (150,000 males a week) and placed in release devices (480 a week). Around 3.3 million adult males are released 3 times a week). The receiving environment is monitored by taking observations on the number of eggs, adults, and larvae with the fluorescent marker. Many trials indicate that a suppression of 80%−99% can be achieved using RIDL mosquitoes.

In the United States, the FDA is reviewing draft environmental assessment prior to field trial in Florida Keys (FDA, 2016). In India, Gangabishan Bhikulal Investment and Trading Ltd. (GBIT, http://gbitindia.com/), a sister company of Maharashtra Hybrid Company (Mahyco), has licensed the technology to breed OX513A strain of mosquitoes from Oxitec in 2011, and has been permitted by the Indian regulatory authorities to conduct small-scale trials (Koshy, 2016).

Oxitec plans to use this technique for control of other species of mosquitoes and agricultural insect pests such as the Pink Bollworm (Glass, 2014). Problems with the adoption of the technology include public suspicion of genetic modification of living organisms.

References

FDA (2016, March 11) *Oxitec mosquito.* Retrieved from http://www.fda.gov/AnimalVeterinary/DevelopmentApprovalProcess/GeneticEngineering/GeneticallyEngineeredAnimals/ucm446529.htm.

Glass, N. (2014, July 4) (Reporter) *Breeding killer mosquitoes out of existence.* Retrieved from http://edition.cnn.com/videos/business/2014/07/04/spc-make-create-innovate-genetic-modification.cnn/video/playlists/intl-make-create-innovate/.

Koshy, J. (2016, February 8) (Reporter) *With Zika, Indian firm scales up trials for 'GM mosquitoes'.* Retrieved from http://www.thehindu.com/news/national/with-zika-indian-firm-scales-up-trials-for-gm-mosquitoes/article8206480.ece.

Phuc, H. K., Andreasen, M. H., Burton, R. S., Vass, C., Epton, M. J., Pape, G., ... Alphey, A. (2007). Late-acting dominant lethal genetic systems and mosquito control. *BMC Biology,* 5, 11 . http://dx.doi.org/10.1186/1741-7007-5-11. Retrieved from http://bmcbiol.biomedcentral.com/articles/10.1186/1741-7007-5-11

4.6.6 HYPOALLERGENIC PETS

In what could possibly have been a scam, a biotechnology company, Allerca/Lifestyle Pets, Delaware, California (http://www.lifestylepets.com/, http://www.allerca.com), claimed to have created hypoallergenic cats (Grant, 2009). For an interesting account of the issue, see Venjara (2013), and watch video *The case of the hypoallergenic cat.*

KEY TAKEAWAYS

Use of transgenic animals:

- Disease models: example, Harvard Oncomouse, Alzheimer's Mouse. Parkinson's Fly
- Increasing food production: example, GM Salmon
- Producing proteins for industry: example, BioSteel (spider silk in goat milk)
- Vector control: example, RIDL mosquitos
- Pets: example, GloFish

4.7 SYNTHETIC ORGANISMS

In an extreme form of recombinant DNA technology, it is possible to splice together gene sequences from different sources and to place it in a cell devoid of genetic material to **create a** new "**organism.**" The first synthetic life form was created in 2003 in Craig Venter's laboratory. From genomic studies, especially of the organism with the smallest known genome *Mycoplasma genitalium*, it is now fairly well understood that there are roughly 256 genes that constitutes a "**minimal genome**" that is essential for survival. By bringing together select genes, novel metabolic pathways could be induced in cells. Prospective applications include the synthesis of hydrocarbons to replace fossil fuels and resolve the energy crisis, create organisms capable of biodegradation and control of environmental pollutants (e.g., plastic waste), novel drugs used in therapy, and newer antibiotics and vaccines (see also Synthetic Biology in Chapter 1: Genes, Genomes, and Genomics).

4.8 CHALLENGES IN APPLICATIONS OF RDNA TECHNOLOGY AND GMO RELEASE

In the over four decades of rDNA technology, applications that have emerged in the health sector including diagnostics and biologics, such as **vaccines and drugs produced in transgenic organisms, have been better accepted by the public than GM foods**, such as GM fish and GM crops. One reason for this disparity may be that the **biologics developed using this technology are often unique and cannot be made by any other technique, whereas for the consumer, GM foods may not offer an advantage over unmodified foods** (any advantage of GM crops may be experienced only by the farmer as it reduces input costs for herbicides and/or pesticides and increases productivity by better yield recovery). With many lifestyle disorders such as diabetes, hypertension, obesity, cancers, and heart disease being on the rise in several populations and studies showing a strong correlation between foods and diseases, people are more particular about what they eat and the manner in which food is grown. Moreover, with a growing trend toward "organic" foods, **GM foods are often viewed with suspicion**. Some of the factors that influence the adoption of GMOs are summarized below, and further elaborated in the section on GMOs and "biopolitics" (Section 4.10).

4.8.1 EFFECT ON ENVIRONMENT

The risk—cost—benefit analysis with respect to GMOs has been controversial with vastly different perceptions between the scientists, the public, and the government, on the consequences of release of GMOs into the environment. This is partly due to a lack of scientific understanding of the complexity of human and environmental systems. Making a decision to release a GMO into the environment is **difficult due to the inability to predict long-term adverse effects on health and biodiversity, and unintended ecological effects**.

4.8.2 EFFECT ON BIODIVERSITY

Several NonGovernmental Organizations (NGOs), such as Greenpeace International (http://www. greenpeace.org), and Navdhanya (http://www.navdanya.org/) in India, have alleged that the introduction of GMOs has resulted in a loss of local seeds and local varieties (Greenpeace, n.d.).

Many who are opposed to GM technology allege "**gene pollution**" due to transfer of genes through pollen to related species and horizontal gene transfer to unrelated species.

4.8.3 EFFECT ON SOCIOCULTURAL NORMS

Agriculture in many communities is considered as a way of life and food has several cultural and societal norms or beliefs attached to it. **New technologies that involve genetic manipulations of plants are often viewed with distrust** and have stirred conflicting ideas and opinions and polarized sectors among stakeholders. At least two religious sectors, the Muslim community and the Roman Catholic Church have expressed opinions on biotechnological developments. For Muslims in Malaysia, there is a "fatwa" (religious decree) that states that **GM foods with DNA from pigs are "haram" (not permissible) for Muslims to eat** (Malasian Biotechnology Information Center, 2004). In 2004, the Pontifical Council for Justice and Peace released the Compendium of the Social Doctrine of the Church which aims to present *a precise but complete overview* of *questions concerning life in society* (Vatican, 2004). The Compendium has three parts with twelve chapters. Chapter Ten in Part Two of the Compendium presents the opinions of the church regarding safeguarding the environment, and has a section on the use of biotechnology. In the Christian vision, *nature is not a sacred or divine reality that man must leave alone,* but *at the same time makes a strong appeal for responsibility.* It emphasizes the necessity to *evaluate accurately the real benefits as well as the possible consequences in terms of risks* and exhorts different stakeholders: scientists, technicians, entrepreneurs and directors of public agencies, politicians, legislators, and public administrators, and leaders, to *make decisions that are best suited to the common good* and to *guide developments in the area of biotechnologies toward very promising ends as far as concerns the fight against hunger, especially in poorer countries, the fight against disease and the fight to safeguard the ecosystem, the common patrimony of all* (Vatican, 2004). **In 2009, the Pontifical Academy of Science**, which is the scientific academy of the Vatican founded in 1936, **issued a 15-page statement in favor of GM crops as "**...*There is nothing intrinsic about the use of GE technologies for crop improvement that would cause the plants themselves or the resulting products to be unsafe***"** (Pontifical Academy of Science, 2009). The statement also called on countries to rationalize approval process for new crop varieties and criticized *excessive, unscientific regulation* of GM crops.

4.8.4 EFFECT ON SOCIOECONOMIC STATUS OF FARMERS

GMOs particularly GM crops have taken a special significance to **food security and poverty alleviation strategies especially in the third world countries**. Most of the GM crops have been developed and distributed by multinational seed companies for large-scale farming and many do not consider these crops suitable for small scale, low input farming prevalent in the third world nations. Also, it is feared that with the gradual erosion of indigenous technologies and local varieties, farmers may become dependent on multinational companies for seed inputs resulting in "**bioserfdom**" (Hund, n.d.). Technologies such as the GURT (dubbed as the "**Terminator technology**," see Chapter 5: Relevance of Ethics in Biotechnology) which were developed by biotech seed companies such as Monsanto and Syngenta, overtly to prevent gene pollution, but covertly to implement property rights, further incite fears of **unethical practices and exploitation of farmers**.

4.9 OBJECTIONS TO GENETIC ENGINEERING AND GMOs

The **validity of major objections to GMOs** can be discussed on the basis of the principle of proportionality, the slippery slope argument, and the principle of subsidiarity.

4.9.1 PROPORTIONALITY

The principle of proportionality examines whether an **important goal can be served by the application of rDNA technology**. Since the development of the technology in the 1970s, this technology has spawned a thriving industry in biologics and improved healthcare globally. Gene therapy which involves replacement of a defective/disease causing gene with a normal one could probably cure several genetic diseases. Crop plants developed using this technology have made significant improvements to farming practices and helped to meet nutrition goals. The technology undoubtedly meets the proportionality requirement.

4.9.2 SLIPPERY SLOPE

This argument supposes that by **accepting this technology, other (potentially undesirable) practices may also be accepted.** One fear is that if genes can be introduced into organisms to alter their characteristics, it would be possible to introduce disease causing genes to nonpathogenic strains or to increase the virulence of pathogenic strains of microbes and use them in warfare/terrorism ("**biowarfare**" or "**bioterrorism**," see Chapter 6: Bioterrorism and Dual Use Research of Concern). Also, if gene therapy to cure diseases (somatic gene therapy) is practiced, it may be a short step to creating "**designer**" **babies** with "desirable" genes introduced into their genome (germ line gene therapy). It does appear that **there is a case for regulating the use of the technology** in order to avoid the slippery slope. Ensuring safety of humans and of the environment, and preventing the misuse of dual use research are the primary goals of the regulation of rDNA technology discussed in other chapters, particularly Chapters 6 and 9, of this book.

4.9.3 SUBSIDIARITY

The argument of subsidiarity relates to whether there are other, simpler technologies that can deliver the same effect. Many of the **biologics** used in healthcare such as vaccines and diagnostics are **unique and impossible to produce by any other technique**. Public acceptance of these products also seems to be in recognition of this fact. However, in the case of **GM crops, establishing the advantage of the technology has been difficult**. For one, most of the initial GM crops had only two traits—herbicide tolerance and insect resistance. Although herbicide tolerance makes large-scale cultivation easier by use of a chemical spray to rid the field of weeds, it is of dubious value in third world nations where farmers typically have small holdings, manual labor is inexpensive, and budgets for seed and chemical inputs are low. Insect resistance in GM crops is due to the expression of a bacterial toxin (Bt toxin), which kills certain classes of insects and reduces pesticide sprays. However, safety concerns regarding the presence of the toxin in feed/food have prevented the adoption of such crops in several countries. Yet another instance is that of "Golden Rice" (see Box 4.3).

BOX 4.3 GOLDEN RICE

Golden Rice: 40 g when cooked per day will save lives and sight.

(Photo courtesy: Golden Rice Humanitarian Board *www.goldenrice.org*)

Golden Rice is a genetically modified rice that contains two genes crucial to beta-carotene biosynthesis:

1. *psy* (encodes phytoene synthase) from maize and
2. *ctr 1* (encodes carotene desaturase) from the soil bacterium, *Erwinia uredovora*.

Golden Rice represents a landmark in GE as genes for an entire biosynthetic pathway were transferred. The presence of these genes causes the production of beta-carotene in rice grains giving them a "golden" color, hence the name. The significance of this modification is that the beta-carotene which is converted in the human body to Vitamin A, will help prevent blindness caused by Vitamin A deficiency in thousands of children, and death from depressed immune response in the poorest of poor rice-eating countries. The rice was first produced in 1999 by Ingo Potrykus of Swiss Federal Institute of Technology, Zurich, and Peter Beyer of the University of Freiburg, German. *Golden Rice 1* was made by transfer of three genes: *psy* from daffodil, *ctr 1* from *Erwinia*, and *lcy* also from daffodil (encodes lycopene cyclase, later found to be unnecessary as already present in rice) in a project that lasted 8 years (Ye et al., 2000). An improved *Golden Rice 2* (which produces 23 times more carotenoids) was produced by a team of researchers at Syngenta in 2005 (Paine et al., 2005). It has subsequently been bred with locally adapted rice varieties in Philippines, Bangladesh, and India. The nutritional trait which is Golden Rice has been donated by its inventors Ingo Potrykus and Peter Beyer, who have licensed it to all the government institutions and the International Rice Research Institute (IRRI). Golden Rice is being developed by IRRI together with PhilRice, Philippine Rice Research Institute. In partnership with collaborating national research agencies in the Philippines, Indonesia, Vietnam, India, and Bangladesh, the rice has been tested under laboratory as well as field conditions. As of now commercial cultivation of Golden Rice is awaiting biosafety evaluation and regulatory approval submission in different countries.

Golden Rice has been heavily criticized by Greenpeace as being of no value as a mix of vegetables and fruits would deliver the recommended dietary allowance (RDA) far more efficiently than the estimated amount of Golden Rice that would have to be eaten to achieve RDA for Vitamin A. Organizations like the Greenpeace allege that it is detrimental to human health and environment. In IRRI, Philippines, and several other places, experimental fields growing Golden Rice have been vandalized by anti-GMO activists. In a letter addressed to the Leaders of Greenpeace, the United Nations, and to Governments around the world, 113 Nobel laureates have come out in support of GMOs. They call upon Greenpeace *to cease and desist in its campaign against Golden Rice specifically, and crops and foods improved*

(Continued)

BOX 4.3 (CONTINUED)

through biotechnology in general, and to governments of the world to reject Greenpeace's campaign against GM crops (Support Precision Agriculture, 2016). Many opponents of GM crops are also changing their minds. For instance, Mark Lynas, who actively participated in the late 1990s protests including vandalism of field trials in the United Kingdom, in a public apology says that opposition to Golden Rice *is immoral and inhumane, depriving the needy of something that would save them*... (Support Precision Agriculture, n.d.). Other well-known Greenpeace members and/or critics who have come out in support of Golden Rice include Patrick Moore, Stephen Tindale, Stewart Brand, Dan Piraro, and Bill Nye.

Proponents of GM technology have hailed the development of Golden Rice as a laudable humanitarian effort. A number of patented technologies were involved in the production of Golden Rice (Krattiger and Potrykus, 2007), but access to these enabled the Golden Rice Humanitarian Board to sublicense the technology free of charge to breeding institutions in developing countries (Dubock, 2013). In 2015, the Golden Rice Project, (in particular Prof. Ingo Potrykus, Prof. Peter Beyer, and Dr. Adrian Dubock), was one of seven recipients of the Patents for Humanity Award 2015. Instituted in 2012 by President Obama, the Patents for Humanity Award is a US Patent and Trademark Office program *that recognizes patent owners and licensees working to improve global health and living standards for underserved populations* (http://www.goldenrice.org/).

The Golden Rice Project has received funding over the last 25 years from a number of sources including the European Union, the Rockefeller Foundation, USAID, Syngenta Foundation, Switzerland, India, Philippines, Bangladesh, and since 2011, the Bill and Melinda Gates Foundation (Dubock, 2013).

References

Dubock, A. (2013). Golden Rice: A long-running story at the watershed of the GM debate—Viewpoints. *Retrieved from* http://www.goldenrice.org/PDFs/GR_A_long-running_story.pdf.

Krattiger, A., & Potrykus, I. (2007). Golden Rice: A product-development partnership in agricultural biotechnology and humanitarian licensing. Case Study 3. In A. Krattiger, R. T. Mahoney, L. Nelson, et al. *Executive guide to intellectual property management in health and agriculture innovation: A handbook of best practices*. Oxford, UK and PIPRA: Davis, USA: MIHR. Retrieved from http://www.iphandbook.org/handbook/case_studies/csPDFs/casestudy03.pdf.

Paine, J. A., Shipton, C. A., Chaggar, S., Howells, R. M., Kennedy, M. J., Vernon, G., ... Adams, J. L. (2005). Improving the nutritional value of Golden Rice through increased pro-vitamin A content. *Nature Biotechnology*, 23(4), 482−487. http://dx.doi.org/10.1038/nbt1082.

Support Precision Agriculture (2016, June 29) Laureates letter supporting precision agriculture (GMOs) [Webpage] Retrieved from http://supportprecisionagriculture.org/nobel-laureate-gmo-letter_rjr.html.

Support Precision Agriculture (n.d.) Former critics of GM have a change of heart [Webpage] Retrieved from http://supportprecisionagriculture.org/former-critics-of-gm-have-a-change-of-heart_rjr.html.

Ye, X., Al-Babili, S., Klöti, A., Zhang, J., Lucca, P., Beyer, P., & Potrykus, I. (2000). Engineering the provitamin A (beta-carotene) biosynthetic pathway into (carotenoid-free) rice endosperm. *Science (New York, N.Y.)*, 287(5451), 303−305. http://dx.doi.org/10.1126/science.287.5451.303.

Further resources

Beyer P., & Potrykus I. (2011). How much Vitamin A rice must one eat? AgBioWorld. Retrieved from http://www.agbioworld.org/biotech-info/topics/goldenrice/how_much.html.

These rice varieties have a higher content of β-carotene due to the transfer of genes in the carotenoid biosynthetic pathway. In humans β-carotene is converted to Vitamin A and helps to prevent blindness and immune disorders especially in children. Golden Rice was developed as a humanitarian effort aimed at the poorest of poor rice eating nations and involved the collaborative effort of several laboratories around the world. However, critics of the technology point out that the amount of Vitamin A per serving of rice is below the recommended daily requirement for children. Encouraging a diet containing a variety of foods and leafy vegetables, would probably be a far better way of solving Vitamin A deficiency, than spending millions of dollars on genetically engineering rice. It is therefore not clear whether rDNA technology satisfies the criterion of subsidiarity with respect to GM crops, although it possibly satisfies this criterion for biologics. There have however been instances **where GE has been crucial to solving problems in food production**. For example, in Hawaii which has a thriving papaya industry, in the 1950s, the trees became infected with **papaya ring spot virus that threatened to wipe out the Hawaiian papaya by the 1970s**. Efforts of plant pathologist Dennis Gonsalves in **genetically engineering papaya to express the viral coat-protein gene "vaccinated" and saved the papaya industry** (Gonsalves, 2004). Another example is that of banana which in South Africa is a major food and cash crop. Since banana is a vegetatively propagated crop incapable of producing viable seeds, introducing disease resistance genes by hybridization is not an option. **Resistance to Banana Xanthomonas Wilt (BXW) in Cavendish banana** which is staple food in East and Central Africa has been possible by transferring genes from green pepper. The GM banana is currently undergoing field trials and will possibly be approved for commercial cultivation by 2020 (Bafana, 2015). In conclusion, it may be argued **that there are applications which may be possible only through GE**, and therefore satisfy the criteria of subsidiarity.

4.10 GMOs AND "BIOPOLITICS"

GMOs by virtue of the foreign genes introduced into their genomes have been generally regarded to be "**unnatural**" and the effects on human and animal health, as well as on the environment upon release from a laboratory environment, has been largely unknown. Risk assessment to ensure safety is part of the regulatory mechanism for approval for release into the market. In the case of GMOs for production of pharmaceuticals, since the organisms are limited to fermenters or growth facilities, the risks are better managed than with GM crops, since persistence and pollen transmission mean that once released, the GMO can probably not be removed from the receiving environment. **Public perception of the risks, and the regulatory approval for market release of GMOs, especially GM crops, have been extremely diverse and dependent on factors such as the geographical location, the economic status, notions of food safety, and the political will of the region.**

In the United States, the first GM food crop approved for commercial cultivation was the *FlavrSavr* tomato in 1994, and **around 111 crops have since been approved by the FDA and USDA** under the Coordinated Framework for the Regulation of Biotechnology (see Chapter 9: Ensuring Safety in Biotechnology). At present, there are 33 varieties of GM Corn, 20 varieties of GM Soybean, and 16 varieties of GM Cotton approved for cultivation in the United States, in addition to genetically modified potato (6 varieties), two varieties each of squash, papaya, canola, and alfalfa, and one each of sugar beet and apple (Johnson and O'Connor, 2015; see also https://www.aphis.usda.gov/biotechnology/petitions_table_pending.shtml#not_reg). The USDA Economic Research Service notes *"Driven by farmers' expectations of higher crop yields and/or lower production costs, management time savings, and other benefits, the rate at which US farmers adopt genetically engineered (GE) crop varieties appears to have reached a plateau at **high adoption rates (around 92%−94% of planted acres) for corn, soybeans, and cotton**"* (USDA-ERS, 2015).

In contrast with the United States, **GM maize variety MON810** (which is a Bt maize produced by Monsanto) is the **only GM crop being cultivated in the EU** (GMO Compass, 2008). Approved for commercial cultivation in 1997 **for cattle feed**, it is grown mainly in Spain, but is banned in France, Germany, Greece, Austria, Luxemburg, and Hungary. A GM potato, "Amflora" produced by BASF, with high starch content meant for the paper industry had been approved by the EU in 2010 for cultivation in Germany and Sweden, but the approval has since been annulled by the European court in 2013 (GeneWatch UK, n.d.). **As a result of a massive anti-GMO wave, a total of 19 of the 28 EU**

countries in October 2015 decided to ban their farmers from growing GM crops (Coghlan, 2015). Authorization of cultivation of GMOs in the EU is effected by the Directive 2001/18/EC, Regulation (EC) 1829/2003 (see Chapter 9: Ensuring Safety in Biotechnology). Under a new Directive (EU) 2015/412, **Member States are empowered to restrict or prohibit GMO cultivation on their territory**. The countries opting out include: Austria, Belgium for the Wallonia region, Britain for Scotland, Wales and Northern Ireland, Bulgaria, Croatia, Cyprus, Denmark, France, Germany, Greece, Hungary, Italy, Latvia, Lithuania, Luxembourg, Malta, the Netherlands, Poland, and Slovenia (Chow, 2015; GMO-free Europe, n.d.). One reason for this development according to the Greenpeace EU food policy director Franziska Achterberg is: *"They **don't trust EU safety assessments** and are rightly taking action to protect their agriculture and food...The only way to restore trust in the EU system now is for the Commission to hit the pause button on GMO crop approvals and to urgently reform safety testing and the approval system"* (Greenpeace, 2015).

Historically, public trust in the ability of regulatory officials at both the national and EU level to protect health and safety was severely eroded in the 1990s primarily due to the regulatory failure to **prevent the spread of mad-cow disease**. When the disease (technically, Bovine Spongiform Encephalopathy) was first noticed in 1982 in cattle in the United Kingdom, the **British Ministry of Agriculture assured the public that it posed no dangers to humans**. The European Commission accepting this assurance placed no restrictions on the sale of British beef, despite growing public concerns on the safety of meat from infected animals. The epidemic reached **incredible proportions** with more than **160,000 infected cows** and involving more than **50% of the dairy herds in the United Kingdom**. Although in 1989—90, human consumption of meat from affected animals had been banned by the European Commission, by 1996 **at least 10 cases of Creutzfeldt-Jakob disease caused by the same infectious agent (prions) had been diagnosed in humans**, possibly due to exposure to infected cattle. This development triggered a **global ban on export of British beef and culling of animals in the United Kingdom** and some of the other Member States. Although not proven, it is believed that sheep offal which was included in protein supplements to feed cattle was probably infected with prions (these agents cause scrapie in sheep and probably jumped species to move into cattle). Consequently, protein supplements containing sheep and cattle offal were banned in the United Kingdom in 1988, although it was not until 1991—92 that the ban was strictly enforced (Henahan, n.d.). Coming as it did at a time when GM corn was being approved for cattle feed, the European public, understandably wary of the ingredients that went into cattle feed, was **suspicious of claims about the safety of GMOs**. Validating these fears was the controversy triggered by Arpad Pusztai of Rowett Institute, Aberdeen, Scotland, claiming in 1998 that GMO potatoes tested in his laboratory were not safe to eat. The sentiment was echoed by several influential persons such as Prince Charles, who in his advocacy of organic farming holds the view that *"GM crops risk causing the biggest-ever environmental disaster"* (Randall, 2008).

Lack of confidence in scientific evaluation of safety of GM foods has prevented the adoption of GM crops in other parts of the world also. For instance, despite safety data being generated in several laboratories, **Bt Brinjal (eggplant)** engineered to express Bt toxin to control the fruit and shoot borer, which could potentially increase yields by 30% and reduce pesticide sprays by 70%—90%, has not received regulatory approval for commercial cultivation in India. This is in spite of more than 80% of the country's cotton production being from Bt-cotton approved for commercial cultivation in 2002 (see Box 4.4). The vegetable was approved for commercial cultivation in Bangladesh in 2014 (ISAAA, 2014).

BOX 4.4 BT COTTON AND BT BRINJAL IN INDIA

Baskets of many different kinds of Brinjal (aka "Eggplant") put out by protesters during the listening tour of India's environment minister relating to the introduction of BT Brinjal. Spring 2010 in Bangalore, India. One sign singling out Monsanto. (*By Infoeco (Own work) [CC BY-SA 3.0 (*http://creativecommons.org/licenses/by-sa/3.0) *or GFDL (*http://www.gnu.org/copyleft/fdl.html*)], via Wikimedia Commons.*)

For India, cotton is an important fiber crop providing direct livelihood for around 6 million farmers and supporting another 40−50 million in processing and trade. High yielding varieties of cotton have been developed by breeding, but the crop is susceptible to a number of different insect pests and consumes over 40% of total pesticides used in the country. In 2002, a joint venture between an Indian hybrid seed company, Maharashtra Hybrid Seed Company (Mayhco) and Monsanto, Mahyco Monsanto Biotech (India) Private Limited, received approval from the regulatory authorities in India to market Bollgard Bt Cotton which had the Cry1Ac gene conferring resistance to bollworms. The company has sublicensed the Bollgard II and Bollgard technologies to 28 Indian seed companies each of whom has introduced the Bollgard technology into their own germplasm. There are currently around 330 Bt Cotton hybrid seed varieties in the market.

The cost of seed of Bt cotton is higher in comparison to the unmodified seed since Monsanto charges a "trait developer fee" even though the seed cannot be patented in India. Notwithstanding the higher input cost, since introduction of Bt cotton, production has increased from around 140 lakh bales in 2002 to 295 lakh bales in 2010. India has overtaken China to become the world's largest cotton producer in 2014−15 marketing year producing around 352 lakh bales. Bt Cotton constitutes roughly 86% of cotton grown in the country; so much of the increase could be attributed to Biotech varieties.

Adoption of Bt cotton has been tumultuous with anti-GM Non-Governmental Organizations (NGOs) opposing it on grounds of it not living up to its promise of not requiring pesticide sprays. Cotton is affected by several insect pests; bollworm is only one of them. To protect the crop therefore, insecticide sprays are still required (although fewer number of sprays are necessary, reducing pesticide usage by 30%−40%). In the 2015 growing season, in some areas of Punjab where 99% of cotton grown is Bt Cotton, 75% of the crop was destroyed by white fly attack, triggering farmers' protests (Vasudeva, 2015). The high cost of cultivation of Bt Cotton has been blamed for farmer suicides in rain-fed areas (Venkat, 2015). Safety has also been an issue since cattle deaths have been allegedly due to grazing Bt Cotton (The Hindu, 2006).

(*Continued*)

> **BOX 4.4 (CONTINUED)**
>
> Adoption of GM varieties that have been developed in other crops has also been controversial. For example, Bt brinjal (aubergine) had been approved for commercial cultivation in 2009, but has been banned from cultivation due to public outcry and protests from NGOs. The moratorium is in order that safety assessments could be conducted. GM mustard developed by Deepak Pental's laboratory at Delhi University is also being evaluated for safety by the regulatory body, the GE Appraisal Committee (GEAC) (Vasudeva, 2016).
>
> Despite sufficient safety data having been generated for regulatory approval, public objection to Bt brinjal stems from several reasons: consumers of this popular vegetable are not convinced that the toxin will not affect humans; India is the center for biodiversity for the species and long term effects on germ plasm cannot be predicted; that the crop would be used like a "Trojan Horse" to introduce more GM crops paving the way for multinationals to control food production in India, which as the Indian economy is primarily agrarian, would be disastrous to the country's sovereignty (Todhunter, 2016).
>
> There is a strong move by several agencies, especially the NGOs, to eschew technology in favor of traditional cultural practices and "organic cultivation." One of the most vocal has been an NGO "Navdhaya" spearheaded by social activist and biopolitician Vandana Shiva (http://vandanashiva.com). Another is the Coalition for a GM-Free India (http://gmfreeindiacoalition.blogspot.in/).
>
> **References**
>
> The Hindu (2006, June 14). *Greenpeace wants probe into cattle death*. Retrieved from http://www.thehindu.com/todays-paper/tp-national/greenpeace-wants-probe-into-cattle-death/article3119331.ece.
> Todhunter, C. (2016, February 11) (Reporter) *Trojan Horse arguments and the GMO issue: Indian food and agriculture under attack*. Retrieved from http://www.globalresearch.ca/trojan-horse-arguments-and-the-gmo-issue-indian-food-and-agriculture-under-attack/5507102.
> Vasudeva, V. (2015, October 18) (Reporter) *GM cotton: whitefly attack raises anxiety among farmers*. Retrieved from http://www.thehindu.com/sci-tech/agriculture/gm-cotton-whitefly-attack-raises-anxiety-among-farmers/article7775306.ece.
> Vasudeva, V. (2016, February 6) (Reporter) *No nod for GM mustard now*. Retrieved from http://www.thehindu.com/todays-paper/tp-national/no-nod-for-gm-mustard-now/article8200239.ece.
> Venkat, V. (2015, June 24) (Reporter) *Bt cotton responsible for suicides in rain-fed areas, says study*. Retrieved from http://www.thehindu.com/news/national/bt-cotton-responsible-for-suicides-in-rainfed-areas-says-study/article7337684.ece.
>
> **Further resources**
>
> ISAAA Bt Cotton Video "The Story of Bt Cotton in India" http://www.isaaa.org/resources/videos/btcotton/default.asp.
> ISAAA Bt Brinjal Video "The Story of Bt Brinjal in India" http://www.isaaa.org/resources/videos/btbrinjalindia/default.asp.

Another significant factor affecting the adoption of GM crops is the **fear that corporations developing and holding intellectual property rights on the technology would be able to influence the food reserves of nations**. Six corporations, dubbed the "Big 6," namely Monsanto, Dow, BASF, Bayer, Syngenta, and DuPont dominate the world's seed, pesticide and agribiotech industry (Pesticide Action Network, n.d.). The largest of them all is Monsanto Company, St. Louis, Missouri (http://www.monsanto.com), a chemical company founded in 1901 which entered into the biotech business in the 1980s and 1990s having acquired a number of smaller seed companies such as G.D. Searle & Company, Agracetus, and Cargill. Many of Monsanto's GM seeds include those that are resistant to herbicides that it manufactures, for instance "Round-up Ready" seeds tolerant to its herbicide "Round-up" (glyphosate). Other GM crops it has produced have Bt toxin that confers resistance to several insect pests such as bollworm in cotton, stem borer in corn, and fruit and shoot borer in eggplant. The company has been at the receiving end of several anti-GMO campaigns such as the "**Millions Against Monsanto**" (https://www.organicconsumers.org/campaigns/millions-against-monsanto) and has been accused of contaminating the food supply, and environmental destruction. Monsanto has also been **accused of lobbying and influencing US regulators for approvals for GM crops** (HubPages, 2012). The allegations are partly due to many Monsanto employees having been appointed to positions in the FDA and EPA. In a review comparing the regulation of GM food in Europe and the United States published in 2001, Diahanna Lynch and David Vogel express the opinion *"There is no question that public opposition* [in Europe] *to GMOs has assumed an anti-American or anti-globalization flavor. This was largely due to three factors. First, the American firm Monsanto was the first mover: the first GMO crops to arrive in Europe came from the United States, rather than being grown in Europe. In addition, Monsanto chose not to label them, thus prompting widespread antagonism on the part of European consumer groups who claimed that European consumers were being deprived of their freedom of choice. Secondly, Monsanto's purchase of a large number of seed companies as well as the rumors surrounding its introduction of a 'terminator gene' created uneasiness among many European farmers: they regarded the firm's marketing of GMO seeds as part of an American strategy to control European agriculture. Thirdly, the initial American exports arrived in Europe just as the United States was imposing $100 million of punitive tariffs against European exports to the United States, many of which were directed against European agricultural products"* (Lynch and Vogel, 2001). Fears that **multinationals would control the fate of farmers** by controlling access to seeds plays a prominent part in anti-GMO protests in several parts of the world including India (see Box 4.4).

KEY TAKEAWAYS
- Public acceptance of medical applications of rDNA technology higher than that of GM crops and GM foods.
- Factors influencing the adoption of GMOs are public perceptions of
  - Effect on environment
  - Effect on sociocultural norms
  - Effect on socioeconomic status of farmers
  - Effect on biodiversity
- rDNA technology could be supported on the basis of proportionality as significant benefits may be obtained from adopting the technology.
- Regulation and oversight of rDNA technology is necessary to avoid the "slippery slope" of potential misuse.
- Regarding subsidiarity, it is unclear whether simpler techniques would not be as effective if not more effective in developing improved crop varieties, but rDNA technology can deliver unique healthcare products, not possible to produce by any other method.

4.11 SUMMARY

GE introduces foreign genes into microbes, plant, and animals in order to express new characteristics. The technique has been used in breeding crops and livestock to increase food production, as well as to manufacture pharmaceuticals and industrial chemicals. However, since the technique involves tinkering with nature and the effects on human and animal health and on the environment cannot be accurately predicted, regulatory oversight is essential. Although biopharma products have gained public acceptance, food produced from GMOs are yet to find universal acceptance. It is significant that even though GM foods have been grown in different parts of the globe for over 20 years, in the EU and several countries including India, the public remains unconvinced about the safety of GM foods.

REFERENCES

Technique

Cohen, S. N., Chang, A. C. Y., Boyer, H. W., & Helling, R. B. (1973). Construction of biologically functional bacterial plasmids in *vitro*. *Proceedings of the National Academy of Sciences of the United States of America*, *70*(11), 3240–3244 . Retrieved from http://www.ncbi.nlm.nih.gov/pmc/articles/PMC427208/

Genentech, (n.d.) *A history of firsts*. Retrieved from http://www.gene.com/media/company-information/chronology.

Nobel Media AB (2014) Physiology or medicine 1978—Press release. *Nobelprize.org*. Retrieved from http://www.nobelprize.org/nobel_prizes/medicine/laureates/1978/press.html.

Statista (2016) Statistics and facts about biotech industry. Retrieved from https://www.statista.com/topics/1634/biotechnology-industry/.

Gene therapy

FDA (2015, October 27) FDA approves first-of-its-kind product for the treatment of melanoma. [FDA News Release] Retrieved from http://www.fda.gov/NewsEvents/Newsroom/PressAnnouncements/ucm469571.htm.

Gallagher, J. (2012, November 2) (Reporter) Gene therapy: Glybera approved by European commission. BBC News. Retrieved from http://www.bbc.com/news/health-20179561.

Geddes, L. (2013, October 30) (Reporter) *'Bubble kid' success puts gene therapy back on track*. [New Scientist] Retrieved from https://www.newscientist.com/article/mg22029413-200-bubble-kid-success-puts-gene-therapy-back-on-track/.

Le Page, M. (2015, November 5) (Reporter) Gene editing saves girl dying from leukemia in world first. New Scientist. Retrieved from https://www.newscientist.com/article/dn28454-gene-editing-saves-life-of-girl-dying-from-leukaemia-in-world-first/.

Maeder, M. L., & Gersbach, C. A. (2016). Genome-editing Technologies for Gene and Cell Therapy. *Molecular Therapy*, *24*(3), 430–446 . Retrieved from http://www.nature.com/mt/journal/v24/n3/full/mt201610a.html

Reardon, S. (2014). Gene-editing method tackles HIV in first clinical test. *Nature News*. Retrieved from http://www.nature.com/news/gene-editing-method-tackles-hiv-in-first-clinical-test-1.14813.

Recombinant antibodies

Boder, E. T., & Wittrup, K. D. (1997). Yeast surface display for screening combinatorial polypeptide libraries. *Nature Biotechnology*, *15*(6), 553–557.

Echko, M.M. & Dozier S.K. (2010). *Recombinant antibody technology for the production of antibodies without the use of animals*. Retrieved from http://alttox.org/recombinant-antibody-technology-for-the-production-of-antibodies-without-the-use-of-animals/.

Johns Hopkins Bloomberg School of Public Health (2016) *Recombinant antibodies (rAbs) produced without the use of animals*. Retrieved from http://altweb.jhsph.edu/mabs/rabs.html.

Köhler, G., & Milstein, C. (1975). Continuous cultures of fused cells secreting antibody of predefined specificity. *Nature*, *256*(5517), 495–497.

Mazor, Y., Van Blarcom, T., Iverson, B. L., & Georgiou, G. (2008). E-clonal antibodies: Selection of full-length IgG antibodies using bacterial periplasmic display. *Nature Protocols*, *3*(11), 1766–1777.

McCafferty, J., Griffiths, A. D., Winter, G., & Chiswell, D. J. (1990). Phage antibodies: Filamentous phage displaying antibody variable domains. *Nature*, *348*(6301), 552–554.

Pharming

Barta, A., Sommengruber, K., Thompson, D., Hartmuth, K., Matzke, M. A., & Matzke, A. J. M. (1986). The expression of a napoline synthase human growth hormone chimeric gene in transformed tobacco and sunflower callus tissue. *Plant Molecular Biology*, *6*, 347–357.

Hiatt, A., Cafferkey, R., & Bowdish, K. (1989). Production of antibodies in transgenic plants. *Nature*, *342* (6245), 76–78.

Hood, E. E., Witcher, D. R., Maddock, S., Meyer, T., Baszczynski, C., Bailey, M., et al. (1997). Commercial production of avidin from transgenic maize: Characterization of transformant, production, processing, extraction and purification. *Molecular Breeding*, *3*(4), 291–306.

Liew, P. S., & Hair-Bejo, M. (2015). Farming of plant based veterinary vaccines and their applications for disease prevention in animals. *Advances in Virology.*, *2015*, 936940. Available from http://dx.doi.org/ 10.1155/2015/936940.

Maksimenko, O. G., Deykin, A. V., Khodarovich, Y. M., & Georgiev, P. G. (2013). Use of transgenic animals in biotechnology: Prospects and problems. *Acta Naturae*, *5*(1), 33–46. Retrieved from http://www.ncbi. nlm.nih.gov/pmc/articles/PMC3612824/

Maxmen, A. (2012, May 02) (Reporter) First plant-made drug on the market. *Nature Newsblog*. Retrieved from http://blogs.nature.com/news/2012/05/first-plant-made-drug-on-the-market.html.

Obembe, O. O., Popoola, J. O., Leelavathi, S., & Reddy, S. V. (2011). Advances in plant molecular farming. *Biotechnology Advances*, *29*(2), 210–222. Retrieved from http://www.sciencedirect.com/science/article/pii/ S0734975010001448.

Vedantam, S. (2009, February 7) (Washington Post Staff Writer) Drug made in milk of altered goats is approved. *The Washington Post*. Retrieved from http://www.washingtonpost.com/wp-dyn/content/article/ 2009/02/06/AR2009020603727.html.

Transgenic plants

Bevan, M. W., Flavell, R. B., & Chilton, M. D. (1983). A chimaeric antibiotic resistance gene as a selectable marker for plant cell transformation. *Nature*, *304*, 184–187.

Fraley, R. T., Rogers, S. G., Horsch, R. B., Sanders, P. R., Flick, J. S., Adams, S. P., ... Woo, S. C. (1983). Expression of bacterial genes in plant cells. *Proceedings of the National Academy of Sciences*, *80*, 4803–4807.

Herrera-Estrella, L., Depicker, A., van Montagu, M., & Schell, J. (1983). Expression of chimaeric genes transferred into plant cells using a Ti-plasmid-derived vector. *Nature*, *303*, 209–213.

Murai, N., Sutton, D. W., Murray, M. G., Slightom, J. L., Merlo, D. J., Reichert, N. A., ... Hall, T. C. (1983). Phaseolin gene from bean is expressed after transfer to sunflower via tumor-inducing plasmid vectors. *Science (New York, N.Y.)*, *222*, 476–482.

GM crops

Aldemita, R. R., Reano, I. M. E., Solis, R. O., & Hautea, R. A. (2015). Trends in global approvals of biotech crops (1992–2014). *GM Crops & Food: Biotechnology in Agriculture and the Food Chain*, *6*(3), 150–166. Available from http://dx.doi.org/10.1080/21645698.2015.1056972.

Ewen, S. W., & Pusztai, A. (1999). Effect of diets containing genetically modified potatoes expressing Galanthus nivalis lectin on rat small intestine. *Lancet*, *354*(9187), 1353–1354. http://dx.doi.org/10.1016/ S0140-6736(98)05860-7. PMID 10533866

James, C. (2015) *Global adoption and approval of biotech crops in 2014*. Retrieved from http://www.isaaa. org/resources/videos/globalstatusreport2014/default.asp.

Losey, J. E., Rayor, L. S., & Carter, M. E. (1999). Transgenic pollen harms monarch larvae. *Nature*, *399*, 214. (20 May 1999) http://dx.doi.org/10.1038/20338.

Pollack, A. (2000, September 3) *(Reporter)* Kraft recalls taco shells with bioengineered corn. Retrieved from http://www.nytimes.com/2000/09/23/business/kraft-recalls-taco-shells-with-bioengineered-corn.html.

Sears, M. K., Hellmich, R. L., Stanley-Horn, D. E., Oberhauser, K. S., Pleasants, J. M., Mattila, H. R., ... Dively, G. P. (2001). Impact of Bt corn pollen on monarch butterfly populations: A risk assessment. *Proceedings of the National Academy of Sciences United States of America, 98*(21), 11937−11942. Available from http://dx.doi.org/10.1073/pnas.211329998.

Smith, J. (2010) Anniversary of a whistleblowing hero. Retrieved from http://www.huffingtonpost.com/entry/anniversary-of-a-whistleb_b_675817.html?section = india.

Transgenic animals: Alzheimer's Mouse

Elder, G. A., Gama Sosa, M. A., & De Gasperi, R. (2010). Transgenic mouse models of Alzheimer's disease. *The Mount Sinai Journal of Medicine, New York, 77*(1), 69−81. http://dx.doi.org/10.1002/msj.20159.

Games, D., Adams, D., Alessandrini, R., et al. (1995). Alzheimer-type neuropathology in transgenic mice over-expressing V717F beta-amyloid precursor protein. *Nature, 373*, 523−527, [PubMed].

Schenk, D., Barbour, R., Dunn, W., Gordon, G., Grajeda, H., Guido, T., et al. (1999). Immunization with amyloid-beta attenuates Alzheimer-disease-like pathology in the PDAPP mouse. *Nature, 400*(6740), 173−177, [PubMed].

Oncomouse

Stewart, T. A., Pattengale, P. K., & Leder, P. (1984). Spontaneous mammary adenocarcinomas in transgenic mice that carry and express MTV/myc fusion genes. *Cell, 38*, 627−637.

Parkinson's fly

Feany, M. B., & Bender, W. W. (2000). A Drosophila model of Parkinson's disease. *Nature, 404*, 394−398. (23 March 2000) | http://dx.doi.org/10.1038/35006074.

Superpig

Marshall, J. (2006) (Reporter) *Transgenic pigs are rich in healthy fats.*[New Scientist] Retrieved from http://www.newscientist.com/article/dn8900-transgenic-pigs-are-rich-in-healthy-fats.html.

Miller, K., Bolt, D., Pursel, V., Hammer, R., Pinkert, C., Palmiter, R., & Brinster, R. (1989). Expression of human or bovine growth hormone gene with a mouse metallothionein-1 promoter in transgenic swine alters the secretion of porcine growth hormone and insulin-like growth factor-I. *Journal of Endocrinology, 120*(3), 481−488.

Biosteel

Hirsch, J. (2013, September 16) (Reporter) *The silky, milky, totally strange saga of the spider goat.* Retrieved from http://modernfarmer.com/2013/09/saga-spidergoat/.

GloFish

Gong, Z., Wan, H., Tay, T. L., Wang, H., Chen, M., & Yan, T. (2003). Development of transgenic fish for ornamental and bioreactor by strong expression of fluorescent proteins in the skeletal muscle. *Biochemical and Biophysical Research Communications, 308*, 58−63.

RILD mosquitoes

Alphey, L., & Andreasen, M. (2002). Dominant lethality and insect population control. *Molecular and Biochemical Parasitology, 121*, 173−178.

Hypoallergenic pets

Grant, B. (2009, December 1) *The end of hypoallergenic cats?* Retrieved from http://www.the-scientist.com/?articles.view/articleNo/27838/title/The-end-of-hypoallergenic-cats-/.

Venjara, S. (2013, July 17) (Reporter) *World's first hypoallergenic cat: Scientific breakthrough or hype?* Retrieved from http://abcnews.go.com/Business/worlds-hypoallergenic-cat-scientific-breakthrough-hype/story?id = 19692501.

Challenges in applications of rDNA technology and GMO release: Socio-cultural Norms

Malaysian Biotechnology Information Centre (2004). *Biotechnology and religion: Are they compatible?* *BICNews*. Malaysia: Petaling Jaya.

Pontifical Academy of Science (2009) *Transgenic plants for food security in the context of development.* PAS Study Week, Vatican City, 15−19 May 2009. Retrieved from http://www.ask-force.org/web/Vatican-PAS-Statement-FPT-PDF/PAS-Statement-English-FPT.pdf.

Vatican (2004) *Compendium of the Social Doctrine of the Church.* Retrieved from http://www.vatican.va/roman_curia/pontifical_councils/justpeace/documents/rc_pc_justpeace_doc_20060526_compendio-dott-soc_en.html. (For section on Biotechnology, see: http://www.vatican.va/roman_curia/pontifical_councils/justpeace/documents/rc_pc_justpeace_doc_20060526_compendio-dott-soc_en.html#The%20use%20of%20biotechnology.)

Socio-economic status of farmers

Hund, A. (n.d.) Monsanto: Visionary or architect of bioserfdom? A global socio-economic examination of Genetically Modified Organisms. Retrieved from https://www.organicconsumers.org/old_articles/Monsanto/bioserf.php.

Effect on biodiversity

Bafana, B. (2015, December14) (Reporter) *Africa closer to a cure for banana disease.* IPS News Agency. Retrieved from http://www.ipsnews.net/2015/12/africa-closer-to-a-cure-for-banana-disease/.

Gonsalves, D. (2004). Transgenic Papaya in Hawaii and beyond. *AgBioForum*, 7(1−2). Retrieved from http://www.agbioforum.org/v7n12/v7n12a07-gonsalves.htm

Greenpeace (n.d.) *Genetic contamination.* Retrieved from http://www.greenpeace.org/international/en/campaigns/agriculture/problem/genetic-engineering/ge-agriculture-and-genetic-pol/.

GMOs and Biopolitics

Chow, L. (2015, October 5) (Reporter) *It's official: 19 European countries say 'No' to GMOs.* Retrieved from http://ecowatch.com/2015/10/05/european-union-ban-gmos/.

Coghlan, A. (2015, October 5) (Reporter) *More than half of EU officially bans genetically modified crops.* [New Scientist] Retrieved from https://www.newscientist.com/article/dn28283-more-than-half-of-european-union-votes-to-ban-growing-gm-crops/.

GeneWatch UK (n.d.) *GM crops and foods in Britain and Europe.* Retrieved from http://www.genewatch.org/sub-568547.

GMO Compass (2008, December 3) *Crops.* Retrieved from http://www.gmo-compass.org/eng/grocery_shopping/crops/18.genetically_modified_maize_eu.html.

GMO-free Europe (n.d.) *Opt-out monitor.* Retrieved from http://www.gmo-free-regions.org/gmo-free-regions/opt-out-monitor.html.

Greenpeace (2015, October 1) *Distrust over EU GM crop approvals grows as 17 countries move towards national bans.* [Press release] Retrieved from http://www.greenpeace.org/eu-unit/en/News/2015/Distrust-over-EU-GM-crop-approvals-grows-as-at-least-13-countries-move-towards-national-bans/.

Henahan, S. (n.d.) *Mad cow disease: The BSE epidemic in Great Britain.* [Interview with Dr. Frederick. A. Murphy, Dean of the School of Veterinary Medicine, University of California Davis] Retrieved from http://mad-cow.org/~tom/vet_interview.html.

HubPages (2012, May 9) *The revolving door between Monsanto, the FDA, and the EPA: Your safety in peril.* Retrieved from http://hubpages.com/politics/The-Revolving-Door-Between-Monsanto-the-FDA-and-the-EPA-Your-Safety-in-Peril.

ISAAA (2014) *ISAAA Brief 49-2014: Executive Summary: Global status of commercialized Biotech/GM Crops: 2014.* Retrieved from http://www.isaaa.org/resources/publications/briefs/49/executivesummary/default.asp.

Johnson, D., & O'Connor, S. (2015, April 30) *These charts show every genetically modified food people already eat in the U.S.* Retrieved from http://time.com/3840073/gmo-food-charts/.

Lynch, D., & Vogel, D. (2001). *The regulation of GMOs in Europe and the United States: A case-study of contemporary European regulatory politics.* Council on Foreign Relations Press. Retrieved from http://www.cfr.org/agricultural-policy/regulation-gmos-europe-united-states-case-study-contemporary-european-regulatory-politics/p8688.

Pesticide Action Network (n.d.) *GMOs, pesticides & profit.* Retrieved from http://www.panna.org/key-issues/gmos-pesticides-profit.

Randall, J. (2008, August 12) (Reporter) *Prince Charles warns GM crops risk causing the biggest-ever environmental disaster.* Retrieved from http://www.telegraph.co.uk/news/earth/earthnews/3349308/Prince-Charles-warns-GM-crops-risk-causing-the-biggest-ever-environmental-disaster.html.

USDA-ERS (2015, July 30) *Biotechnology.* Retrieved from http://www.ers.usda.gov/topics/farm-practices-management/biotechnology.aspx.

FURTHER READING

Reviews

Lynch, D. & Vogel, D. (2001). *The regulation of GMOs in Europe and the United States: A case-study of contemporary European regulatory politics.* Council on Foreign Relations Press Retrieved from http://www.cfr.org/agricultural-policy/regulation-gmos-europe-united-states-case-study-contemporary-european-regulatory-politics/p8688#.

Ma, J. K.-C., Drake, P. M. W., & Christou, P. (2003). Genetic modification: The production of recombinant pharmaceutical proteins in plants. *Nature Reviews. Genetics, 4,* 794–805. http://dx.doi.org/10.1038/nrg1177.

Maksimenko, O. G., Deykin, A. V., Khodarovich, Y. M., & Georgiev, P. G. (2013). Use of transgenic animals in biotechnology: Prospects and problems. *Acta Naturae, 5*(1), 33–46. http://www.ncbi.nlm.nih.gov/pmc/articles/PMC3612824/#R10.

Aldemita, R. R., Reaño, I. M. E., Solis, R. O., & Hautea, R. A. (2015). Trends in global approvals of biotech crops (1992–2014). *GM Crops & Food, 6*(3), 150–166. Available from http://dx.doi.org/10.1080/21645698.2015.1056972.

Web resources

GM Crops: http://www.isaaa.org/resources/videos/globalstatusreport2014/default.asp

  RIDL mosquitoes: http://edition.cnn.com/videos/business/2014/07/04/spc-make-create-innovate-genetic-modification.cnn/video/playlists/intl-make-create-innovate/

  The case of the hypoallergenic cat: http://abcnews.go.com/Nightline/video/case-hypoallergenic-cat-19695208

  Golden rice: Listen to a BBC interview with Prof Hans-Jörg Jacobsen and Vandana Shiva, 20 April 2015: http://www.goldenrice.org/audio/BBC-Apr2015.mp3

  GM Papaya: "Hawaii Snapshot—a Legendary Scientist and His Papaya Dreams." YouTube. https://www.youtube.com/watch?t = 95&v = fn_0KdbTlR8

RELEVANCE OF ETHICS IN BIOTECHNOLOGY

Ethics is knowing the difference between what you have a right to do and what is right to do.

-Potter Stewart, Associate Justice, United States Supreme Court

CHAPTER OUTLINE

5.1 Introduction 128
    5.1.1 Ethics and Bioethics 129
    5.1.2 "Values," "Morals," and "Ethics" 129
5.2 Theories in Ethics 129
5.3 Promoting Ethically Sound Science 131
5.4 Ethical Issues in Biotechnology 131
    5.4.1 Environmental Ethics 131
    5.4.2 Ethical Issues in Plant Biotechnology 132
    5.4.3 Animal Rights and the Use of Animals in Medical Research 134
    5.4.4 Ethical Issues in Human Clinical Trials 138
5.5 Medical Ethics to Biomedical Ethics 140
5.6 Good Clinical Practices 141
5.7 Bioethics 143
5.8 Summary 146
References 146
Further Reading 148

An Introduction to Ethical, Safety and Intellectual Property Rights Issues in Biotechnology.
DOI: http://dx.doi.org/10.1016/B978-0-12-809231-6.00005-3

Nuremberg Trials.
Defendants in their dock, c. 1945–1946 (in front row, from left to right): Hermann Göring, Rudolf Heß, Joachim von Ribbentrop, Wilhelm Keitel (in second row, from left to right): Karl Dönitz, Erich Raeder, Baldur von Schirach, Fritz Sauckel (*By English: Work of the United States Government [Public domain], via Wikimedia Commons*)

5.1 INTRODUCTION

During the Second World War, in Nazi Germany, a number of experiments in the name of furthering science and knowledge were conducted on humans and animals. Details of the experiments which emerged in the aftermath of the war, shocked and disgusted people around the world. What appalled many was that the experiments were conducted by doctors, persons who were expected to save and protect, not harm, human subjects. Immediately after the war, concerted efforts were made to define ethical conduct by doctors and scientists doing experiments on living systems especially human beings, laying the foundation for the discipline of bioethics.

Science experiments are admittedly empirical resulting in often unforeseen outcomes, and consequently deserve careful consideration to ensure safety and minimize risk to human and/or animal subjects. In some types of medical or translational research, it may be impossible to eradicate all risk of harm, but adhering to widely accepted principles of ethics could prevent the abuse of animal or human subjects. These principles of ethics have evolved from theological traditions, from codes, regulations, and rules. **Abiding by the rules that define the framework of** *ethically sound science* **encourages public confidence in the research and leads to better public acceptance of the application.** Ethical issues typically arise because of concerns regarding fairness and in human subjects, notions of autonomy, dignity, and respect for persons. This chapter discusses the basic principles of ethics and its relevance in areas of animal experimentation, clinical trials on human subjects, care for the environment, and agriculture.

5.1.1 ETHICS AND BIOETHICS

Although defined in many different ways, **"ethics" is a philosophical study of morality or the rules that guide our conduct**. It also includes the values and principles shaping the rules of conduct. "Bioethics" literally refers to the ethics of life, but the term has often been restricted to areas in which medicine or biomedical sciences affect human life and well-being.

5.1.2 "VALUES," "MORALS," AND "ETHICS"

These terms have taken on distinctive and arguably narrow meanings in discussions on ethics. *Values* **refer to the relative importance that one would attribute to a quality or behavior that would influence our decision of good or bad/right or wrong.** (For instance, a scientist interested in using embryonic stem cells for finding new ways to treat a disease might place a lesser value to embryos, compared to a practicing Christian for whom the embryo as a potential human being is more valuable than a putative therapy.) **Values are *intrinsic*** to a person or a community and can therefore differ from one person/community to another, and reflect attitudes or intentions. *Morals*, **on the other hand, are imposed on members of a society.** They are **based on religious, cultural, and social norms** of that society, often handed down from one generation to the next. Morals **define socially acceptable behaviors** in a given community and can therefore differ between communities. (For instance, polygamy would be acceptable in some communities but considered immoral in others). Morals embody the rules (principles, codes, or laws) that define "rights" in a society. *Ethics* **refers to both the study as well as critique of human actions** with respect to whether it is good or bad/right or wrong. Ethics define the goals (mission and vision) of a society.

5.2 THEORIES IN ETHICS

In the study of ethics, three "levels" have been distinguished. At the first level is *descriptive ethics* which may be regarded as a specialized form of social sciences and is primarily an observation of moral opinions in different communities or populations. At the second level is *normative ethics* (or "*prescriptive* ethics") that seeks to analyze moral issues with respect to right/wrong, benefit/harm, duties/rights, or fairness, in order to evolve a morally acceptable action or policy. At the next level is *critical ethics* or *metaethics* that seeks to examine the basis of normative arguments. Moral philosophers caution that a failure to distinguish between these levels of ethical inquiry can result in considerable errors and confusion—failure to distinguish between descriptive and normative ethics can result in naïve comparisons between cultures ("cultural relativism") or subjects ("subjective relativism"); while not being able to distinguish between normative ethics and critical ethics can lead to hasty moral conclusions.

Science policies have been evaluated primarily on the basis of one or more normative ethical theories typically based on *consequences* or *duties*. **Consequentialist ethical theories** consider a policy morally acceptable based on the consequences of implementing the policy. As popularized by **Jeremy Bentham** (1748−1832) (Bentham, 1789) and **John Stuart Mill** (1806−1873), **an action is morally acceptable if it promotes the best consequences.** One form of consequentialism is *utilitarianism* (Mill,1861) which **promotes those policies which produce the greatest good** (utility) to the greatest number of individuals. ***Duty based or deontological ethics*** (from Greek, "*deon*" meaning duty, obligation), popularized by **Immanuel Kant** (1724—1804), rejects this criterion for the acceptability of a policy, as a policy that yields desirable consequences for the greatest number of people could still be morally wrong if it is at the expense of a minority. This theory maintains that **an action is right only if it is in accordance with a moral rule or principle**. In Kantian ethics, all individuals affected by a policy are to be given equal consideration and it is not morally acceptable for some individuals to be used as mere means to an end. Kant's version of duty based ethics was based on a ***categorical imperative***, a rule that is true in all circumstances (Kant, 1797). Respect for human autonomy and dignity are to be upheld under all circumstances. While most ethical theories are *anthropocentric* (considering humans as being most significant), environmental ethicists believe that animals and the environment too deserve the same consideration as human beings.

Both consequentialist ethics and duty based ethics have been criticized as being **extreme positions**: in promoting the happiness of the majority consequentialism (utilitarianism) sacrifices the importance of justice and fairness for each individual; while the emphasis on the morally right action in deontological ethics sometimes prevents the adoption of policies that may benefit many. Forging **a middle path in normative ethics** is the theory of ***common morality***, advocated by **Bernard Gert**. According to his definition:

> "*Morality is an informal public system applying to all rational persons, governing behavior that affects others, and includes what are commonly known as the moral rules, ideals, and virtues and has the lessening of evil or harm as its goal*".
>
> **(Gert, 1998. Pg.13)**

Gert believed that common sense and rationality dictate moral rules that prohibit and discourage harm/evil. He proposed a moral system with 10 rules that served to avoid the five harms/evils to be avoided namely: *death, pain, disability, loss of freedom, and loss of pleasure*. The rules are:

1. *Do not kill*
2. *Do not cause pain*
3. *Do not disable*
4. *Do not deprive of freedom*
5. *Do not deprive of pleasure*
6. *Do not deceive*
7. *Keep your promises*
8. *Do not cheat*
9. *Obey the law*
10. *Do your duty.*

(Gert, 1998. Pg.165, 166)

The first five rules are intended to conform to the moral obligations of *do no harm*, while the rest, if broken, could lead to further harm, and essentially recommend *do not violate trust*.

Gert's theories have found favor with bioethicists, as entirely avoiding "harm" is not possible in experiments in biomedical science. Common morality provides a normative framework that scientists could use to ensure that their research is ethically sound.

5.3 PROMOTING ETHICALLY SOUND SCIENCE

When confronted with an ethical dilemma, scientists and policy makers have resorted to several different approaches to resolve the issue such that harm/risks are minimized. One approach has been to **pose questions that help in a critical evaluation of risks and benefits** and in understanding the ethical issues involved, for example:

- Does this research and/or products of the research meet a social need?
- Will anyone be harmed or benefited by this activity?
- How will the costs and benefits of the research and/or products be distributed?
- Will this be fair to all concerned?
- Is there an issue of trust?
- Is the research/product acceptable to the public?

Another approach has been to **evaluate the issue on the basis of arguments or justifications for pursuing the science in the face of potential harm/risk**. These include arguments based on:

- *The principle of proportionality*: This argument seeks to establish the greater goal that can be achieved by the practice and effectively prevents "frivolous" science.
- *The slippery slope argument*: This argument examines whether an acceptance or moral justification of a practice X will lead inevitably to an acceptance of an "undesirable" practice Y.
- *The principle of subsidiarity:* This argument looks at whether alternative practices exist which are simpler and less likely to cause harm, and serves to justify the practice under question.

5.4 ETHICAL ISSUES IN BIOTECHNOLOGY

Several of the ethical issues in biotechnology have resulted from **divergent interests of stakeholders**. Modern biotechnology is expensive as it requires considerable investment in terms of equipment, chemicals, and trained manpower. Examples include manufacture of pharmaceuticals using rDNA technology, development of GM crops, kits for disease diagnosis, stem cell therapy, and gene therapy, to name a few. In many countries, investment in research and development (R&D) and production facilities is made by industries belonging to the private sector. Given the corporation's obligation to its shareholders, making a profit from the enterprise is of tantamount importance to biotech industries (**shareholder primacy**). Government funding, using revenue generated from taxes, in most cases (both in developed and third world countries), is limited to areas of national importance. **Increasingly, the products of biotechnology research are entering the marketplace, and public awareness through media participation is significantly high, making it necessary for biotech industries to conform to ethical standards.** In the following sections, ethical issues in biotechnology in the environment, agriculture, and healthcare sectors are discussed with a view to understand the issue.

5.4.1 ENVIRONMENTAL ETHICS

Scientists working on environmental issues recognize three positions in discussions on environmental ethics: (1) only humans (*anthropocentrism*), (2) animals (and plants) (*biocentrism*), and (3) ecosystems (*ecocentrism*). While most of the early science policies were considered to be primarily anthropocentric, promoting human welfare with scant regard to the effect on nonhuman entities or the ecosystem, contemporary environmental ethicists strive to unify the three positions of the triangle with the aim of making sustainable development a reality (Regan, 1981). With increasing knowledge of ecosystems and the effect of anthropogenic activities on the environment, it is imperative that we **assume moral responsibility to nature and to its conservation for the future** (Partridge, 1980).

In biotechnology research and application, environmental ethical issues arise in part because the technology may have international effects resulting from **ownership issues** over source material and the potentially high mobility of products. In the planning and funding of research, ethicists have often mooted the idea of **a moral limit to certain scientific procedures in order to protect the intrinsic value of biodiversity**. (The idea of a moral limit is not foreign to ethics as we already accept a moral limit to experiments contrary to human dignity as explained in later sections.) An expression of this view is seen in international agreements such as the Convention on Biological Diversity (Chapter 8: Biodiversity and Sharing of Biological Resources) and it's supplementary, the Cartagena Protocol of Biosafety (Chapter 9: Ensuring Safety in Biotechnology).

An ethical problem arising from ownership issues is that of **fairness and equitable sharing of natural resources**. Plants and animals in biodiversity hotspots (mostly in technology poor third world countries) often yield products of value in healthcare, agriculture, and environment remediation. The term *biopiracy* has been coined to express the **exploitative relationship, in the absence of sharing, which arises when technology rich countries use the local knowledge regarding biota in the third world countries for deriving commercially viable products** (see Chapter 16: Protection of Traditional Knowledge Associated With Genetic Resources). Concomitant issues are those of Intellectual Property Rights—who owns natural resources and who should benefit from products derived from them. An illustrative example is that of the development of PCR (Polymerase Chain Reaction) made possible by the discovery of an enzyme, Taq polymerase, from a bacterium found in the hot springs of the Yellowstone National Park. In the spirit of fairness, the question is whether the Park should not have received a share of the profits accruing from the commercialization of PCR (see Chapter 13: Relevance of Intellectual Property Rights in Biotechnology).

Another major ethical problem results from accidental or deliberate release of the products of biotechnology, such as genetically modified organisms, which can persist in an environment due to their ability to reproduce. Contamination of the gene-pool (*gene pollution*) due to vertical (to members of the same species) and horizontal (to members of unrelated species) gene transfer, **loss of biodiversity** due to a possibly invasive nature of the introduced species, and **probable habitat destruction** are some of the risks. The problem gains significance due to the outcome of environmental release being largely unpredictable and risk assessment difficulty (see Chapter 10: Risk Analysis). In recognition of the importance of environmental ethics, Environment Impact Assessment is mandatory before the adoption of any technology, including biotechnology (see Chapter 10: Risk Analysis).

5.4.2 ETHICAL ISSUES IN PLANT BIOTECHNOLOGY

Since the domestication of plant species for food, fodder, fiber, fuel, and pharmaceutical purposes (the 5Fs) in the early phase of human civilization, plant breeding has changed genetic characteristics so as to improve yield, quality, and agronomic performance. Improvement of crop plants have been traditionally achieved by *introduction* (picking useful plants from their natural habitats and growing them in new areas/fields), *selection* (picking and propagating the best performers in the field), and *hybridization* (combining useful characters by making crosses between selected plants). Modern plant breeding has used *chemical and physical mutagens* and *tissue culture* (somaclonal variation) to create new agronomic traits not found in natural populations of the crop plant and incorporated them into breeding strategies. The most recent technique in the arsenal of the plant breeder is *recombinant DNA technology* or genetic engineering which (as discussed in Chapter 4: Recombinant DNA Technology and Genetically Modified Organisms) has generated crops that are herbicide tolerant, resistant to insect pests and viruses, and have improved quality. Some of the concerns regarding GM crops and the challenges with respect to the adoption of GM foods have been discussed in Chapter 4: Recombinant DNA Technology and Genetically Modified Organisms.

5.4.2.1 "Terminator" technologies

In the 1990s, a method of restricting the use of GM crops by causing the second generation seed to be sterile was developed under a cooperative R&D agreement between a private company, Delta and Pine Land Company of Mississippi, and the Agriculture Research Service of the United States Department of Agriculture (USDA). The patent application for the technology called it *genetic use restriction technology (GURT)*. Two types of GURT were developed: *V-GURT (varietal)* in which the seeds of the second generation were sterile unless treated with a chemical, and *T-GURT (trait)* in which the seeds were fertile, but the modified trait could be expressed only if the seeds were treated. Delta and Pine Land Company was bought in 2006 by Monsanto, a company in the forefront of the agribiotech industry. Patent applications for similar GURTs were also submitted by other major seed companies DuPont (Pioneer Hi-Bred), Syngenta (Zeneca), and Novartis (Syngenta) (Shi, 2006). The purpose of developing the technology was to prevent the spread of introduced genes through pollen transmission to unmodified crop plants and to be able to control the persistence of GM traits in farmers' fields. Many farmer forums and Non-Governmental Organizations (NGOs) however insisted that the purpose of the technology was to strengthen enforcement of patent rights on the modified seeds since patent infringement in plant seeds is often difficult to prove (see Box-13.5 in Chapter 13: Relevance of Intellectual Property Rights in Biotechnology, the case of Monsanto Canada vs Schmeiser). The technology was denounced as being unethical by farmers, indigenous people, NGOs, and some governments. Most vocal in denouncing the technology was an NGO organization—Rural Advancement Foundation International (http://rafiusa.org/) which dubbed the V-GURT technology as the *Terminator technology* and the T-GURT as *Traitor technology*, terms that have been widely used in discussions of the technology. Seeds have a special value to farmers and restricting the use of seed through terminator technologies is contrary to notions of fairness and sharing of natural resources. It was moreover viewed as an exploitation of farmers by rich multinational companies (especially in the developing nations). Under the UN Convention on Biological Diversity, there has been a *de facto worldwide moratorium on field testing and commercial sale of "terminator seeds" since 2001* (Ban Terminator, n.d.). Countries such as India and Brazil have passed national laws prohibiting the use of the technology.

5.4.2.2 *"Bioserfdom"*

Many of the GM crops currently in the market have been developed and patented by multinational agri-biotech companies. As explained in Chapter 4: Recombinant DNA Technology and Genetically Modified Organisms, many are afraid that this would lead to dependencies and unethical exploitation of farmers, dubbed as "**bioserfdom**," in the event of the erosion of local seeds and varieties.

5.4.3 ANIMAL RIGHTS AND THE USE OF ANIMALS IN MEDICAL RESEARCH

Animals are used in various biotechnological applications: for genetic engineering, production of pharmaceuticals, as models of human diseases, and in medical research. The **use of experimental animals has become a part of** *preclinical trials* **for new medicines, therapies, or establishing safety of chemicals** (such as cosmetics, drugs, and food additives) meant for human use. Statistics published by the USDA record the number of animals covered by the Animal Welfare Act (AWA) (which excludes mice, rats, and fish) used in research as 834,453 in 2014 (Speaking of Research, 2015). Many **experiments conducted using animals cause pain and suffering, reduce the quality of life, or even cause death of the animals involved.**

There have been **two polarized positions** with regard to animal experimentation: **those in favor of animal experiments,** who maintain that such **experiments are morally acceptable if the information sought could not be obtained through any other means,** and if animal suffering is minimized to the extent possible; and those **who are opposed to animal experiments on account of the suffering caused to animals and the fact that in many cases the information obtained from animals may not be true in humans.** The ethical dilemma stems from the *moral status to be accorded to animals*, and whether it is **ethical to use animals as a mere means of furthering human interests.** Animal rights activists maintain that experimentation on animals is in violation of animal rights and cannot be justified on any grounds, even if it means that valuable information cannot be obtained.

Use of animals in medical research draws on *consequentialist ethics* according to which **harm to animals may be acceptable in view of the greater benefit to humans.** Scientists should however strive to minimize the suffering to animals and treat experimental animals as humanely as possible. Scientists are encouraged to follow the principle of *three Rs* proposed by Russell & Burch (1959). The consequentialist justification also draws on the *moral consequences of doing or not doing an experiment—* the harm to animals by doing an experiment is weighed against the **harm to humans by not doing the experiment.** The ramification of not doing drug trials (preclinical trials) in animals is that either no new drugs can be developed, or human beings would have to be used for all safety trials.

5.4.3.1 *The three Rs of animal experimentation*

The concept of the three Rs to achieve the goal of ethical animal experimentation was developed in a report of a study commissioned by the Universities Federation for Animal Welfare in 1954 prepared by zoologist **William Russell** and microbiologist **Rex Burch**. The report was published as "The Principles of Humane Experimental Techniques" in 1959. The three Rs are:

Replacement: This tenet refers to methods that seek to use alternatives to animals in scientific experiments. It includes both **absolute replacement** in which animals are replaced by computer models or artificial cells; and **relative replacement**, where animals provide cells or tissues, but in vitro techniques such as cell/tissue cultures, perfused organs, or tissue slices are used; or a non-sentient or less sentient material is used to replace vertebrates with higher consciousness. Russell and Burch proposed an arbitrary classification of living organisms on the basis of consciousness—microorganisms and plants were considered nonsentient, while invertebrates and lesser vertebrates were less sentient, and primates and human beings the most sentient. Thus, **replacement could be in the form of a *comparative substitution* of an animal with one of lesser sentience**.

Reduction: This tenet seeks to **optimize the number of animals required to get a statistically meaningful answer to a research question** and to maximize the information obtained from a single animal without compromising animal welfare.

Refinement: This tenet attempts to ensure humane treatment of experimental animals by **adoption of animal husbandry or experimental protocols that minimized pain, stress, and discomfort**. It advocated use of anesthesia for surgical procedures and euthanasia.

The three Rs have been hugely successful in providing a framework for the ethical treatment of animals in science experimentation and have been incorporated into national and international guidelines and legislation regarding use of animals in science. It has proved to be useful not only in research but also in education by encouraging the use of computer models instead of animal dissections to teach anatomy in graduate science programs (e.g., Froguts—http://www.froguts.com/, or virtual cadaver dissection tool from McGraw Hill—http://www.mhhe.com/sem/apr3/). Other areas which have benefitted from the three Rs of animal welfare include rearing animals for food, and maintaining them in captivity such as zoos and circuses. The concept has thus evolved as a common ethic of animal use.

The three Rs have however also been criticized on various counts (Fenwick, Griffin, & Gauthier, 2009). To begin with, **the concept has been criticized for its underlying premise that the use of animals in experimentation is acceptable**. For another, there could be **conflicts in the aim of the experiment and the tenets of the three Rs**—for instance, it could be necessary to induce a disease in an animal model or subject the animal to stress in order to test for a putative remedy, which would be in contradiction with the goal of refinement.

5.4.3.2 Regulation for the protection of animals used for scientific purposes

Protection of animals used in research has been promoted by specific guidelines as well as through legislation regarding their use in experiments. Use of animals such as amphibians, birds, reptiles, fishes, nonhuman primates, and especially rodents, in fundamental research and applied medical research started in late 19th century and became part of translational research in the 20th century. Criticisms regarding their use began in 1850 in France and soon after in the United Kingdom (UK), resulting in the UK parliament adopting the *Cruelty to Animal Act* **in 1876** aimed at the protection of experimental animals. This is the first ever law to protect animals used in experiments, but it took over a century for other national and international legislation to come into existence.

- *European Union*:
    In 1949, the Treaty of London created the Council of Europe initially signed by ten states, but eventually expanded to include 47 member countries. These member states in 1986 agreed on the *Convention for the Protection of Vertebrate Animals used for Experimental or other Scientific Purposes* (European Treaty Series 123). Members of the Convention were legally bound to abide by rules regarding the use of animals in research and agreed on restricting the use to specified areas of research and only in the absence of alternate methods being available.

In order to harmonize national regulations, the European Commission published a ***Directive for the Protection of Vertebrate Animals used for Experimental and other Scientific Purposes (86/609/EEC)*** in 1986 (Belot, n.d.). In 2010, the European Union adopted ***Directive 2010/63/EU*** replacing Directive 86/609/EEC on the protection of animals used for scientific purposes. The **directive is based on the 3Rs** and lays down minimum standards for housing and care of experimental animals. The directive seeks to regulate the use of animals by a systematic project evaluation which takes into account the assessment of pain, distress, and lasting harm caused to the animals. The development, validation, and implementation of alternative methods are promoted among member states by measures including the establishment of a Union reference laboratory for the purpose. A ***Consolidated Commission Implementing Decision 2012/70/EU as corrected by Decision 2014/11/EU*** sets out a common format for submitting information on the use of animals for scientific purposes as referred to in Article 54 of Directive 2010/63/EU, which would allow the commission to assess the effectiveness of the legislation and ensure consistency in its application.

- ***United States***:

    In the United States (US), the only federal law that protects animals in research is the **Animal Welfare Act (AWA) of 1966** (amended in **1976** and **1980**). The Act excludes cold-blooded animals (amphibians, reptiles, and fish), farmed animals such as cows and pigs, and rats, mice and birds bred for research and affords minimal protection for certain animals such as dogs, cats, guinea pigs, hamsters, rabbits, and nonhuman primates such as the chimpanzees and monkeys. For the animals covered under the AWA, the law specifies minimum standards for housing, feeding, handling, and care. **The law is enforced by the USDA, more specifically by its Animal and Plant Health Inspection Service (APHIS)**. Under this law, research organizations are required to establish an ***Institutional Animal Care and Use Committee*** (IUCUC) that provides oversight to all aspects of the institution's animal care and use in experimentation. Facilities using live vertebrates funded by the Public Health Service (PHS) such as the National Institutes of Health (NIH), the Centers for Disease Control and Prevention (CDC), and the Food and Drug Administration (FDA), must also conform to the ***PHS Policy on Humane Care and Use of Laboratory Animals (the PHS Policy)***. These laboratories must also implement the ***Guide for the Care and Use of Laboratory Animals*** specified by the **Institute for Laboratory Animal Research (ILAR)**. The **Office of Laboratory Animal Welfare of the NIH** is responsible for the administration and coordination of the PHS Policy. To improve the possibility of attracting government funding, research facilities can apply for accreditation from the ***Association for Assessment and Accreditation of Laboratory Care International*** which follows the guidelines of the ILAR (New England Anti-Vivisection Society, n.d.).

- ***Australia***:

    Victoria was the first state in Australia to introduce legislation to protect animals in 1883, modeled on the Cruelty to Animals Act of UK. Currently, each state has laws governing the use of animals in research, teaching and product testing which conform to the ***Australian Code of Practice for the Care and Use of Animals for Scientific Purposes*** (Animal Ethics Infolink, n.d.). The ***Animal Research Act 1985*** requires that organizations or individuals carrying our research using animals are authorized to do so, and to comply with the ***Animal Research Regulation 2010***. Institutions using animals for research must constitute an ***Animal Experimentation Ethics Committee***. Promoting standards for animal care and use in research and teaching in Australia

and New Zealand is the *Australian and New Zealand Council for the Care of Animals in Research and Teaching.*
- *Canada*:

    At the federal level, animal welfare is safeguarded by the Criminal Code, while provincial bills for protection of animals in research have been enacted in some provinces, for instance Alberta, Saskatchewan, and Ontario. The Council of Medical Research in 1968 established the *Canadian Council of Animal Care* which has formulated and published guidelines for the care and use of animals in research. Oversight is provided by the *Animal Care Committee*.
- *India*:

    In India, animal welfare is legislatively ensured by the *Prevention of Cruelty to Animals Act, 1960* (revised several times, the latest being in **2008**). Chapter 4 of the Act addresses experimentation on animals. It establishes an **Animal Welfare Board** which is required to constitute a *Committee for the Purpose of Control and Supervision of Experiments on Animals*. This Committee is empowered to handle all legal and ethical aspects of animal experimentation. In discharge of its duties, the Committee has published **guidelines** for good laboratory practices for animal facilities (CPCSEA, n.d.). The Committee also monitors the activities of *Institutional Ethical Committees (IEC)*. Well established institutions in India also follow the *Guidelines for Care and Use of Animals in Scientific Research* drafted by the **Indian National Science Academy** (INSA, 2000).

KEY TAKEAWAYS

Examples of Regulation for the Protection of Animals Used for Scientific Purposes:

Region/Country	Legislation	Implementation
United Kingdom	1876—Cruelty to Animal Act	
European Union	1986—Directive for the Protection of Vertebrate Animals used for Experimental and other Scientific Purposes (86/609/EEC) 2010—Directive 2010/63/EU	Consolidated Commission Implementing Decision 2012/70/EU as corrected by Decision 2014/11/EU
United States	1966—Animal Welfare Act (AWA) (amended in 1976 and 1980)	Institutional Animal Care and Use Committee (IACUC) Public Health Service Policy on Humane Care and Use of Laboratory Animals (the PHS Policy) Accreditation from the Association for Assessment and Accreditation of Laboratory Care International (AAALAC)
Australia	1985—Animal Research Act 2010—Animal Research Regulation	Australian Code of Practice for the Care and Use of Animals for Scientific Purposes
Canada	Criminal Code	Canadian Council of Animal Care (CCAC) Animal Care Committee (ACC)
India	1960—Prevention of Cruelty to Animals Act, 1960 (revised 2008)	Committee for the Purpose of Control and Supervision of Experiments on Animals (CPCSEA) Guidelines for Care and Use of Animals in Scientific Research (INSA)

5.4.4 ETHICAL ISSUES IN HUMAN CLINICAL TRIALS

Testing of novel chemicals for use in therapy or diagnosis is an essential part of medical biotechnology. Although the initial part of the evaluation (preclinical trials) for safety and efficacy, are done in experimental animals, if the drugs are being developed for human use, they have to be tested on human volunteers. Clinical trials involve different phases:

- **Phase I** require the drug to be tested in a **small number of healthy volunteers**, usually at low, ascending doses, mainly to determine whether the drug is **safe**;
- **Phase II** is done on a **small number of patients (100–200)** to test both **safety** as well as **efficacy**;
- **Phase III** is done on a **large sample of patients (1000–2000)** to test for **therapeutic efficiency** and for **manufacture and marketing licensing purposes**; and finally
- **Phase IV** which is the **postmarketing surveillance** to determine the **long-term effects** of the drug.

For a new drug to go from discovery phase to final distribution in the market it takes about 15 years, and since only about one in 10,000 putative chemical entities in the discovery phase enter the market as medicines, the process is extremely capital intensive and risky. Vested interests of drug manufacturers could result in unethical practices in the recruitment of volunteers in clinical trials and experimentation on humans. The following are a few examples of unethical trials conducted on human beings.

5.4.4.1 Tuskegee syphilis trials

Despite international efforts to prevent "crimes against humanity" after the Second World War, in the 1960s and 1970s reports emerged of a number of unethical medical research and drug trials funded by the US government. The most shocking of these trials was the Tuskegee Syphilis Study. **Between 1932 and 1972** the **US PHS**, conducted a study in Tuskegee, Alabama, on the **effects of syphilis in humans** (U.S. Public Health Service Syphilis Study at Tuskegee, 2015). In this study **400 African-American males from lower income groups**, identified through free medical examinations as being infected with syphilis, were **monitored for 40 years**. The subjects **were not informed** about their disease and were **denied treatment** even though proven cure in the form of penicillin was available by the 1950s. Researchers even intervened to prevent treatment when participants were diagnosed by other physicians. During the course of the study many of the subjects died of syphilis. The **study was stopped in 1973**, but only after its existence was publicized by the media and the **issue became a political embarrassment**. In another similar US government sponsored study in **1946–1948**, the **US PHS and the Pan American Sanitary Bureau worked with several Guatemalan government agencies** to study the effect of the then newly discovered penicillin by **deliberately exposing 1300 persons to gonorrhea, syphilis, or chancroid**. The test population included prostitutes, prisoners, mental patients, and soldiers.

5.4.4.2 Manhattan Project

In the 1940s–1960s, US officials of the Manhattan Project authorized experiments to study the **effects of radiation** in order to establish health and safety standards for thousands of workers in atomic bomb plants (ushistory.org, n.d.). The experiments involved subjecting participants to radioactive substances. The **participants included pregnant women, disabled children, hospital patients, and enlisted military personnel**. While some of the participants gave informed consent, most of them had **no knowledge of the experiments**.

5.4.4.3 Thalidomide disaster

Thalidomide was prescribed in the 1950s to control nausea in the early stages of pregnancy. **Patients were however not informed that the drug was investigational**, in the testing phase of the regulatory process. Unfortunately, after it had been used by women in Europe, Canada, and the United States, the **drug was discovered to have teratogenic effects**. The drug was soon banned, but had **caused severe deformities in about 12,000 babies** due to its use (Dove, 2011). The obligation to obtain informed consent from potential subjects before administering investigational medication was made mandatory by the US FDA as a result of the thalidomide disaster. The Food Drug and Cosmetic Act was altered in 1962 to include the Kefauver Amendment which required drug manufacturers to prove safety and effectiveness of their products, and physicians to obtain informed consent from patients, before administering investigational drugs (FDA, 2012).

5.4.4.4 Gene therapy

Indicative of ethical issues that may arise in clinical translation of recombinant DNA technology was the trial conducted in the University of Pennsylvania in 1999, which resulted in the **death of 18-year-old Jesse Gelsinger**. Having enrolled in a clinical trial to test the safety of using gene therapy for Ornithine Trans Carbamylase, Gelsinger was injected with an adenoviral vector carrying a corrected gene. He died four days later apparently of a massive immune reaction to the viral vector (BBC Two, 2014). The University and the investigators including James Wilson, Director of the Institute for Human Gene Therapy, were criticized on several counts including ethical and moral grounds for **failure to disclose in the informed consent documentation the risks involved in the use of adenoviral vectors**. These vectors had earlier shown adverse side effects in animal studies and in at least two other human patients.

5.4.4.5 TeGenero clinical trial

In 2006, after testing for safety and efficacy in preclinical trials in rodents and nonhuman primates, TeGenero Immuno Therapeutics, a London based company, conducted a Phase I clinical trial for an antibody TGN1412 intended to treat rheumatoid arthritis and B cell chronic lymphocyte leukemia. In the trial **six healthy human volunteers received the antibody** at a dose 500 times smaller than that found to be safe in animal studies. All six volunteers however **faced multiorgan failure and life threatening conditions within 12 hours** of receiving the first dose, requiring hospitalization in the intensive care unit (Attarwala, 2010). Although press reports indicated that five of the men were discharged after a month and the sixth a little later, doctors suggested that they may never fully recover. Several ethical issues emerged from investigations into the incident, for one—contrary to the Good Clinical Practice (GCP) guideline which prohibits payment to Phase I volunteers on the basis of risk, Paraxel, the independent organization that conducted the trial, recruited the volunteers with a £2000 fee. For another, **deviations from the approved protocol were alleged**—the dose was to be administered at 2-hour intervals, but was apparently done at 10-minute intervals. **It was alleged that the first volunteer had started showing signs of distress but the dose was still administered to the last volunteer**.

5.4.4.6 Bial clinical trial

In January 2016, Bial, a Portuguese company tested a drug aimed at treating anxiety and motor disorders associated with Parkinson's Disease, and chronic pain associated with cancer. The Phase I clinical trial was conducted by Biotrial, a French contract research organization (CRO) at its facilities in Rennes. The trial recruited 128 healthy volunteers between 18 and 55 years, of whom ninety received the experimental drug, and the rest a placebo. Unfortunately, **the trial left one dead and five others had to be hospitalized with brain injuries**. Trial data secrecy has made it difficult to establish the identity of the drug; Bial said the drug was a Fatty Acid Amide Hydrolase inhibitor code named BIA 10-2474 (Butler & Callaway, 2016). Allegations of **unethical conduct of the trial** include the administration of the experimental drug to all six volunteers simultaneously, instead of one receiving a test dose and being checked for adverse reactions, before giving it to the others. The incident has prompted the European Medicines Agency to consider changes in current guidance on first-in-human clinical trials to identify and mitigate risks to trial participants (CenterWatch, 2016).

5.5 MEDICAL ETHICS TO BIOMEDICAL ETHICS

The history of medical ethics possibly traces back to the *Hippocratic Oath* taken by medical professionals. Although the ethical principles of beneficence and nonmaleficence are enshrined in the Hippocratic Oath, after the Nuremberg Doctor's trials in 1946, it became apparent that the Oath is insufficient to ensure ethical treatment of patients at the hands of their doctors. The **World Medical Association** (WMA, http://www.wma.net/) was founded in 1947 and it developed the *Declaration of Geneva* with the aim to prevent further *crimes against humanity* (WMA Declaration of Geneva, 1948). The Declaration of Geneva requires that the medical practitioner maintains respect for human life and even under threat will not use his/her knowledge contrary to the laws of humanity. (First adopted in 1948, it has been amended in 1968, 1983, and 1994 and editorially revised in 2005 and 2006). The WMA in June 1964 developed a **set of ethical principles for human experimentation** known as the *Declaration of Helsinki* which represents the first significant effort to regulate research on human subjects. (This Declaration has since been revised seven times, the latest being in 2013). The Declaration of Helsinki (Box 5.1) represents a higher standard of protection for humans as it is morally binding on physicians and overrides any national or local laws. The declaration formed the basis for current *GCPs*.

Since the 1970s, there has been a growing interest in **medical ethics** from a focus on health care and clinical relationship; the field has expanded to include the **study of the ethical implications of new treatment and diagnostic methods as afforded by vaccines, pharmaceuticals, stem cells, new surgical techniques, and perhaps most significantly, genome information revealed by the Human Genome Project**. While *medical ethics* in its true sense dealt primarily with the relationship between the doctor and patient, the field gradually expanded to include issues of how new medical technologies would impact humankind, thereby evolving into *biomedical ethics*.

In the absence of a universal framework for analysis of medical ethics, **a guide to morality in health professions** is provided by the *four principles* approach (often referred to as the "Georgetown mantra") advocated by **Beauchamp and Childress** in their 1979 textbook *Principles*

BOX 5.1 THE DECLARATION OF HELSINKI

Although the Declaration of Helsinki was drafted by the World Medical Association in 1964 to address the issues of clinical experimentation, it has evolved to become a model document on human research ethics. In keeping with its mandate, it has been revised seven times, the latest being in October 2013 and has expanded in its treatment from the original 11 paragraphs to 37 in the current version. The general principles elaborated in Articles 3 to 15 establish the role of physicians in ensuring beneficence and justice in research involving their patients. As risks and burdens are inevitable in medical practice and medical research, Articles 16—18, emphasize the role of the physician to ensure that risks involved have been adequately assessed, satisfactorily managed and monitored to minimize harm to human subjects. Of especial importance is the need for informed consent from research participants. Articles 25 to 32 elaborate on the modalities of obtaining informed consent. Article 33 recommends that new interventions may be tested against those of proven interventions and limits the use of placebos to situations where no proven interventions exist. Unproven interventions in clinical practice is addressed in Article 37, which calls for experiments to be carefully designed to evaluate its safety and efficacy and information to be documented and made publicly available. The privacy and confidentiality of personal information of the patients should be protected as recommended in Article 24. Research on human subjects should be registered in publicly available databases and results, including negative findings, made publicly available in keeping with accepted guidelines for ethical reporting (Articles 35 and 36). Article 34 requires post trial access to beneficial interventions to be provided to the participants

The Declaration is morally binding on physicians, although investigators are also required to abide by local laws or regulations, and this has led to apparent conflict with other codes and guidelines such as the guidelines prepared by the Council for International Organizations of Medical Sciences (CIOMS, 2002), or the Canadian Tri-Council Policy Statement (Canadian Institute of Health Research, Natural Sciences, & Engineering Research Council of Canada & Social Sciences and Humanities Research Council of Canada, 2010). Although doubts regarding its universal adoption have been raised, principles prescribed in this document form an integral part of Good Clinical Practices being followed internationally.

References

Canadian Institutes of Health Research, Natural Sciences and Engineering Research Council of Canada, and Social Sciences and Humanities Research Council of Canada (2010) *Tri-Council Policy Statement: Ethical conduct for research involving humans*, December 2010. Retrieved from http://www.pre.ethics.gc.ca/pdf/eng/tcps2/TCPS_2_FINAL_Web.pdf.

CIOMS (2002). *International ethical guidelines for biomedical research involving human subjects*. [Prepared by the Council for International Organizations of Medical Sciences (CIOMS) in collaboration with the World Health Organization (WHO)]. Retrieved from http://www.cioms.ch/publications/layout_guide2002.pdf.

---

of Biomedical Ethics. **The four principles offer a way of structuring the kind of questions that must be asked when confronting a moral problem in biomedical ethics** (Beauchamp & Childress, 2012). The four principles are:

- *Respect for autonomy*: patients have the right to choose or to refuse treatment.
- *Beneficence*: the treatment should be in the best interest of the patient.
- *Nonmaleficence:* the treatment should not harm the patient.
- *Justice*: access to treatment should be fair.

5.6 GOOD CLINICAL PRACTICES

The fundamental tenet of GCPs is that *in human experimentation the well-being of the subject is not sacrificed in the interest of science and society*. By the 1960s and 1970s, global efforts to develop new medicinal products had resulted in a number of such items being marketed. In order to ensure the safety, quality, and efficacy of medicinal products, several countries put in place guidelines, regulations, and laws to evaluate data on preclinical and clinical trials prior to market approval. As many of the companies developing these products were international, divergence in technical requirements from one country to the next meant that expensive and time-consuming tests had to be repeated in order to obtain market approvals. Industries in Europe, the United States, and Japan were especially interested in rationalization and harmonization of regulatory requirements so that medicines could be delivered to patients with minimum delay and at lesser costs. Discussions at the **WHO International Conference of Drug Regulatory Authorities** in Paris, in 1989, resulted in the creation of the *International Conference on Harmonisation (ICH) of Technical Requirements for Registration of Pharmaceuticals for Human Use* in April 1990 (ICH, n.d.). The immediate task of ICH was to develop the Tripartite ICH Guidelines on Safety, Quality and Efficacy topics. In April 1996, the ICH released the *Guidance for Industry E6 GCP* (ICH, 1996).

The purpose of this international quality standard is to enforce tight guidelines on the ethical aspects of a clinical study. It delineates the moral responsibilities of the investigators, the sponsor of the trial, and the trial monitors. It also specifies protocols for the clinical trial and reporting of trial results including adverse reactions. *Compliance with this standard provides public assurance that the rights, safety, and well-being of trial subjects are protected, consistent with the principles that have their origin in the Declaration of Helsinki, and that the clinical trial data are credible* (ICH, 1996).

Oversight to clinical trials in the GCP guidelines is affected by *the Institutional Review Board/Independent Ethics Committee (IRB/IEC)*. The duties of this committee include examination and approval of documents relating to the trial protocol, informed consent forms, recruitment of subjects, and also the competence of the principal investigator to conduct the trial. The investigator should comply with GCP and applicable regulatory requirements and is **responsible for the care and well-being of test subjects**. The investigator is also responsible for compliance with the approved protocol, for advance informed consent from participants, and reporting to the IRB/IEC on the progress and safety (serious adverse events (SAEs) are to be immediately reported), and the final results upon completion (or premature termination/suspension) of the trial. The GCP places the responsibility for implementing and maintaining quality assurance and quality control with written **Standard Operating Procedures** (SOP) on the sponsor. All or many of the duties of the sponsor may be assigned to a Contract Research Organization (**CRO**), but the **responsibility of the trial would be incumbent on the sponsor**. This includes designating appropriately qualified medical personnel, designing the trial and managing the data from the trial, supplying and handling the investigational product (including the manufacture, packaging, labeling and coding the investigational product and placebos to be used in the trial), and maintaining records and reports of the trial. The GCP has elaborate specifications for the clinical trial protocol addressing details such as randomization and blinding, selection of trial subjects, assessment of efficacy, assessment of safety, and handling of data. The GCP also has instructions for compilation of the **Investigator's Brochure** of nonclinical and clinical data relevant to the investigational product used in the trial.

5.7 BIOETHICS

The field of "bioethics" developed out of concerns regarding the **morality of doctors and health care workers,** but after the Second World War, has developed to include other areas of human health and welfare. The Nazi War Crime Tribunals held in Nuremberg (1945–49) revealed many ruthless and lethal experiments conducted on prisoners in concentration camps by doctors. At the Doctor's trial in 1946, the Nazi human experimentation was declared to be *crimes against humanity*. A 10-point *Nuremberg Code* (Box 5.2) was established by the **Counsel for War Crimes in 1947** that sought to define the boundaries of legitimate medical experimentation. Although the code itself was **not legally binding**, it formed the basis for regulations regarding experimentation on human subjects in many nations. The code is especially significant in that it established the *principle of autonomy of the patient*: **informed and voluntary consent is mandatory, and the experiment itself is to be designed keeping in mind beneficence toward the participants**.

In the United States, in 1974, the US government passed the *National Research Act* in order to protect human participants in medical trials. In implementing this Act, the **National Commission for the Protection of Human Subjects of Biomedical and Behavioral Research**

BOX 5.2 THE NUREMBERG CODE

1. *Voluntary informed consent of the human subject*: The subject should be given all information with respect to the risks and benefits of the intervention and consent to participate should be obtained without threat, fraud, deceit or any form of coercion. In case the subject does not have the legal capacity to give consent, it should be obtained from next-of-kin/legal guardian.
2. *The experiment should yield results for the benefit of society that cannot be obtained in any other way*: This code seeks to prevent frivolous experiments with no apparent benefit from being conducted.
3. *Justification for the experiment must come from previous studies*: Nonclinical or preclinical trials on animals should have been conducted and sufficient indication of possible benefit present for the trial to be done on human subjects.
4. *Unnecessary physical and mental suffering or injury to human subjects should be avoided*: This code requires careful planning of the trial protocol.
5. *No trial that could result in death or disabling injury should be conducted*: The original code made an exception for cases in which the physicians also serve as subjects.
6. *The risks of the experiment should not exceed the anticipated benefit to society*: This code ensures that the interest of science or society does not take precedence over the welfare of the subject.
7. *Adequate precautions to minimize risk to the subject should be in place before the start of the trial*: Safety of the subject is more important than the trial.
8. *The experiment is to be performed by qualified personnel*: The highest degree of skill and care is required in all aspects of the trial.
9. *The subject should be at liberty to end the experiment at any point during the trial*: Participation in the trial should be voluntary and this includes being able to terminate the experiment at any stage.
10. *The scientist(s)/trial monitors should terminate the trial if probable cause for death or injury exists*: Trials once started should be abandoned if in the scientists' superior skill or judgement there is no possibility of benefit or if continuation of the experiment would result in harm.

was created. The Commission was given the charge of identifying the basic ethical principles and to develop guidelines to be followed to ensure that research is conducted in accordance with those principles. In 1979, the Commission prepared the ***Belmont Report*** (Box 5.3) to provide *an analytical framework to guide the resolution of ethical problems arising from research with human subjects.*

An attempt to provide a universal framework of principles and procedures in the field of bioethics was made by the **United Nations Educational, Scientific and Cultural Organization** which in 2005, adopted the ***Universal Declaration on Bioethics and Human Rights*** (UNESCO, 2005). This declaration aims to provide norms for human experimentation which confer protection to vulnerable individuals and groups while demanding respect and protection for human rights and human dignity (see also Chapter 7: Genetic Testing, Genetic Discrimination and Human Rights).

BOX 5.3 THE BELMONT REPORT

Ethical Principles and Guidelines for the Protection of Human Subjects of Research

The Belmont Report prepared by the National Commission for the Protection of Human Subjects of Biomedical and Behavioral Research is a statement of basic ethical principles and guidelines that provide an analytical framework to guide the resolution of ethical problems that arise from research with human subjects. The basic ethical principles delineated in the report include:

- **Respect for Persons**:
    It entails treating individuals as autonomous persons capable of choosing for themselves. In the case of persons with limited autonomy, additional protection even to the extent of excluding them from activities that may harm them should be advocated. The extent of protection would depend on the nature of potential risk of harm and the likelihood of benefit. The application of this principle involves an **informed consent process** during which subjects are provided all information (in a comprehensible form) necessary for an individual to make a decision to voluntarily participate in a study.
- **Beneficence**:
    This requires an assessment of the potential **risks** (probable harm) to the anticipated **benefits** (promotion of health, well-being, or welfare). Investigators are required to devise mechanisms that maximize the benefits and reduce the risk that may be involved in the research. The public too need to take cognizance of the risks and benefits that may result from novel medical, psychological, and social processes and procedures.
- **Justice**:
    This principle advocates fair treatment for all and a fair distribution of the risks and benefits of the research. It forbids exploitation of vulnerable people (for instance, economically disadvantaged or those with limited cognitive capacity) or those who are easily manipulated as a result of their situation. It also requires that the researcher verifies that the potential subject pool is appropriate for the research and that the recruitment of volunteers is fair and impartial.
    Although never officially adopted by the US Congress or the Department of Health Education and Welfare (now Department of Health and Human Services), the Belmont Report has served as an ethical framework for protecting human subjects and its recommendations incorporated into other guidelines. It is an essential reference document for Institutional Review Boards (IRBs) that review and ensure that research proposals involving human subjects conducted or supported by the Human & Health Services (HHS) meet the ethical standards of the regulations.

KEY TAKEAWAYS

Medical Ethics and Bioethics:

- Counsel for War Crimes, 1947—Nuremberg Code—10-point code defining medical experimentation on human subjects.
- WMA founded in 1947 to prevent further "crimes against humanity"
  - Declaration of Geneva, 1948—medical practitioners maintain respect for human life
  - Declaration of Helsinki, 1964—regulate research on human subjects, basis for GCPs
- Four Principles of biomedical ethics (Beauchamp & Childress, 2012):
  - *Respect for autonomy*: patients have the right to choose or to refuse treatment.
  - *Beneficence*: the treatment should be in the best interest of the patient.
  - *Nonmaleficence:* the treatment should not harm the patient.
  - *Justice:* access to treatment should be fair.
- Belmont Report, 1979—prepared by the US National Commission for the Protection of Human Subjects of Biomedical and Behavioral Research provides an analytical framework for resolution of ethical problems in research with human subjects.
- Universal Declaration on Bioethics and Human Rights (UNESCO, 2005)—provides norms for human experimentation.

Table 5.1 Bioethics Advisory Committees

Organization	Bodies	Activities
United Nations Educational, Scientific and Cultural Organization (UNESCO)	International Bioethics Committee (IBC) (created in 1993)Body of 36 independent members appointed by the Director-General of UNESCO	Promote reflection on the ethical and legal issues raised by research in the life sciences and their applications
UNESCO	Intergovernmental Bioethics Committee (IGBC) (created in 1998 under Article 11 of the Statutes of the IBC)	Contributes to the dissemination of the principles set out in the UNESCO Declarations in the field of bioethics, and to the further examination of issues raised by their applications and by the evolution of the technologies in question
International Association of Bioethics (IAB)	Global, multicultural perspective with 21-member board (with no more than three members from any one country) drawn from all regions of the world	Involved in a multicultural dialogue that helps to resolve bioethical issues since there are currently no international laws to prevent abuse of human subjects in medical experimentation
World Health Organization (WHO)	Global Health Ethics Unit provides the permanent secretariat for the Global Summit of National Bioethics Advisory Bodies	The Global Health Ethics Unit maintains a database containing *Opinions submitted by National Ethics Committees (ONEC)* which provides easy access to information and opinions submitted by **National Ethics Committees** to other such committees and to the general public

Providing guidance and oversight on ethical and legal issues arising from research in the life sciences and their application are several international advisory committees, summarized in Table 5.1.

5.8 SUMMARY

Advances in biotechnology are rapidly determining the future of mankind, touching all aspects of life, from food and healthcare to life style and environment. There is therefore a pressing need to examine the ethical ramifications of biotechnology research. The moral approach to science policy demands that we justify our actions, account for our intentions, and avoid being at the mercy of the "technological imperative" (employing a technique merely because we can). A critical examination of the moral value of the technology, is necessary to determine if it serves to protect individual rights and dignity, is fair to all concerned so that the risks and benefits are shared equally, and is for the common good.

REFERENCES
Theories in Ethics
Bentham, J. (1789). An introduction to the principles of morals and legislation. In J. H. Burns, & H. L. A. Hart (Eds.), *Reprinted in 1996 in collected works of Jeremy Bentham*. Oxford: Clarendon Press.
Gert, B. (1998). *Morality: Its nature and justification*. Oxford: Oxford University Press.
Kant, I. (1797). In Mary J. Gregor (Ed.), *The metaphysics of morals* (p. 1996). Cambridge University Press.
Mill, J.S. (1861) *Utilitarianism*. Reprinted in 1998 (R. Crisp, ed.) Oxford: Oxford University Press.

Environmental Ethics
Partridge, E. (1980) *Environmental ethics: An introduction*. Retrieved from http://gadfly.igc.org/e-ethics/Intro-ee.htm.
Regan, T. (1981). The nature and possibility of an environmental ethic. *Environmental Ethics*, *3*, 19—34.

Terminator Technologies:
Ban Terminator (n.d.) *Moratorium*. Retrieved from http://www.banterminator.org/content/view/full/153.
Shi, G. (2006) *Intellectual Property Rights, Genetic Use Restriction Technologies (GURTs), and strategic behavior*. [Paper No. 156059 presented at the American Agricultural Economics Association Annual Meeting, Long Beach, California, July 23—26, 2006]. Retrieved from http://ageconsearch.umn.edu/bit-stream/21434/1/sp06sh04.pdf.

Use of Animal in Research
Fenwick, N., Griffin, G., & Gauthier, C. (2009). The welfare of animals used in science: How the "Three Rs" ethic guides improvements. *Canadian Veterinary Journal*, *50*(5), 523—530 . Retrieved from http://www.ncbi.nlm.nih.gov/pmc/articles/PMC2671878/.
Russell, W. M. S., & Burch, R. L. (1959). *The principles of humane experimental technique*. London, UK: Universities Federation for Animal Welfare.

Speaking of research (2015, July 9) *USDA publishes 2014 animal research statistics.* Retrieved from http://speakingofresearch.com/2015/07/09/usda-publishes-2014-animal-research-statistics/.

Regulation for the Protection of Animals in Research

Animal Ethics Infolink (n.d.) *Legislation.* Retrieved from http://www.animalethics.org.au/legislation.

Belot, J. (n.d.) *Animal experimentation—Legislation and protection.* Retrieved from http://www.ecopa.eu/wp-content/uploads/belgian_legislation.pdf.

CPCSEA (n.d.) CPCSEA guidelines for laboratory animal facility. Retrieved from http://icmr.nic.in/bioethics/final_cpcsea.pdf.

INSA (2000) *Guidelines for care and use of animals in scientific research.* Retrieved from http://icmr.nic.in/bioethics/INSA_Guidelines.pdf.

New England Anti-Vivisection Society (n.d.) *Animals in research.* Retrieved from http://www.neavs.org/research/laws.

Ethical issues in Human Clinical Trials

Attarwala, H. (2010). TGN1412: From discovery to disaster. *Journal of Young Pharmacists, 2*(3), 332–336 . http://dx.doi.org/10.4103/0975-1483.66810.

BBC Two (September 24, 2014) *Trial and error.* Retrieved from: http://www.bbc.co.uk/science/horizon/2003/trialerror.shtml.

Butler, D., & Callaway, E (January 18, 2016) (Reporters) Scientists in dark after French clinical trial proves fatal. *Nature News.* Retrieved from http://www.nature.com/news/scientists-in-the-dark-after-french-clinical-trial-proves-fatal-1.19189.

CenterWatch (July 25, 2016) *EMA to revise guidance on first-in-human clinical trials after Bial trial deaths.* Retrieved from http://www.centerwatch.com/news-online/2016/07/25/ema-revise-guidance-first-human-clinical-trials-bial-trial-deaths/.

Dove, F. (November 3, 2011) (Reporter) *What's happened to Thalidomide babies?* Retrieved from http://www.bbc.com/news/magazine-15536544.

FDA (October 10, 2012) *Kefauver-Harris amendments revolutionized drug development.* Retrieved from http://www.fda.gov/ForConsumers/ConsumerUpdates/ucm322856.htm#top.

ushistory.org (n.d.) *The Manhattan Project.* Retrieved from http://www.ushistory.org/us/51f.asp.

U.S. Public Health Service Syphilis Study at Tuskegee (2015) *The Tuskegee timeline.* Retrieved from http://www.cdc.gov/tuskegee/timeline.htm.

Good Clinical Practices

ICH (n.d.) *History.* Retrieved from http://www.ich.org/about/history.html.

ICH (1996) *Guidance for Industry E6 Good Clinical Practice: Consolidated guidance.* Retrieved from http://www.fda.gov/downloads/Drugs/.../Guidances/ucm073122.pdf.

Medical Ethics

WMA Declaration of Geneva (1948) Retrieved from http://www.wma.net/en/30publications/10policies/g1/.

Beauchamp, T. L., & Childress, J. F. (2012). *Principles of biomedical ethics* (7 EditionNew York: Oxford University Press.

Bioethics

UNESCO (October 19, 2005) Universal declaration on bioethics and human rights. Retrieved from http://portal.unesco.org/en/ev.php-URL_ID = 31058&URL_DO = DO_TOPIC&URL_SECTION = 201.html.

FURTHER READING

Directive 2010/63/EU of the European Parliament and of the Council of 22 September 2010 on the protection of animals used for scientific purposes. Text with EEA relevance. Retrieved from http://eur-lex.europa.eu/legal-content/EN/TXT/?qid = 1458717696374&uri = CELEX:32010L0063.

HHS.gov (n.d.) *The Nuremberg Code*. Retrieved from http://www.hhs.gov/ohrp/archive/nurcode.html.

HHS.gov (April 18, 1979). *The belmont report: Ethical principles and guidelines for the protection of human subjects of research*. Retrieved from http://www.hhs.gov/ohrp/humansubjects/guidance/belmont.html.

WMA Declaration of Helsinki (1964) Retrieved from http://www.wma.net/en/30publications/10policies/b3/.

Reviews and Books

Beauchamp, T. L., & Childress, J. F. (2012). *Principles of biomedical ethics* (7 EditionNew York: Oxford University Press.

Web Resources

Learn about Clinical trials: https://clinicaltrials.gov/ct2/about-studies/learn

International Bioethics Commissions and Committees: https://bioethics.georgetown.edu/explore-bioethics/international-bioethics-commissions-and-committees/

BIOTERRORISM AND DUAL-USE RESEARCH OF CONCERN

Armed with a single vial of a biological agent, small groups of fanatics, or failing states, could gain the power to threaten great nations, threaten the world peace. America, and the entire civilized world, will face this threat for decades to come. We must confront the danger with open eyes, and unbending purpose.

-George W. Bush, 43rd President of the United States of America (2001−09) February 11, 2004.

CHAPTER OUTLINE

6.1 Introduction 150
6.2 Bioterrorism.. 151
    6.2.1 Weaponizing Microbes 151
    6.2.2 Biological Weapons in History... 152
6.3 The 1972 Biological Weapons Convention... 153
6.4 Biosecurity Measures for Preventing Bioterrorism 155
6.5 United States Approach to Bioterrorism 157
6.6 European Union Approach to Bioterrorism... 161
6.7 Biodefense Programs 162
6.8 Dual-Use Research of Concern.. 163
    6.8.1 National Science Advisory Board for Biosecurity... 163
    6.8.2 Core Responsibilities of Life Scientists in Regard to Dual-Use Research of Concern 165
6.9 Summary 165
References 167
Further Reading 169

An Introduction to Ethical, Safety and Intellectual Property Rights Issues in Biotechnology.
DOI: http://dx.doi.org/10.1016/B978-0-12-809231-6.00006-5

Terrorist letter with anthrax addressed to Democrat Senator Tom Daschle during the 2001 anthrax attacks. The return address given in the top left says: "4th Grade, Greendale School, Franklin Park, New Jersey, 08852." There is no Greendale School at that address, though there is a Greenbrook School in the locality. Tests conducted at USAMRIID confirmed the presence of fine, "energetic," powdered anthrax within this prestamped 34 cent transmittal envelope. Also present was a one-page handwritten letter, clues from which enabled the FBI to create a profile of the sender. (*By anonymous FBI's employment* (http://www.fbi.gov/pressrel/pressrel01/102301.htm) *[Public domain], via Wikimedia Commons.*)

6.1 INTRODUCTION

On September 11, 2001, as the world watched in horror, two passenger airliners were hijacked by al-Qaeda terrorists and deliberately crashed into the twin towers of the World Trade Centre in New York City, forcing terrorism to public consciousness. One week later, beginning on September 18, 2001, letters filled with spores of Anthrax bacteria that causes respiratory distress were mailed to several news media offices and two Democratic US senators, killing 5 persons and infecting 17 others. It became evident that terrorists did not necessarily have to use explosive devices, but could use biological entities to intimidate people.

Advances in genetic engineering have made it possible for persons with even a rudimentary understanding of biology to mix genetic material from unrelated organisms and create a novel one. Although the process is more complicated in higher order organisms, creating simple microbes with altered characteristics has become "child's play." It is therefore not inconceivable that the process in the hand of malevolent individuals could be used as a weapon for terrorism. The activity raises ethical issues on several counts. For one, it goes against the basic tenet of "**do no harm.**" For another, research into pathogenesis of microbes is crucial for the development of disease control (for instance vaccines) and therapy. **Restricting access to information to prevent misuse** as in dual-use research (DUR) of concern (or DURC) could **impede the development of medical research** and bona fide commercial activities, which would be **unfair to humanity** as also to scientists and researchers. This chapter looks at the possibility of creating such "biological weapons of mass destruction" and discusses some of the current strategies that have been adopted by societies and governments to prevent it.

6.2 BIOTERRORISM

Terrorism by definition is the **unlawful use of force or violence in order to intimidate or coerce a government or civilian population in the furtherance of social or political objectives**. Terrorist weapons include bombs, missiles, poisons, and radioactive substances. Also in their arsenal are diseases causing organisms capable of damaging populations, economies, and food supplies. Such biological agents are effective weapons as they are relatively inexpensive to make and can be directed at either a small group of people or an entire population causing panic and social disruption. Creating microbes for use as weapons can be achieved in many different ways (see Table 6.1). Biological agents used in warfare are not always lethal; "incapacitating" agents with low fatality can still be used to weaken the enemy. A common misconception is that biological weapons are used only to cause disease in humans. As a matter of fact, they can also have a devastating effect on other life forms, affecting plants (e.g., the rice blast virus can destroy rice crops), and animals (e.g., the avian influenza virus). Also, biological weapons are not necessarily microbes—insect pests, such as thrips, which can destroy crops, could also be used. **Bioterrorism is defined by the US Centers for Diseases Control and Prevention as** *"the deliberate release of viruses, bacteria, toxins or other harmful agents used to cause illness or death in people, animals, or plants"* (Centers for Disease Control, n.d.).

6.2.1 WEAPONIZING MICROBES

Biological agents that can be weaponized include **bacteria** (e.g., those that cause diseases such as anthrax, plague, typhoid, cholera, and tularemia), **rickettsia** (such as the causal agents of typhus,

Table 6.1 Weaponizing Microbes

Method	Example	Reference
Transferring genes for pathogenicity or antibiotic resistance to nonpathogenic bacteria	Transfer of the gene for the lethal factor of anthrax to harmless gut bacteria *Escherichia coli* created a superbug that could attack human cells	Robertson and Leppla (1986)
"Gene shuffling:" rearranging gene sequences or so that new traits are created by sequences being expressed in altered gene contexts or altered metabolic pathways	Researchers in Germany found that when a part of a gene was eliminated, Ebola viruses became more toxic to human cells apparently due to elimination of the part of the virus that downregulates toxicity	Volchkov et al. (2001)
"Synthetic" microbes: Using gene sequences as "building blocks" and splicing together such blocks	Reconstruction of a live polio virus from synthetic oligonucleotides that were linked together and transfected into cells	Cello, Paul, and Wimmer (2002)
	Deadly mousepox virus created by adding an interleukin gene IL-4 to pox genome, engineered virus was able to kill mice that had been vaccinated against mousepox	Jackson et al. (2001)
Hybrid viruses: Created by recombining related strains	Hybrid "Dengatitis" virus created by combining the viruses causing hepatitis C and dengue in an attempt to find a vaccine for hepatitis C	Arthur (2001)

rocky mountain spotted fever, Q fever, India tick fever, and Mediterranean tick fever), **viruses** (such as Ebola, influenza, smallpox, Lassa fever, viral hepatitis, and viral hemorrhagic fevers), and **biotoxins** (e.g., botulinum, ricin, and staphylococcal enterotoxin B).

Based on the principles of microbial pathogenesis (particularly, the interaction between host and pathogen) attempts have been made to determine the **weapon potential of microbes**. To be an effective weapon, the biological entity should be **capable of causing maximum damage in the least possible time**. This depends on the **virulence** of the pathogen, which in turn, depends on the **communicability**, the **stability**, and **period of the attack**. Also influencing the weapon potential of a microbe is the technological competence of the aggressor in terms of **deliverability** and the **amount of panic or fear that can be induced** (Zimmerman & Zimmerman, 2003).

The development of biological weapons is a growing concern partly because of the relative ease with which they can be manufactured on a small scale, and also, because scientific literature exists to provide vital information on potential genes involved in pathogenicity and virulence, colonization of host cells, immune response, and antibiotic resistance.

6.2.2 BIOLOGICAL WEAPONS IN HISTORY

Use of biological entities as weapons predates developments in genetic engineering technology. History is replete with instances of the use of disease causing organisms in warfare (reviewed by Riedel, 2004). One of the earliest recorded examples is the use of the **cadavers of plague victims by the Tartars in 1346**. After a 3-year siege of the city of Kaffa, the Tartars are believed to have catapulted the infected corpses over the city walls, in order to weaken the city's resistance. The last known incident using the same strategy of using plague corpses was in 1710, when Russian forces attacked the Swedish city of Reval. In the 17th century, the **British army distributed blankets infected with smallpox virus among Native Americans** in an attempt to get rid of the natives. Instances of biowarfare tactics occurred during the two World Wars. During the **World War I, enemy horses and cattle were infected with glanders and anthrax by German saboteurs**. During the **World War II, anthrax bombs were tested** by military researchers of United Kingdom on the Scottish island of Gruinard, contaminating it so severely that the island was off-limits for 50 years. Also during this war, the Japanese Army Unit 731 unleashed what is considered to be the largest use of biological weapons, attacking Chinese villages in Manchuria with **bombs laced with plague and other diseases** and subjecting prisoners of war to biowarfare experiments. Spurred by intelligence reports that Nazi Germany was involved in a biowarfare program, counter offensive preparations were made by the Allies. United Kingdom produced **millions of cattle-feed cakes laced with anthrax** designed to be air dropped into Germany in order to destroy livestock and food supply. In the **United States, a biological bomb production plant was built in 1944** capable of producing fifty thousand four-pound anthrax bombs every month. Later in the 1960s, United States stockpiled **36,000 kg of wheat stem rust** and nearly a **ton of rice blast** which could be dispersed by coated feathers dropped from airplanes. In the mid-1970s, genetic engineering techniques began to be used for the creation of bioweapons. Apparently, both the United States and the Soviet Union collected samples of Ebola and Marburg viruses from Africa for bioweapons research. The Soviet Union's biological weapons directorate, **Biopreparat**, commissioned thousands of scientists to genetically alter microbes and to study the most efficient and effective form of application of pathogenic microbes in the battlefield. By the **late 1960s, the United States military had developed a biological arsenal** that included several pathogens causing diseases in animals and humans, toxins, as well as fungal pathogens capable of destroying crops. Studies to determine vulnerability of humans to certain aerosolized pathogens were conducted at **Fort Detrick** inside 1-million-liter, metallic, spherical aerosolization chambers known as the **"eight ball"** where volunteers were exposed to *Francisella tularensis* (**tularemia** or rabbit fever) and *Coxiella burnetii* (**Q fever**). The more recent example of bioterrorism is the **anthrax attacks of September 11, 2001.** Envelopes filled with highly refined *Bacillus anthracis* (anthrax) spores were sent through the US Postal Service to several prominent US senators and media personnel. The attack resulted in **5 deaths, 22 unknown infections, and millions of dollars' worth of decontamination efforts**. The spores used in the attack were later found to be genetically identical to strains of the bacteria stored in the research laboratories at the Centers for Disease Control and Prevention (CDC) and the United States Army Medical Research Institute of Infectious Diseases (Leitenberg, 2002). Despite a long history of research, the use of biological weapons in warfare and terrorism has been limited by the lack of specificity of the infectious agent and the inability to contain an infection within the target population.

KEY TAKEAWAYS

- Bioterrorism is the deliberate release of disease causing or other harmful biological agents to cause disability or death in people, animals, or plants.
- Historical examples of bioterrorism include:
  - 1356—siege of Kaffa, Tartars catapulted cadavers of plague victims over city walls
  - 1710—Russian attack on Swedish city of Reval, used plague corpses in attack
  - 17th Century—British army distributed blankets infected with smallpox virus to native Americans
  - World War I—German saboteurs infected enemy hoses and cattle with glanders and anthrax
  - World War II—Japanese Army Unit 731 attacked Chinese villages in Manchuria with bombs laced with plague and other diseases, subjected prisoners to biowarfare experiments; Intelligence reports on biowarfare program in Nazi Germany; UK produced millions of cattle-feed cakes containing anthrax spores to be air dropped; the United States builds a facility capable of producing 50,000 four-pound anthrax bombs every month
  - 1960s—the United States stockpiled 36,000 kg of wheat stem rust and about a ton of rice blast to be dispersed by coated feathers from airplanes
  - 1970s—Bioweapon research using rDNA technology is both the United States (Fort Derrick "eight ball") and Soviet Union (Biopreparat)
  - 2001—anthrax attacks in the United States

6.3 **THE 1972 BIOLOGICAL WEAPONS CONVENTION**

Although there were several allegations on the use of biological warfare agents by the Germans during World War I, no hard evidence for the same could be found by the subcommittees of the League of Nations. However, the horror of chemical warfare used extensively during the war led to international diplomatic efforts to limit the proliferation and use of weapons of mass destruction. On June 17, 1925, the *Protocol for the Prohibition of the Use in War of Asphyxiating, Poisonous or Other Gases and of Bacteriological Methods of Warfare*, commonly called the **Geneva Protocol of 1925**, was signed by a total of 108 nations, including the five permanent members of the United Nations Security Council. The Geneva Protocol however did not have a method of ensuring compliance, hence soon after its ratification, several signatory countries such as Canada, Belgium, Japan, and the Soviet Union began developing biological weapons. The United States ratified the Geneva Protocol only in 1975 after the termination of its own biological weapons program in 1970.

Under the auspices of the United Nations Office for Disarmament Affairs (UNODA), the *Convention on the Prohibition of the Development, Production and Stockpiling of Bacteriological (Biological) and Toxin Weapons and on their Destruction*, referred to as the **Biological Weapons Convention (BWC)** was opened for signature on April 10, 1972, and entered into force on March 26, 1975 (United Nations Office for Disarmament Affairs, n.d.). Currently, 165 states are party to it with 12 signatory states. The BWC is the first multilateral disarmament treaty banning the "development, production and stockpiling" of an entire category of weapons of mass destruction. As per the terms of the treaty, the BWC bans:

- The development, stockpiling, acquisition, retention, and production of:
  1. Biological agents and toxins "of types and in quantities that have no justification for prophylactic, protective, or other peaceful purposes"
  2. Weapons, equipment, and delivery vehicles "designed to use such agents or toxins for hostile purposes or in armed conflict."
- The transfer of or assistance with acquiring the agents, toxins, weapons, equipment, and delivery vehicles described above.

(Archy, 2010)

The treaty also requires that within 9 months of coming into force, the state parties should **destroy or divert to peaceful purposes** biological agents, toxins, weapons or equipment, and means of delivery. **The BWC does not ban biodefense programs**.

Implementation and ensuring compliance to the treaty has been difficult, and several instances of violation have been documented, resulting in efforts to review and improve the treaty's implementation about every 5 years. In the Second Review Conference held in 1986, and the Third Review Conference held in 1991, the States Parties attempted to reduce ambiguities by several **confidence building measures** (CBMs) and to improve international cooperation. This included the **submission of annual reports** by the States Parties on specific activities such as information on biological research and development facilities, biodefense programs, disease outbreaks and control, regulations, legislation, or other measures. However, only a minority of member states submitted annual reports and the efforts remained mostly unsuccessful. In 1994, the States Parties held a Special Conference and agreed to establish an **Ad Hoc Group of States Parties** in order **to develop a legally binding verification regimen**, which was welcomed at the Fourth Review Conference held in 1996. The draft legal document was to be finalized in the Fifth Review Conference to be held in 2001. However, negotiations on the inspection protocol collapsed due to the **United States rejecting the draft text**. The Bush administration in the United States considered the acquisition and use of biological weapons by terrorist outfits as being more of a threat to security than state-level proliferation, and wanted that under the BWC, States should implement national legislation to thwart access to select agents. The Fifth Review Conference held in November 2002, on US insistence, adopted a resolution to hold three annual meetings of state parties and expert groups with the objective of evolving a **common program to implement measure at the national level to strengthen the BWC**. The measures included: penal legislation, pathogen security measures, enhanced international procedures to investigate and mitigate the alleged use of biological weapons or suspicious outbreaks of infectious disease, improved mechanisms for global disease surveillance and response, and scientific codes of conduct. This was finally **adopted by consensus in the Sixth Review Conference in 2006**. The States parties also adopted a detailed plan for promoting universal adherence including submission and distribution of CBMs. A comprehensive inter-sessional program from 2007 to 2010 and the establishment of Implementation Support Unit to assist State parties in implementing the BWC were also set up.

6.4 BIOSECURITY MEASURES FOR PREVENTING BIOTERRORISM

By definition, *"biosecurity is the effective implementation of measures that prevent would be terrorists, criminals and spies from gaining access to dangerous pathogens and toxins"* (Barletta, 2002). "Biosecurity" is distinguished from a similar term "biosafety" in that the former refers to measures to thwart deliberate release of disease causing agents, whereas the latter addresses accidental or unintended infections. In practice, some of the measures taken for ensuring biosafety, such as posting signs in areas where dangerous pathogens are handled in order to avoid accidental exposure, may sometimes be in conflict with measures taken for biosecurity, such as removing laboratory labels to hide the presence of infectious agents (Atlas, 2005). Nevertheless, **restricting access to biothreat agents**, which is crucial to good biosafety practices, is also an effective biosecurity measure. According to the World Health Organization (2004) Laboratory Biosafety Manual *"effective biosafety practices are the very foundation of laboratory biosecurity activities"*. This is because basic biosafety procedures for handling of microorganisms in laboratories mandate the use of physical and biological containment for infectious agents and restricts entry to such facilities to authorized, trained personnel (see Chapter 11: Laboratory Biosafety and Good Laboratory Practices). Restricting access to dangerous agents is also in keeping with Article IV of the 1972 Biological and Toxin Weapons Convention that requires State parties to take adequate measures to prevent the development, production and stockpiling of biological, and toxin weapons in its territory. National laws for controlling access to potential biothreat agents were first enacted in the United States. Similar laws were adopted in several nations such as Canada, Germany, France, the United Kingdom, Japan, and Israel. For instance, in the United Kingdom, access to pathogens and toxins in research facilities is regulated under the *Anti-terrorism, Crime and Securities Act (ATCSA) of 2001*. Under Part 7 (Security of Pathogens and Toxins) of this Act, research facilities as in universities must register with the government (notify the Secretary of State) before keeping or using dangerous substances, and must furnish on request, details of personnel with access to the agents. The Act also confers powers to the police to inspect the premises

(Queen's Printer of Acts of Parliament, 2001). Many other nations rely on existing regulation for ensuring biosafety to also address biosecurity by including specific provisions (Tucker, 2003).

Typically, biosecurity measures are applied to BSL-3 or 4 laboratory settings (see Chapter 11: Laboratory Biosafety and Good Laboratory Practices). These include:

1. *Physical security measures:* restricting access to authorized personnel and deploying security services to enforce it. Such measures would prevent theft of potential biothreat agents by outsiders.
2. *Personnel security:* including identity verification and background checks to prevent threats/thefts of biological materials by staff.
3. *Material accountability and security:* including the handling as well as transport of infectious agents.

These measures are specified in biosafety and biosecurity manuals of the WHO (World Health Organization, 2004, 2006; Zaki, 2010). However, as there are thousands of companies and laboratories around the world that possess or work with dangerous pathogens, having biosecurity strategies that vary from one country to the next could potentially result in terrorists indulging in "venue shopping." The World Federation for Culture Collections, which is an international association of germ banks, has urged its members to establish tighter controls on access to biothreat agents; however, it has no means of ensuring compliance (Tucker, 2002). Some of the international biosecurity initiatives are summarized in Table 6.2.

Table 6.2 International Biosecurity Initiatives

Organization	Instrument	Activity
Organization for Economic Cooperation and Development (OECD) (30 advanced industrial countries)	Biological Research Centres (BRCs)	An expert Task Force set up by the OECD Working Party on Biotechnology developed a series of best practices for BRCs in 2007 (OECD, 2007). The section *Best Practice Guidelines on Biosecurity for BRCs* (pages 45 to 57 of OECD, 2007) specifies methods and protocols to be followed by BRCs entrusted with the maintenance and exchange of hazardous biological materials
Australia Group (http://www.australiagroup.net/en/) (the United States and 32 other countries)	States parties to the BWC	Harmonize national controls on exports of dangerous pathogens
Global Health Security Initiative (GHSI) (http://www.ghsi.ca/english/index.asp) launched in November 2001 by Canada, the European Union, France, Germany, Italy, Japan, Mexico, the United Kingdom, and the United States	Support the World Health Organization's disease surveillance network	Cooperate in procuring vaccines and antibiotics, and in the development of vaccines especially smallpox vaccines

KEY TAKEAWAYS

- **Geneva Protocol** of 1925 (the *Protocol for the Prohibition of the Use in War of Asphyxiating, Poisonous or Other Gases and of Bacteriological Methods of Warfare*) signed by 108 nations.
- **BWC** of 1972 (the *Convention on the Prohibition of the Development, Production and Stockpiling of Bacteriological (Biological) and Toxin Weapons and on their Destruction*) signed by 165 nations under the auspices of UNODA—first multilateral disarmament treaty banning the "development, production, and stockpiling" of an entire category of weapons of mass destruction. The BWC does not ban biodefense programs.
- **Biosecurity** is the effective implementation of measures that prevent would be terrorists, criminals, and spies from gaining access to dangerous pathogens and toxins.
- Biosecurity measures include:
  - *Physical security measures*: Restricting access prevent thefts.
  - *Personnel security*: including identity verification and background checks to prevent threats/thefts of biological materials by staff.
  - *Material accountability and security*: including the handling as well as transport of infectious agents.
- Specified in WHO Laboratory Biosafety Manual, 2004, 2006; Article IV of the 1972 Biological and Toxin Weapons Convention; National laws.

International Biosecurity Initiatives:

- OECD—*Best Practice Guidelines on Biosecurity for Biological Research Centres, 2007*.
- Australia Group—United States and 43 members resolve to strengthen BWC.
- Global Health Security Initiative—brings together like-minded countries to strengthen preparedness to CBRN threats and support WHO in disease surveillance.

6.5 UNITED STATES APPROACH TO BIOTERRORISM

The United States Congress felt the need to introduce legislation in order to regulate access to biothreat agents in 1995 after a licensed microbiologist Larry Wayne Harris in Columbus, Ohio, ordered three vials of freeze–dried plague bacteria from the American Type Culture Collection (ATCC) using a forged letter head and identification number of the laboratory where he worked. His action aroused suspicion because he made repeated calls to the ATCC to check on the status of his order. Harris was a known neo-Nazi sympathizer, but could be arrested and convicted only for mail fraud on account of the forged letter head, as there was no law preventing ordinary citizens from ordering microbes for own use (Tucker, 2007). In 1996, the US Congress passed the *Anti-Terrorism and Effective Death Penalty Act* containing a section regulating facilities that handled dangerous microbes and toxins. This federal regulation, which came into effect on April 15, 1997, established the **Select Agent Rule** according to which **hazardous agents (microbes and toxins) were listed by the CDC of the Department of Health and Human Services** (DHHS) (http://www.selectagents.gov/), and anyone wanting to send or receive listed agents had to **register and**

document the transaction with the CDC. Failure to do so could attract prison terms and fines up to $500,000. One major flaw with this legislation was that facilities which possessed such agents or which were working with them did not have to register, only transfers required to be registered (Tucker, 2002).

Biosecurity in the United States became an important issue consequent to the anthrax attacks of 2001. As an immediate response to the attack, the US government passed the ***Uniting and Strengthening America by Providing Appropriate Tools Required to Intercept and Obstruct Terrorism (USA PATRIOT) Act of 2001*** (Public Law 107-56). The Act defines biological agents as *"microorganisms, or any recombinant or synthesized component thereof, capable of causing death, disease, or other biological malfunction in a human, animal, plant or other living organism; deterioration of food, water, equipment, supplies, or material of any kind; or deleterious alteration of the environment."* The USA PATRIOT Act included many of the measures conceptualized to ensure biosecurity. For instance, the Act:

- *Makes it a felony to possess a type or quantity of a biological agent that cannot be justified for prophylactic, protective, or peaceful purposes;*
- *Makes it a federal offense for convicted felons, illegal aliens or fugitives to possess or transport biological agents or toxins, in any quantity or for any reason.*

The following year, the US government also passed the ***Public Health Security and Bioterrorism Preparedness and Response Act of 2002***, commonly referred to as the ***Bioterrorism Preparedness Act***. This Act requires that the United States improves its ability to prevent, prepare for, and respond to acts of bioterrorism and other public health emergencies that could threaten either public health and safety or agriculture. Also, the heads of the DHHS and the US Department of Agriculture (USDA) were charged with the responsibility of determining what should be listed as a select agent. The Act requires that:

- *Certain federal agencies must be informed of research, possession, and transport of select agents;*
- *FBI background checks must be performed for anyone accessing, transporting, or receiving these agents;*
- *Defined security procedures must be followed for facilities within which these agents are contained.*

Acting in accordance with this Act, both the DHHS and the USDA passed rules governing the possession, use, and transfer of "select agents," or those agents which are deemed to be the most dangerous to human, plant, or animal health. The two rules passed were the ***HHS Select Agent Final Rule (HHS Select Agent Rule 2005)*** and the ***USDA Select Agent Final Rule (USDA Select Agent Rule 2005)***. The Bioterrorism Preparedness Act and the Select Agent Rule was used to prosecute Thomas Butler, a microbiologist at Texas Tech University for failing to report 30 missing vials of plague bacteria (he was acquitted of this charge, but was found guilty of 47 other counts mostly involving fraud) (Enserink & Malakoff, 2003).

Concomitant development of biodefense strategies was also made. In 2004, the US Congress passed the ***Project BioShield Act*** which was a 10-year program designed to acquire medical countermeasures to terrorist attacks. The legislation authorized the use of a Special Reserve Fund which made available US$5.6 billion till the fiscal year of 2013 for development/purchase of medical countermeasures.

BOX 6.1 PROJECT BIOSHIELD

In addition to authorizing funds for stockpiling *medical countermeasures (MCM)* against CBRN threat agents, the Project BioShield (PBS) Act of 2004 also amended the Public Health Services (PHS) Act and the Federal Food, Drug and Cosmetic (FD&C) Act in order to provide flexibility in development, procurement, and to authorize the government to permit their use during an emergency. Further refinement of these authorities was incorporated in the Pandemic and All-Hazards Preparedness Act (PAHPA) of 2006 and the Pandemic and All-Hazards Preparedness Reauthorization Act (PAHPRA) of 2013.

- *Research and development*: Section 2 of the PBS Act (enacted Section 319 F-1 of the PHS Act) authorizes the use of various streamlined procedures for awarding grants, contracts, and cooperative agreements for research and development of MCMs.
- *Security countermeasure procurements*: Section 3 of PBS Act (enacted section 510 of Homeland Security Act) authorizes the original US$ 5.6 billion for the period FY2004 through FY 2013 in a Special Reserve Fund for procuring security countermeasures for placing in the Strategic National Stockpile (SNS). The fund is received by the BARDA through an annual appropriation process. Section 3 of the PBS Act also enacted section 319F-2 of the PHS Act that authorizes the use and reporting of simplified acquisition procedures.
- *Emergency Use Authorization for medical countermeasures*: Section 4 of the PBS Act (enacted section 564 of the FD&C Act) allows the Secretary of the Health and Human Services (HHS) to issue an Emergency Use Authorization (EUA) when justified, or if deemed necessary by the Secretaries of Defense, Homeland Security or HHS. This authority has been delegated to the Food and Drug Administration (FDA) Commissioner. The EUA declaration allows the use of a FDA approved or licensed product for an unapproved indication, or an unapproved product for an indication pending approval, licensure, clearance, or ceasing of the emergency.

The Project BioShield was deemed necessary to incentivize pharmaceutical companies to produce medical countermeasures such as vaccines, antitoxins, and antibiotics as part of the biodefense strategy. Drug development is an expensive and failure prone enterprise with only around 8% of entities entering Phase I clinical trials progressing to market licensure. This is a risk that drug companies are willing to take when the markets are large, as profits margins could be substantial for successful products. However, for medical products to be stockpiled, with the government being the prime customer, the expected profit is too low to be attractive to large pharma companies. The creation of the permanent Special Reserve Fund was to encourage pharma companies to manufacture MCMs for meeting CBRN threats (Russell, 2007). Table 6.3 summarizes the acquisition programs between 2004 and 2006, during which the emphasis was on countermeasures to meet the threat from anthrax, botulinum toxins, smallpox, radiological and nuclear threats. Since the setting up of BARDA in 2007, MCMs added to the SNS include those against anthrax, botulism, plague, tularemia, glanders, melioidosis, typhus, viral hemorrhagic fever, radionuclide exposure, ionizing irradiation, nuclear detonation, and chemical agent exposure. BARDA has invested just over US$2.2 billion of the SRF through FY2013 and US$415 million in FY2014 for development of nearly 90 CBRN vaccine, therapeutic, and diagnostic candidates (United States Department of Health and Human Services, n.d.). A comprehensive list of BARDA's current CBRN advanced research and development projects is available as a regularly updated version on https://www.medicalcountermeasures.gov/.

In addition to funding development of MCMs for the SNS, BARDA has also supported MCMs for epidemics. As part of the US response to the Ebola outbreak in West Africa, BARDA launched a new Advance Research and Development (ARD) program in August 2014 in which it transitioned 10 Ebola vaccine, monoclonal antibodies, and antiviral candidates from early development supported by the National Institutes of Health and Department of Defense, into advanced development using funds from the FY2014 and FY2015 supplemental appropriations. This included supporting manufacturing scale up, optimizing product formulation, animal challenge studies, and human clinical trials (United States Department of Health and Human Services, n.d.).

References

Russell, P. K. (2007). Project BioShield: What it is, why it is needed, and its accomplishments so far. *Clinical Infectious Diseases*, *45*(Supplement 1), S68–S72 . Available from http://dx.doi.org/10.1086/518151. Retrieved from http://cid.oxfordjournals.org/content/45/Supplement_1/S68.full
United States Department of Health and Human Services (n.d.) *Project BioShield Annual Report January 2015-December 2014*. Retrieved from https://www.medicalcountermeasures.gov/media/36816/pbs-report-2014.pdf.

(Continued)

BOX 6.1 (CONTINUED)

Table 6.3 Summary of Project BioShield Medical Countermeasure Acquisition Programs Between 2004 and 2006

Threat Agent, Medical Countermeasure	Quantity	Funds Obligated/Status of Contract
Anthrax		
Recombinant protective antigen anthrax vaccine	75 million doses	Contract awarded in Now 2004 to VaxGen: $879.2 million (contract terminated for default Dec 2006)
Anthrax vaccine adsorbed	10 million doses	Contract awarded in May 2005 to BipPort: $122.1 million for 5 million doses; contract options for an additional 5 million doses ($120 million) exercised in May 2006; delivery to SNS completed in early 2007
	10.4 million doses	Presolicitation notice issued in Apr 2007, stating intent to purchase 10.4 million doses, with an option for an additional 8.4 million doses
Anthrax therapeutics	20,001 treatment courses	Contract awarded in Jun 2006 to Human Genome Sciences for 20,001 treatment courses for ABthrax: $165.2 million
	10,000 treatment courses	Contract awarded in Jul 2006 to Cangene for 9900 treatment coursed of human anthrax immune globulin: $144 million
Botulinum		
Botulinum anitioxin	200,000 doses	Contract awarded Jun 2006 to Cangene for 200,000 doses of heptavalent botulism antitoxin: $362.6 million
Smallpox		
Modified vaccinia ankara	10–20 million doses	Request for proposals closed Oct 2005; contract anticipated in fiscal year 2007
Radiological/nuclear		
Pediatric (liquid) potassium iodide	1.7 million 1-ounce bottles	Contract awarded Mar 2005 to Fleming Pharmaceuticals for 1.7 million bottles: $5.7 million; delivery to the SNS was completed in Sep 2005
	Additional 3.1 million bottles	Contract options exercised on 1 Feb and May 2006, including additional 3.1 million bottles: $11.8 million; total obligation, $17.5 million; delivery to the SNS started in May 2006
Medical countermeasures to treat/mitigate neutropenia associated with acute radiation syndrome	Up to 100,000 treatment courses	Request for proposals closed in Feb 2006 and canceled in Mar 2007
Chelating agents Zn- and Ca-DTPA	∼475,000 doses	Contract awarded Feb 2006 to Akron for delivery of >390,00 doses of Ca-DTPA and >60,000 doses of Zn-DTPA: total obligation, $22 million; delivery to the SNS completed in Apr 2006

Note: DTPA, *diethylenetriaminepentaacetate*; SNS, *Strategic National Stockpile*.
Russell, P.K. (2007) Clin Infect Dis. (2007) 45 (Supplement 1): S68–S72. *(Reproduced with permission of Oxford University Press).*

The Project BioShield (see Box 6.1) implemented by the US DHHS and its component agencies was aimed at expediting R&D, and acquisition and availability of medical countermeasures to improve the government's preparedness for and ability to counter chemical, biological, radiological, and nuclear (CBRN) threat agents. Initial efforts focused on vaccines for category A threats, anthrax and small pox. By 2006, nine additional agents were added.

Since passage of the Project BioShield Act, the US Congress has considered additional measures to further encourage countermeasure development. This includes the *Pandemic and All-Hazards Preparedness Act, 2006*, under which the **Biomedical Advanced Research and Development Authority (BARDA)** was created in HHS. This Act also modified the Project BioShield procurement process. The *Pandemic and All-Hazards Preparedness Reauthorization Act of 2013* authorized Project BioShield appropriations of $2.8 billion for FY2014 through FY2018. The *Consolidated Appropriations Act, 2014* provided $255 million for Project BioShield procurements to remain available until expended (Gotton, 2014).

KEY TAKEAWAYS

United States approach to Bioterrorism:

- *Anti-Terrorism and Effective Death Penalty Act of 1996*
  - *Select Agent Rule*—hazardous agents listed by the CDC of the DHHS; permits required for sending or receiving biological agents on the list
- *USA PATRIOT Act of 2001 (Uniting and Strengthening America by Providing Appropriate Tools Required to Intercept and Obstruct Terrorism)*
  - makes it a felony to possess select agents without proper justification
- *Bioterrorism Preparedness Act of 2002 (Public Health Security and Bioterrorism Preparedness and Response Act)*—rules governing possession, use, and transfer of select agents:
  - *HHS Select Agent Final Rule (HHS Select Agent Rule 2005)*
  - *USDA Select Agent Final Rule (USDA Select Agent Rule 2005)*
- *Project BioShield Act of 2004*—authorized use of Special Reserve Fund to acquire medical countermeasures to terrorist attack
- *Pandemic and All-Hazards Preparedness Act of 2006*—created the BARDA under HHS
- *Pandemic and All-Hazards Preparedness Reauthorization Act of 2013 and Consolidated Appropriations Act, 2014*—provides finance for Project BioShield

---

6.6 EUROPEAN UNION APPROACH TO BIOTERRORISM

Possibly because Member States of the European Union (EU) have not had any recent bioterrorism attacks, the level of public concern is lower than in the United States. However, after the 2001 anthrax attacks, recognition that a biological attack could simultaneously affect several Member States, prompted efforts to put in place a comprehensive approach to prevent and protect against possible biothreats. To begin with a **Health Security Committee** was established in November 2001, to coordinate and exchange information among Member States of possible release of agents threatening public health. This was achieved through a series of Communications (Communication

2003/320 in June 2003, Communication 2004/701 in October 2004, and Communication 2005/605 in November 2005), which although are not legally binding outline policies and forward guidelines for effective surveillance systems and systems for prompt notifications of information, aimed at coordinating health preparedness and emergency response. Community programs for preparedness for biological and chemical agent attacks (**BICHAT program**) and rapid alert systems (**RAS-BICHAT**) were created (Casale, n.d.).

In July 2007, with the intention of stimulating discussions within and between Member States on measures to reduce biological risks and to enhance preparedness and response, the European Commission brought out the **Green Paper on Bio-preparedness**. The Green Paper required all the national authorities responsible for risk prevention and response, human, animal and plant health, customs, civil protection, law enforcement authorities, the military, bioindustry, epidemiological and health communities, academic institutions, and bioresearch institutes, to be involved, to contribute, and to improve the ability of the EU to prevent, respond to, and recover from a biological incident or deliberate criminal activity (Sirbu, 2010).

KEY TAKEAWAYS

EU approach to bioterrorism:

- Health Security Committee established in 2001—to coordinate and exchange information among Member States of possible release of select agents.
- Community programs for preparedness for biological and chemical agent attacks (BICHAT program) and rapid alert systems (RAS-BICHAT) were created.

6.7 BIODEFENSE PROGRAMS

The **BWC** 1972 does not ban biodefense programs. Research into dangerous pathogens with a view to understanding modes of transmission, pathogenesis, means of control, and treatment strategies, is necessary to protect soldiers and citizens from bioterrorism. However, biodefense research programs need to be sufficiently transparent as the distinction between defensive and offensive research and development is suspiciously thin (Tucker, 2007). In the United States, the biodefense strategy is essentially described by the document *Biodefense for the 21st Century* released by the Bush administration in 2004 (The White House, 2004). The document describes the pillars of the biodefense program as being; "*Threat Awareness, Prevention and Protection, Surveillance and Detection, and Response and Recovery.*" For successful implementation, it requires the involvement of a wide range of Federal departments and agencies with the responsibility of coordinating *domestic Federal operations to prepare for, respond to, and recover from biological weapons attacks* resting with the Secretary of Homeland Security.

In a three-part strategy to develop countermeasures such as diagnostics, vaccines and drugs, mainly against small pox and anthrax, the US government spent around US$60 billion between 2001 and 2011 on basic research at the National Institutes of Health (NIH), on further development and testing at the BARDA, and to purchase finished drugs and vaccines under Project BioShield. The funds helped modernize the nation's public health system, and **through BioShield created a stockpile of 20 million doses of smallpox vaccine, 28.75 million doses of anthrax vaccine, and**

1.98 million doses of four medicines to treat complications of smallpox, anthrax, and botulism (Hayden, 2011a). However, questions have been raised about the effectiveness of the US biodefense research and development. In one particularly telling incident that occurred in January 2009, 20-year-old Lance Corporal Cory Belken of the US Marine Corps received a routine smallpox vaccination, but unfortunately was diagnosed with leukemia 2 weeks after the vaccination. As a result of reduced immunity consequent to chemotherapy, he developed vaccinia viral infection. Three counterattack experimental smallpox drugs developed as part of the biodefense program were used to treat the infection. The first drug, an approved antibody had no effect even at 30 times the standard dose; the second called STS-246, had only been tested on one person earlier; and the third, called CMX001, had to be used before he started to recover, although it is still not known which of the three, if any, eventually helped (Hayden, 2011a). Supporters of the biodefense program insist that drug discovery is a long, expensive, and failure prone activity, and that sustained efforts would be required for preparing the nation against bioterrorist attack.

From a relatively small number of facilities capable of handling biological weapons agents, the **United States now has over 400 laboratories with around 15,000 workers engaged in biodefense programs.** Some members of the Congress and members of the Government Accountability Office are worried that this could pose a greater danger by continuing to proliferate the number of places and people that are handling these agents. The CDC and USDA which oversee these types of facilities release minimal information to the public due to the sensitive nature of the information, but several **allegations of insufficient oversight have been made** (Young & Penzenstadler, 2015). For instance, in January 2016, a Pentagon accountability review revealed that a biodefense laboratory, the Army's Dugway Proving Ground in Utah, had **mistakenly shipped live anthrax specimens for more than a decade**. The issue came to light when a private biotech firm found that some of the radiation killed specimens it received were still alive (Young & Vanden Brook, 2016).

6.8 DUAL-USE RESEARCH OF CONCERN

A major issue of concern to scientific communities, government authorities, and the public is that of the potential for **misuse of life sciences research** conducted for legitimate scientific purposes. For instance, research on pathogenicity genes which would help in developing vaccines could also potentially be used to develop a bioweapon. One of the first documented examples of legitimate research accidently creating a potential bioweapon happened in 2001 when Ron Jackson of CSIRO's wildlife division and Ian Ramshaw of Australian National University in Canberra, Australia, found that inserting an Interleukin 4 (IL-4) gene in mousepox virus killed the animals (including many that had been vaccinated for mousepox) within 9 days of inoculation. The scientists had been trying to create a contraceptive vaccine and were trying to stimulate antibodies against mouse eggs to make the animals infertile. The mousepox virus was only intended to be a vector to carry egg proteins to trigger an immune response, and the IL-4 was to boost antibody production (Nowak, 2001).

6.8.1 NATIONAL SCIENCE ADVISORY BOARD FOR BIOSECURITY

Several scientific and professional societies have advocated the development of codes of conduct to guide scientists work in DUR. In the United States, in 2004, the National Institutes of Health established the **National Science Advisory Board for Biosecurity (NSABB)** as a federal advisory committee to address issues related to DUR (National Institutes of Health, n.d.). Also, a US$1.5 billion initiative to sequence the genomes of key microorganisms with bioterrorism potential, known as the **Transformational Medical Technologies Initiative**, was launched by the Department of Defense in 2006. It was made into a permanent program in 2009, although it ceased to exist as a stand-alone program in 2011 (Hayden, 2011b).

From its inception, the NSABB has recognized the importance of **responsible conduct of research (RCR)** and that scientists and researchers are critical to any effort to address and mitigate the risks of DURC. Because of their specialized knowledge, scientists are obviously in a position to anticipate the types of knowledge, products, or technologies that their work might generate and the potential for their misuse. The mandate of the NSABB is to sensitize scientists to the dual-use potential of their work and to **strengthen their moral and ethical values to prevent malevolent dual-use**.

In 2007, the NSABB brought out its *Proposed Framework for the Oversight of Dual-Use Life Sciences Research: Strategies for Minimizing the Potential Misuse of Research Information* (National Science Advisory Board for Biosecurity, 2007), salient excerpts from which are included below. This document proposed the following criterion for identifying DURC:

"Research that, based on current understanding, can be reasonably anticipated to provide knowledge, products, or technologies that could be directly misapplied by others to pose a threat to public health and safety, agricultural crops and other plans, animals, the environment, or material".

Seven categories of research that might satisfy this criterion included knowledge, products, or technologies that would:

1. *Enhance the harmful consequences of a biological agent or toxin.*
2. *Disrupt immunity or the effectiveness of immunization without clinical and/or agricultural justification.*
3. *Confer to a biological agent or toxin, resistance to clinically and/or agriculturally useful prophylactic or therapeutic interventions against that agent or toxin, or facilitate their ability to evade detection methodologies.*
4. *Increase the stability, transmissibility, or the ability to disseminate a biological agent or toxin.*
5. *Alter the host range or tropism of a biological agent or toxin.*
6. *Enhance the susceptibility of a host population.*
7. *Generate a novel pathogenic agent or toxin, or reconstitute an eradicated or extinct biological agent.*

The NSABB also suggested codes of conduct including core responsibilities and a delineation of responsibilities in DURC in life sciences.

KEY TAKEAWAYS

DURC defined as research that *can be reasonably anticipated to provide knowledge, products, or technologies that could be directly misapplied by others to pose a threat to public health and safety, agricultural crops and other plans, animals, the environment, or material.*

6.8.2 CORE RESPONSIBILITIES OF LIFE SCIENTISTS IN REGARD TO DUAL-USE RESEARCH OF CONCERN

Although institutional and regulatory oversight is necessary for DUR, it is evident that biosecurity cannot be achieved by any single law or program. A **"bottom-up" approach with scientific self-governance has been suggested as being crucial** for preventing beneficently intended research being applied for evil ends (Kwik, Fitzgerald, Inglesby, & O'Toole, 2003). This "bottom-up" approach requires scientists to be trained in ethics, to be aware of potential misuse of own research, and to be committed to the task of using one's knowledge and skill for beneficial purposes.

The code of conduct advocated by the NSABB emphasizes that individuals involved in life sciences research have an *ethical obligation to avoid or minimize the risks and harm that could result from malevolent use of research outcomes*. In order to achieve this, scientists should:

1. *Assess their research for dual-use potential*
2. *Stay informed regarding relevant literature, guidance, and requirements*
3. *Train others to identify and appropriately manage and communicate DURC*
4. *Serve as role models of responsible behavior*
5. *Be alert to potential misuse of research*

(National Science Advisory Board for Biosecurity, 2007)

Responsible communication of research with dual-use potential should take into account decisions regarding the content, the timing, as well as the extent of distribution of the information. Depending on the potential for misuse, a redacted version of the content alone could be published (see Box 6.2). As to timing, the communication could be deferred till a time that the communication does not present the same level of risk. Also, the extent of distribution could be restricted to a select group rather than make it accessible to the public.

KEY TAKEAWAYS

Responsibilities of Life scientists in regard to DURC:

- "Bottom-up" approach—scientific self-governance—requires scientists to be trained in ethics, to be aware of potential misuse of own research, and to be committed to the task of using one's knowledge and skill for beneficial purposes.
- Responsible communication of research with DURC potential:
  - publish redacted version
  - timing of publication—defers publication till a time that it does not present the same risk
  - restrict access to select group

6.9 SUMMARY

A sound grounding in ethics alone can prevent the use of research involving dangerous microbes with infective capability to damage or destroy crops, livestock, and human beings, from being used for malevolent purposes. International efforts such as the **BWC** of 1972, as well as national laws,

BOX 6.2 BALANCING NATIONAL SECURITY WITH SCIENTIFIC OPENNESS

In the words of Gerald Fink, *"Biotechnology represents a 'dual-use' dilemma in which the same technologies can be used legitimately for human betterment and misused for bioterrorism"* (National Research Council, 2004). The risk of potential misuse can come from two sources: (1) the risk of dangerous agents being stolen from research facilities and (2) the risk of information regarding dangerous agents such as research results, knowledge, or techniques that would allow the weaponizing or creation of novel microbes for use in terrorism. Restricting access to and limiting the handling of dangerous agents to authorized personnel in licensed facilities have been the strategies adopted to address the risk of theft. However, proposals to limit the publication of research results have been met with expressions of concern from both scientists and publishers. The norm of open publication, with sufficient details of materials and methods to allow the results to be challenged or reproduced in peer laboratories, is critical to uphold the veracity of scientific research. Also, as the progress of science is incremental, open communication is necessary for researchers to build on the results of others. In 2003, the editors of major life sciences journals, including *Science, Nature, Cell,* and the *Proceedings of the National Academy of Sciences (PNAS)*, met to discuss the issue of security and scientific publication (Journal Editors and Authors Group, 2003). The statement issued by the group expressed the opinion that if research information *presents enough risk of use by terrorists . . . it should not be published.* The National Research Council's Committee of Research Standards and Practice to Prevent the Destructive Application of Biotechnology in 2004 recommended the establishment of a **National Science Advisory Board for Biodefense (NSABB)** under the DHHS, *in order to provide advice, guidance, and leadership* to the review and oversight system proposed by the committee. In keeping with its mandate, the NSABB in June 2007 brought out the ***Proposed Framework for the Oversight of Dual-Use Life Sciences Research: Strategies for Minimizing the Potential Misuse of Research Information***. The goal of the framework is to implement reasonable precautions to minimize the risk of misuse while maintaining a robust research environment so that oversight does not impede legitimate life science research. The NSABB has in exercising its function taken decisions in December 2011 to "recommend" that *Science* and *Nature* publish only the redacted (censored) versions of two studies on H5N1 influenza virus created in the labs for fears that the information can be misused by some if published in full (National Institutes of Health, 2011).

1. Yoshihiro Kawaoka and his colleagues at University of Wisconsin-Madison produced a mutant H5N1 influenza virus that had the viability to become transmissible by air. The mutant was created by combining H5 hemagglutinin with genes from the 2009 pandemic H1N1 influenza virus. The NSABB decided to recommend that only the censored version be published in *Nature* as the authors had demonstrated the compatibility of segments of the 2009 pandemic influenza backbone with H5 hemagglutinin to produce a transmissible virus.
2. Ron Fouchier's team of the Erasmus Medical Center in Rotterdam created a mutant H5N1 virus by inserting three mutations and then passing the virus from one infected ferret to another till it became transmissible by air. The lab-created viruses had apparently bypassed the barriers to evolution in the wild, since in nature, pigs are the hosts in which influenza viruses re-assort (or combine) to produce strains that are pathogenic or transmissible by air.

However, a meeting of experts convened by the World Health Organization (WHO) on February 17, 2012, came to a different conclusion (Butler, 2012), although it was apparent at the meeting that there was a need for some consensus as to what parts of such studies should be published and who might qualify for access to the full papers. In the case of the two H5N1 papers, after several months, the NSABB was informed that publishing the redacted versions was not feasible because of a number of factors. Both papers were published in 2012 after the NSABB voted for the publication of the revised manuscript containing all the genetic data.

Submissions for publication of information of DURC has led to a *vexing problem for journal editors* due to the absence of a practical mechanism or process of redaction of critical data by journals (Casadevall et al., 2013). There have, however, been instances of redacted versions of papers having been published after the authors of the study voiced their concerns about the DURC. For example, redacted versions of two papers in the *Journal of Infectious Diseases* (JID) reporting a new type of botulinum toxin were published in October 2013 by the journal after the authors voiced concerns, waiving the requirement of depositing the genomic sequence information in public databases. Redaction thus does seem to be a feasible option for publication of information of DURC.

(Continued)

BOX 6.2 (CONTINUED)

References

Butler, D. (2012, February 17) (Reporter) *Updated: Avian flu controversy comes to roost at WHO- International meeting seeks to chart a way forward for mutant flu research.* Retrieved from http://www.nature.com/news/updated-avian-flu-controversy-comes-to-roost-at-who-1.10055.

Casadevall, A., Enquist, L., Imperiale, M. J., Keim, P., Osterholm, M. T., & Relman, D. A. (2013). Redaction of sensitive data in the publication of dual use research of concern. *mBio, 5*(1). e00991-13. http://dx.doi.org/10.1128/mBio.00991-13.

Journal Editors and Authors Group (2003). Statement on scientific publication and safety. *Science, 299*(5610), 1149. Retrieved from http://www.sciencemag.org/site/feature/data/security/statement.pdf (This statement also appeared in the 18 February 2003 issue of the Proceedings of the National Academy of Sciences and the 20 February 2003 issue of Nature.)

National Institutes of Health (2011, December 20) *Press statement on the NSABB review of H5N1 research.* Retrieved from http://www.nih.gov/news-events/news-releases/press-statement-nsabb-review-h5n1-research.

National Research Council (2004). Biotechnology research in an age of terrorism—Committee of research standards and practice to prevent the destructive application of biotechnology. The National Academies Press. Retrieved from http://www.nap.edu/download.php?record_id = 10827.

National Science Advisory Board for Biosecurity (2007). Proposed framework for the oversight of dual use life sciences research: Strategies for minimizing the potential misuse of research information. Retrieved from <http://osp.od.nih.gov/sites/default/files/resources/Framework%20for%20transmittal%20duplex%209-10-07.pdf>.

for example, the USA PATRIOT Act in the United States, ban the production, development, and stockpiling of dangerous microorganisms and toxins that could be used in terror attacks. Biosecurity measures require background checks on scientific personnel handling select agents, and oversight in facilities handling or transferring these agents. Development of biodefense strategies in order to control the effects and protect soldiers and civilians from biothreats is essential. This however, leads to a problem of DUR, since information on pathogenesis necessary for finding means of controlling infection, could simplistically, be misused in the hands of a terrorist. Mandatory laboratory best practices and educating graduate students and science personnel on ethics could possibly prevent malevolent use of DUR.

REFERENCES

Archy, W. (2010, October 4, updated September 2012). *The Biological Weapons Convention (BWC) at a glance.* Retrieved from https://www.armscontrol.org/factsheets/bwc.

Arthur, C. (2001) (Reporter) Scientists made virus 'more lethal than HIV'. *The Independent, 24 July 2001.*

Atlas, R. M. (2005). Ensuring biosecurity and biosafety through biopolicy mechanisms: Addressing threats of bioterrorism and biowarfare. *Asian Biotechnology and Development Review, 8*(1), 121—137. Retrieved from http://www.ris.org.in/images/RIS_images/pdf/article5_v8n1.pdf.

Barletta, M. (2002). *Biosecurity measures for preventing bioterrorism.* USA: Center for Nonproliferation Studies.

Casale, D. (n.d.) EU approach to bio-terrorism. *CBW magazine, 3.* Retrieved from http://www.idsa.in/cbwmagazine/EUApproachtoBioTerrorism_dcasale_0409.

Cello, J., Paul, A. V., & Wimmer, E. (2002). Chemical synthesis of poliovirus cDNA: Generation of infectious virus in the absence of natural template. *Science, 297*(5583), 1016—1018, Epub 2002 Jul 11.

Centers for Disease Control (n.d.). *Bioterrorism overview.* Retrieved from http://emergency.cdc.gov/bioterrorism/overview.asp.

Enserink, M., & Malakoff, D. (2003). The trials of Thomas Butler. *Science,* 2054—2063, 19 Dec 2003

Gotton, F. (2014, June 18). The Project BioShield Act: Issues for the 113th Congress. *Congressional Research Service Report.* Retrieved from https://www.fas.org/sgp/crs/terror/R43607.pdf.

Hayden, E.C. (2011a) (Reporter) *Biodefense since 9/11: The price of protection.* Published online 7 September 2011 I Nature 477, 150−152 (2011) I http://dx.doi.org/10.1038/477150a. Retrieved from http://www.nature.com/news/2011/110907/full/477150a.html.

Hayden, E.C. (2011b) (Reporter) *Pentagon rethinks bioterror effort.* Published online 21 September 2011 I Nature 477, 380−381 (2011) I http://dx.doi.org/10.1038/477380a. Retrieved from http://www.nature.com/news/2011/110921/full/477380a.html.

Jackson, R. J., Ramsay, A. J., Christensen, C. D., Beaton, S., Hall, D. F., & Ramshaw, I. A. (2001). Expression of mouse interleukin-4 by a recombinant ectromelia virus suppresses cytolytic lymphocyte responses and overcomes genetic resistance to mousepox. *Journal of Virology, 75*(3), 1205−1210. Retrieved from http://jvi.asm.org/content/75/3/1205.full.

Kwik, G., Fitzgerald, J., Inglesby, T. V., & O'Toole, T. (2003). Biosecurity: Responsible stewardship of bioscience in an age of catastrophic terrorism. *Biosecurity and Bioterrorism: Biodefense Strategy, Practice, and Science, 1*(1), 27−35.

Leitenberg, M. (2002). Biological weapons and bioterrorism in the first years of the twenty-first century. *Politics and the Life Sciences, 21,* 3−27.

National Institutes of Health (n.d) *Biosecurity—National Science Advisory Board for Biosecurity (NSABB).* Retrieved from http://osp.od.nih.gov/office-biotechnology-activities/biosecurity/nsabb.

National Science Advisory Board for Biosecurity (2007). Proposed framework for the oversight of dual use life sciences research: Strategies for minimizing the potential misuse of research information. Retrieved from http://osp.od.nih.gov/sites/default/files/resources/Framework%20for%20transmittal%20duplex%209-10-07.pdf.

Nowak, R. (2001, January10) (Reporter) *Killer mousepox virus raises bioterror fears.* Retrieved from https://www.newscientist.com/article/dn311-killer-mousepox-virus-raises-bioterror-fears/.

OECD (2007) *OECD Best practice guidelines for biological resource centres.* Retrieved from http://www.oecd.org/sti/biotech/38777417.pdf.

Queen's Printer of Acts of Parliament (2001). Anti-terrorism, crime and security act. Retrieved from http://www.legislation.gov.uk/ukpga/2001/24/contents.

Riedel, S. (2004). Biological warfare and bioterrorism: a historical review. *Proceedings (Baylor University. Medical Center), 17*(4), 400−406. Retrieved from http://www.ncbi.nlm.nih.gov/pmc/articles/PMC1200679/

Robertson, D. L., & Leppla, S. H. (1986). Molecular cloning and expression in *Escherichia coli* of the lethal factor gene of *Bacillus anthracis. Gene, 44*(1), 71−78.

Sirbu, M. (2010). Green Paper on Bio-preparedness—general comments. *Journal of Medicine and Life, 3*(4), 430−432 . Retrieved from http://www.ncbi.nlm.nih.gov/pmc/articles/PMC3019068/.

The White House (2004, April 28). *Biodefense for the 21st Century.* Retrieved from http://fas.org/irp/offdocs/nspd/hspd-10.html.

Tucker, J. B. (2002). Preventing terrorist access to dangerous pathogens: the need for international biosecurity standards. *Disarmament Diplomacy, 66.* Retrieved from http://www.acronym.org.uk/dd/dd66/66op2.htm.

Tucker, J. B. (2003, June 1). *Preventing the misuse of pathogens: the need for global biosecurity.* Retrieved from http://www.armscontrol.org/act/2003_06/tucker_june03.

Tucker, J.B. (2007, April 30). *Strategies to prevent bioterrorism: Biosecurity policies in the United States and Germany.* Retrieved from http://www.acronym.org.uk/dd/dd84/84jt.htm#en09.

United Nations Office for Disarmament Affairs (n.d.) *The Biological Weapons Convention.* Retrieved from http://www.un.org/disarmament/WMD/Bio/.

Volchkov, V. E., Volchkova, V. A., Muhlberger, E., Kolesnikova, L. V., Weik, M., Dolnik, O., & Klenk, H. D. (2001). Recovery of infectious Ebola virus from complementary DNA: RNA editing of the GP gene and viral cytotoxicity. *Science, 291*(5510), 1965−1969.

World Health Organization (2004). *Laboratory safety manual* (3rd Edition, Retrieved from http://www.who.int/csr/resources/publications/biosafety/WHO_CDS_CSR_LYO_2004_11/en/). Geneva: World Health Organization?

World Health Organization (2006). *Biorisk management. Laboratory biosecurity guidance*. Geneva: World Health Organization. Retrieved from http://www.who.int/csr/resources/publications/biosafety/WHO_CDS_EPR_2006_6.pdf.

Young, A., & Penzenstadler, N. (2015) (Reporters) *Inside America's secretive biolabs*. Retrieved from http://www.usatoday.com/longform/news/2015/05/28/biolabs-pathogens-location-incidents/26587505/.

Young, A., & Vanden Brook, T. (2016, January 19) (Reporters) *Army anthrax revelations raise oversight concerns*. Retrieved from http://www.usatoday.com/story/news/2016/01/15/army-lab-anthrax-accounability-report-reaction/78845202/.

Zaki, A. N. (2010). Biosafety and biosecurity measures: management of biosafety level 3 facilities. *International Journal of Antimicrobial Agents, 36*(Suppl1), S70–S74.

Zimmerman, B. E., & Zimmerman, D. J. (2003). *Killer germs: Microbes and diseases that threaten humanity*. New York, NY: McGraw-Hill.

FURTHER READING

Reviews and Books

Miller, J., Engelberg, S., & Broad, W. (2002). *Germs: Biological weapons and America's secret war*. New York: Simon and Schuster.

Riedel, S. (2004). Biological warfare and bioterrorism: A historical review. *Proceedings (Baylor University. Medical Center), 17*(4), 400–406. Retrieved from http://www.ncbi.nlm.nih.gov/pmc/articles/PMC1200679/pdf/bumc0017-0400.pdf.

Ryan, J. (2016). *Biosecurity and bioterrorism: Containing and preventing biological threats*. Butterworth-Heinemann, 392 pp.

The Sunshine Project (2002, April 1) An introduction to biological weapons, their prohibition, and the relationship to biosafety. Retrieved from http://www.biosafety-info.net/article.php?aid = 25.

Web Resources

Interview with **Edward Hammond**, Director of the U.S. Office of *The Sunshine Project*, an organization focusing on oversight of research involving biological weapons agents. Uploaded on April 20, 2008. Retrieved from https://www.youtube.com/watch?v = q29mKJr7_Xk.

Resources for Research Ethics Education—http://research-ethics.net/topics/biosecurity/#resources.

GENETIC TESTING, GENETIC DISCRIMINATION AND HUMAN RIGHTS

7

A cardinal principle that we must not stray from -no exceptions- is that your genetic information is your business in terms of who sees it. Nobody should be gaining access to that information without your explicit permission, and nobody should be requiring you to take a genetic test unless you decide that that's what you want to do.

-Francis Collins, Director, National Institute of Health.

CHAPTER OUTLINE

7.1 Introduction 172
7.2 Genetic Testing 172
7.3 Genetic Exceptionalism 175
7.4 Genetic Discrimination 175
7.5 Ethical, Legal, Social Implication (ELSI) 177
    7.5.1 The ELSI of Human Genome Project 178
    7.5.2 International and National Programs on ELSI 179
7.6 Mechanisms for Preventing Genetic Discrimination 180
    7.6.1 Rights-Based Advocacy 180
    7.6.2 Rights-Based Policy and Tools 182
    7.6.3 National Legislation 183
7.7 Summary 186
References 186
Further Reading 187

An Introduction to Ethical, Safety and Intellectual Property Rights Issues in Biotechnology.
DOI: http://dx.doi.org/10.1016/B978-0-12-809231-6.00007-7

Angelina Jolie
Hollywood actor and director, made history in May 2013 by going public about her double mastectomy after genetic tests revealed that she had an abnormal breast cancer gene (*By Philipp von Ostau - Own work, CC BY-SA 3.0*, https://commons.wikimedia.org/w/index.php?curid = 17825680)

7.1 INTRODUCTION

An important objective of the Human Genome Project (HGP) was to study diseases that had a genetic basis with a view to improve diagnostic, preventive and therapeutic strategies. Several diseases including Huntington's disease, Parkinson's disease, Duchenne's Muscular Dystrophy, cystic fibrosis, multiple sclerosis, and several types of cancers are known to be heritable, while many others such as diabetes, hypertension, heart disease, and mental disorders such as schizophrenia and Alzheimer's, have been established to have strong familial etiology. **The field of "*pharmacogenomics*" aims to use genome information to identify disease causing gene sequences and to use this information for diagnosis, for designing new drugs, or gene therapy** which replaces a diseased gene with a normal one. One important issue that has emerged is that personal genetic information is unlike other types of information and can result in various types of psychological and social harm to persons. The Ethical, Legal and Social Implication (ELSI) program of the Human Genome Project articulated this concern and deliberated on methods to ensure privacy and fairness of use of genome information. This chapter examines the ethical issues surrounding genetic information and reviews current legislation aimed at protecting personal privacy and preventing discrimination based on genetic information.

7.2 GENETIC TESTING

Genetic disorders could result from many different causes. They could be due to **mutations in single genes** (for example, sickle cell anemia is due to a single amino acid substitution in hemoglobin in red blood corpuscles); or **chromosomal aberrations** that changes the number or structure of chromosomes (for example, Down's syndrome is caused by the presence of an extra chromosome); or due to **mutations in several genes** (for example, colon cancer). These diseases can be detected through tests on blood or other tissues. Genetic tests may be done for a variety of reasons:

- for **diagnosis** (or confirmation of a diagnosis) when disease symptoms are presented,
- to **prevent, or to start early treatment** before disease symptoms are manifested,
- to **screen pre-implantation embryos**,
- to find **genetic diseases in unborn babies**,
- to determine **dose of a medicine** best suited to a patient, and
- to determine whether a person **has a heritable gene for a disease** (this is especially useful in certain late onset diseases such as Huntington's disease).

Results of genetic tests could help in **deciding a treatment regimen** when available. For conditions for which treatment is currently unavailable, the genetic tests can help make **decisions on lifestyles, reproductive choices** and family **planning**, or insurance coverage. Genetic tests can help people take a more proactive role in their medical state as exemplified by Angelina Jolie and her double mastectomy (see Box 7.1). **Genetic counselling** is usually provided to persons taking genetic tests to understand the pros and cons of the testing.

Genetic testing has been useful in studying, screening and treatment of **genetic diseases in select populations** such as the **Ashkenazi Jews**, in which it has been especially useful in reducing the number of cases of hereditary diseases such as Tay-Sachs disease. Other diseases common among these Jews include Gaucher's disease, familial dysautonomia (unique to this group),

---

BOX 7.1 PREVENTIVE MEDICINE AND THE BRCA GENES

Angelina Jolie, Hollywood actor and director, made history in May 2013 by going public about her double mastectomy in a bid to encourage women at risk of developing breast cancer to take a more proactive role in exercising options to prevent the disease (Jolie, 2013). Although she is not the first woman to have asked for a double breast removal after doctors found an abnormal breast cancer gene, being a Hollywood star celebrated as one of the world's most beautiful women, Angelina Jolie drew public attention to the role of genetic screening and preventive medicine.

The incidence of breast cancer has been linked to several causes such as age, race, reproductive and menstrual history, obesity, lifestyle (for example, excessive alcohol intake, smoking and high fat diets) and family predisposition. Ten percent of all breast cancer cases have been attributed to heritable mutations in two tumor suppressor genes, BRCA1 (BReast-CAncer gene 1) and BRCA2. Women with mutations in either of these genes have an increased risk of 56−85 percent of developing breast cancer (depending on the type of mutation), and a risk of 10−66 percent of developing ovarian cancer. Specific breast cancer mutations are common in certain ethnic groups. For instance, about 2.6 percent of Ashkenazi Jews with hereditary breast cancer have three specific mutations (referred to as "founder mutations") which occur in only 0.2 percent of the general American population. Genetic testing for the presence of mutations in BRCA1/2 have been available since 1996, patented as BRACAnalysis, a predictive medicine product for hereditary breast and ovarian cancer by Myriad Genetics, Inc. (https://www.myriad.com), a Utah-based company co-founded by Nobel Laureate, Walter Gilbert. Celebrity endorsement of pre-emptive testing could fuel a rush for these options as seen by the jump in the share price of the company (the share price rose by nearly four percent to a three-year high) on the day that Angelina Jolie published her story in the *New York Times*.

Reference

Jolie, A. (2013, May14) *My medical choice*. [*The New York Times*]. Retrieved from http://www.nytimes.com/2013/05/14/opinion/my-medical-choice.html?_r = 0.

Crohn's disease, and a number of cancers such as colorectal cancer, Kaposi's sarcoma and breast cancer (Jewish Genetic Disease Consortium, n.d.).

Genetic testing has become **common for certain disease for which cures are available**. For instance, in India, the Health Department routinely **screens new-borns** in government hospitals for four metabolic disorders: congenital hypothyroidism, phenylketoneuria (PKU), congenital adrenal hyperplasia, and glucose 6 phosphate dehydrogenase (G6PD) deficiency, which can be treated effectively if detected early (Maya, 2015). With more genes being identified through genome analysis and diagnostic kits being commercially available, DNA-based tests for preventive medicine is becoming increasingly popular. Several companies offer **direct-to-customer genetic testing kits** to customers who aren't necessarily ill or in the high risk category, but are merely curious or worried about the probability of developing various disorders (see Table 7.1). While some of the kits require a prescription from a certified physician, many are sold over the internet. Most kits contain instructions and a vial to collect cells through a cheek swab or saliva, which is to be mailed to the company for analysis. The results are usually delivered to the e-mail account provided by the customer. Typically, the results predict risk for conditions such as heart disease, Alzheimer's disease, or colon cancer, identify gene variants that influence the ability to metabolize alcohol and certain drugs, and in some cases traces ancestry by identifying clusters of mutations often inherited together, indicating common origin. Many companies also provide contact details of counselors to help customers make decisions based on the results. Follow up reports may offer advice on strategies and life style choices that may help minimize the risks. The **merit of these direct-to-customer genetic analysis kits is questionable** due to:

- *Accuracy*—clinical validation for many gene variations is not available; having a mutation may not be predictive of a disease condition.

Table 7.1 Examples of direct-to-consumer genetic testing kits

Company	Products and Services
Pathway Genomics (https://www.pathway.com/)	Founded in 2008, Pathway Genomics based in San Diego, has testing services for a *"variety of conditions including somatic* and *hereditary cancer, cardiac health, carrier screening, diet and weight loss, as well as drug response for specific medications including those used in pain management and mental health."* The company in association with IBM Watson is focused on providing users validated and personalized healthcare information delivered to any mobile device.
Interleukin Genetics (http://ilgenetics.com/about-interleukin/)	Interleukin Genetics Inc. based in Waltham, Massachusetts, launched Inherent Health™ brand of genetic tests in 2009. Other products include the Inherent Health™ Weight Management Test, and PerioPredict Genetic Risk Test. In the pipeline are tests to predict osteoarthritis risk.
23andMe (https://www.23andme.com/en-int/)	The company has a database with over 1,000,000 people and sells an online test that helps you discover your ancestral origins, trace lineage, and gives a personalized analysis of your DNA for US$149.
deCODE Genetics Inc. (http://www.decode.com/company/)	Founded in 1996 by Kari Stefansson and headquartered in Reykjavik, Iceland. The company had offered personal genome scans, DeCODEme in January 2009 as a better way to understand risk of cardiovascular disease (deCODEme Cardio) and common cancers (deCODEme Cancer), but as of April 2013, discontinued selling genetic services from the website. deCODE genetics is now a subsidiary of Amgen.

- *Incomplete information*—the actual risk for developing a disease condition is dependent on a complex interaction of genes and a number of different hereditary and lifestyle factors which are not usually considered in an online test.
- *Knowing the results may not be useful*—identifying a genetic risk could prompt feelings of anxiety or fatalism, while the absence of a genetic risk could create a false sense of security.
- *Unnecessary*—routine laboratory tests coupled with familial information could possibly be as predictive as the DNA tests for several health problems like heart disease, diabetes and hypertension.
- *Expensive*—the tests are often several hundred dollars, not covered by insurance.

7.3 GENETIC EXCEPTIONALISM

The genetic information pertaining to an individual is **intrinsically different** from other personal and medical information. This is because genetic testing **provides information not only about the person from whom the sample is taken, but also about related individuals**. One of the issues resulting from this is that family members, who may not even be carriers of the marker, could be stigmatized, and be forced to face issues of employability and insurability. The issue thus **raises ethical concerns regarding privacy and access to information**. The ethics of clinical practise does recognize confidentiality of medical records, but these records pertain to the individual patient and are usually protected by doctor-patient privileges. Genetic information however contains implied information pertaining to the family members of the tested individual. Insurance companies/ employers for instance, could access sensitive information merely by asking clients whether their relatives have tested positive for listed genetic disorders.

7.4 GENETIC DISCRIMINATION

Using genetic information to make decisions about hiring/firing, advancement opportunities, denying insurance coverage or changing insurance terms and paying for medical care constitutes genetic discrimination. One example of genetic discrimination is that which resulted from the **screening of African American children and young adults for sickle cell anemia in the 1970 s**. At a time when screening for the disease was unable to distinguish between *carriers* (sickle cell trait, who are healthy; heterozygous for the disease gene) and *affected individuals* (sickle cell disease; homozygous for the disease gene), in some areas in the United States, 30–40% of the African American population tested positive. Also, in 1969, four African American army recruits died unexpectedly during extreme exercise, and the deaths were attributed to probable sickle cell disease. This resulted in blacks being excluded from athletics and "high risk" jobs in aviation and several industries. School children were stigmatized, excluded from sports and parents forced to sign waivers (Markel, 1992). The situation changed only after a DNA test for the disease became available in 1978, and it was found that other ethnic groups and white Americans too had the trait.

Several instances of **discrimination by employers and insurance companies on the basis of genetic information** have been documented by the Council for Responsible Genetics, a bioethics advocacy organization based in Cambridge, Massachusetts (Council for Responsible Genetics, 2001).

In the United States, health cover is usually provided by the employer. With rising health care costs, it is natural that **employers would see an economic advantage in employing people with lesser health risks**. Genetic screening of employees has been documented as having been conducted by at least five Fortune 500 companies in a 1989 survey commissioned by the Congressional Office of Technology (U.S. Congress, Office of Technology Assessment, 1990). In some cases, the **tests were being done secretly** as in the case of the Burlington Northern Santa Fe (BNSF) Railroad employees (see Box 7.2). Another instance of secret testing is that of the **use of pre-employment genetic screening** at Lawrence Berkeley Laboratory (see Box 7.2).

BOX 7.2 COERCIVE EMPLOYEE GENETIC TESTING PROGRAMS AND THE LAW

The Burlington Northern and Santa Fe Railway Company (BNSF, http://www.bnsf.com/) headquartered at Fort Worth, Texas, operates one of the largest railroad networks in Northern America. In 2001, lawsuits were filed against BNSF when it appeared that the company had secretly conducted genetic tests for a rare genetic condition (hereditary neuropathy with liability to pressure palsies — HNPP, one of the symptoms of which is carpal tunnel syndrome) on several employees. The tests were conducted in connection with settlement of insurance claims for the treatment of carpel tunnel syndrome, purportedly to determine whether the injuries were work-related, and were pursuant to the BSNF Safety Rule 26.3 which required employees to undergo any medical examinations as deemed necessary by the Medical Department. The problem was that in addition to the usual X-ray and nerve conduction tests, workers were being subjected to genetic tests for carpel tunnel syndrome without their knowledge or informed consent. The law suits were filed by the Brotherhood of Maintenance of Way Employees (BMWE) and the US Equal Employment Opportunity Commission (EEOC). In February 2002, BNSF agreed to halt its genetic testing program (bmwe.org, n.d.), and later that year, conceded to a mediated settlement of US$2.2 million for genetically testing or seeking to test 36 of its employees (U.S. Equal Employment Opportunity Commission, 2002). This was the first case against a corporation involving the misuse of genetic screening in the workplace and garnered sufficient media attention and public awareness for the Americans with Disability Act of 1990, which allows medical examination of employees, but limits it to collecting information related to the employee's ability to do the job.

Norman-Bloodsaw v. Lawrence Berkeley Laboratory

The *Norman-Bloodsaw v. Lawrence Berkeley Laboratory* case tried in 1998, was the first to establish the right of an employee to be free from non-consensual genetic testing in preplacement medical examination. The plaintiffs were seven administrative and clerical employees of the Lawrence Berkeley Laboratory, a national laboratory of the University of California under contract with the United States Department of Energy. The plaintiffs claimed that as part of the preplacement medical examination, blood and urine samples collected from them was used to test for sickle cell anemia, syphilis, and even pregnancy without their knowledge or consent, which constituted an invasion of their privacy under Title VII of the Civil Rights Act of 1964, and the Americans with Disability Act (ADA) of 1990. The claims were dismissed by the district court, but the United States Court of Appeals for the Ninth Circuit held that employers who conduct non-consensual genetic tests may be liable for invasion of privacy under the United States and California Constitutions, and the Title VII of the Civil Rights Act. The court however held that ADA does not limit the scope of testing after the job is offered and prior to actual employment (Echevarria, 1999).

References

bmwe.org (n.d.) *Derailing genetic testing at BNSF*. Retrieved from https://www.bmwe.org/journal/2001/04apr/A1.htm.
Echevarria, C.E. (1999) Employment Law- Norman-Bloodsaw v. Lawrence Berkeley Laboratory, *Golden Gate U.L.Rev. 29(1)* Retrieved from http://digitalcommons.law.ggu.edu/ggulrev/vol29/iss1/10.
Norman-Bloodsaw v. Lawrence Berkeley Laboratory 135 F.3d 1260 (9th Cir. 1998). The appeal from United States District Court for the Northern District of California, was argued and submitted on June 10, 1997 before Circuit Judge Reinhart, Judge T.G. Nelson, and Judge Hawkins. The decision was filed February 3, 1998. Judge Reinhardt authored the opinion.
US Equal Employment Opportunity Commission (2002) (Press release 5-8-02) *EEOC and BNSF settle genetic testing case under Americans with Disabilities Act*. Retrieved from https://www.eeoc.gov/eeoc/newsroom/release/5-8-02.cfm.

The fear of genetic discrimination **could have an adverse effect on choice of medical treatment**—tests may be avoided over apprehensions of loss of jobs or insurance cover, and individuals may hesitate to enroll in clinical trials.

7.5 ETHICAL, LEGAL, SOCIAL IMPLICATION (ELSI)

Genetic information from human patients and their family members is **crucial for drug discovery**. An illustrative example is the attempt to develop a remedy for a genetic disorder known as Huntington's Disease (HD) that affects around one in ten thousand human beings. The disease is known to be caused by a single autosomal dominant gene which means that both men and women are equally affected and there is a 50% probability that the progeny of an affected individual will also inherit the disease. Unfortunately, the disease has a late onset, with symptoms appearing only in the late forties and fifties, often after the reproductive period of the affected person. The symptoms include chorea (dance-like involuntary movements of body and limbs) which gets progressively worse and affects normal body functions including swallowing, and is associated with depression and dementia. The disease is known to be caused by a defective protein (called huntingtin protein, htt) produced by the mutant gene. Since the cloning of the gene in 1993, considerable progress has been made in identifying targets for HD therapeutics (Morse et al., 2011). Concerted efforts in this direction have been made by the European Huntington's Disease Network, the CHDI Foundation Inc., and the Huntington Study Group, which are non-profit groups focused on drug discovery. HD is a neurological disease that affects only human beings. Consequently, invertebrate animal models such as *Drosophila* and *Caenorhabditis* have limited potential for generating clinically relevant information. The Chief Scientific Officer of CHDI, Robert Pacifici, says that **human data is "precious" for HD drug hunters**, more so because variations in disease manifestations, particularly the severity of the symptoms, and the age at which they appear, differ from one patient to the next. Researchers not only have to study the *htt* gene itself but also need a genome wide search for modifiers of the gene action (also known as **Genome-Wide Association Studies, GWAS**). This is **possible only if human subjects share their genetic information with the drug hunters** (Gene Veritas, 2015).

Ethical issues can however arise if adequate care is not exercised in **informing volunteers as to the manner in which their genetic data would be used**, in **obtaining their consent for use**, in **failing to maintain their privacy**, or if **equitable sharing of the benefits of use** does not happen. Illustrative of this is the case of deCODE Genetics Inc., that sought to use the genealogical data, medical health records, and DNA samples of around 100,000 volunteers from Iceland to identify the underlying genes responsible for common diseases such as osteoporosis, osteoarthritis, psoriasis and schizophrenia (see Box 7.3).

The Human Genome Project (HGP) represents the formal global effort in unraveling the genetic make-up of human beings primarily with the view of using this information to diagnose diseases, if possible before the onset of disease symptoms, and to develop new therapies for curing or alleviating the disease symptoms. The potential of this information to be misused in ways that are unfair, or cause harm, was recognized very early in the development of the HGP, prompting efforts to understand the ethical and social implications of the information and to put in place a legal system to prevent misuse.

BOX 7.3 ELSI CONCERNS IN THE DECODE GENETICS CASE

deCODE Genetics Inc. was founded in 1996 by Kari Stefansson in order to identify genes causing chronic diseases such as osteoporosis, osteoarthritis, psoriasis, heart disease, stroke, and allergies, by mining genetic data from populations. Of special interest was the homogenous population of Stefansson's native Iceland. Residents of Iceland share a common ancestry (Norse and Celts) and has remained virtually unchanged since the ninth century, and is thus an ideal resource for population genomics. Stefansson already had access to genealogical data for over 1000 years and he negotiated with the Icelandic government to make available for the study the nation's health care records dating back to 1915. By an Act in 1998, the Iceland parliament authorized the creation of a Health Sector Database, the Icelandic Healthcare Database (IHD), that would link together the medical records, genealogical records, and genetic information, from over 100,000 volunteers. In exchange for the investment and the work involved in setting up the database, deCODE Genetics was given exclusive research access for a period of 12 years (Tavani, 2006). The program came under intense criticism on several counts:

- *Informed consent:* Many who consented to donate DNA samples for the study did not know that it would be cross-referenced to other information about them and mined for data.
- *Confidentiality:* It was unclear how confidentiality would be protected as the data was to be cross referenced and aggregated with other data. A physicians-led citizen's group, Association of Icelanders for Ethics in Science and Medicine, had raised objections based on violation of the Icelandic constitution (Tavani, 2006).
- *Fairness and access:* The exclusive rights given to deCODE Genetics meant that researchers not associated with the company, now had to pay to access medical records that were earlier free to use. Also, as deCODE Genetics had signed contracts with pharmaceutical companies and with IBM for combining deCODE's proprietary data-mining software with IBM's servers and database software, critics feared that Iceland's genetic data was being reduced to a marketable commodity.

Reference

Tavani, H. T. (2006). Ethics at the intersection of computing and genomics. In H. T. Tavani (Ed.), *Ethics, Computing and Genomics* (pp. 5–26). USA: Jones and Bartlett.

7.5.1 THE ELSI OF HUMAN GENOME PROJECT

The issues related to the fair use of genetic information requires the involvement of not only scientists and developers of the technology, but the general public, and requires inputs from philosophers, ethicists, lawyers and policy makers. The ELSI Research Program of the National Human Genome Research Institute (NHGRI) funds and manages studies, supports workshops, research consortia and policy conferences related to the following issues:

- *Privacy and confidentiality of genetic information*
- *Fairness in the use of genetic information by insurers, employers, courts, schools, adoption agencies, military, among others*
- *Psychological impact, possibly stigmatization and discrimination due to an individual's genetic differences*
- *Reproduction issues including adequate and informed use of genetic information in reproductive decision making*
- *Clinical issues in the use of information obtained by genetic testing by doctors and healthcare providers*

- *Fairness in access to advanced genomic technologies*
- *Commercialization of products and accessibility of data and materials.*

In 2003, the NHGRI in collaboration with the US Department of Energy and the National Institute of Child Health and Human Development created the **Centers of Excellence in ELSI Research (CEER)** which are interdisciplinary centers intended to encourage innovative solutions to important new or persistent ethical, social and legal issues related to advances in genetics and genomics (National Human Genome Research Institute, n.d.).

The impact of the ELSI program is that it has **helped to provide a foundation** on which applied studies can be built and policies enacted. Over the past couple of decades, there has been a change in the cultural milieu in which genomic research is conducted—a **greater emphasis on informed consent, better models of assessing risks and benefits, of sharing of data while safeguarding privacy, autonomy, and related interests of genome research participants, and a more precise and ethically sensitive nomenclature for communities** participating in the International HapMap Project (https://hapmap.ncbi.nlm.nih.gov/) and 1000Genome project (http://www.1000genomes.org/). The ELSI research relating to risk of genetic discrimination and genetic privacy helped to create significant momentum that **led to federal enactments in the United States and similar legislation in other countries**.

7.5.2 INTERNATIONAL AND NATIONAL PROGRAMS ON ELSI

The United Nations and organizations such as the United Nations Educational, Scientific and Cultural Organization (UNESCO) and World Health Organization (WHO) have been instrumental in establishing international policies in bioethics. The WHO is committed to make ethics integral to all its activities. In 2002, WHO's *Human Genetics (HGN)*, in recognition of the importance of the development of genomics and the ethical, legal and social issues arising from its use, commissioned the *Advisory Committee on Health Research report, Genomics and World Health*. Priority areas identified by the HGN on consultation with geneticists include: "*genetic testing and screening, genetic patents, genetic databanks and pharmacogenomics*" (World Health Organization, 2016). The ELSI program of the HGN aims to:

- Promote just and equitable access to affordable genetic tests, screening, diagnostics, and other technologies
- Promote the development and safe application of new technologies for health burdens borne by women, children, and disadvantaged groups
- Reduce health risk and ensure ethical conduct by providing guidance for safety standards, monitoring and evaluation of genetic databanks, genetic tests, and screening
- Empower and advocate for women, children, and disadvantaged groups in the development of genomics health research priorities, and in achieving affordable access to research products
- Develop tools for genomics capacity building, including educational modules and mechanisms for accessing bioinformatics, especially in developing countries
- Promote health through sound and ethical regulation of the research, development and use of genomics based technologies, and regulation of genetic service delivery in developing countries. (http://www.who.int/genomics/elsi/elsiatwho/en/)

Table 7.2 Overview of examples of ELSI/A programs

Country/Region	Acronym	Program	Funding Agency	Year
US	ELSI	Ethical, Legal and Social Implications	NIH/ NHGRI	1990
Canada	GE³LS	Genomics-related Ethical, Environmental, Economic, Legal and Social Aspects	Genome Canada	2000
South Korea	ELSI	Ethical, Legal and Social Implications	Government of S.Korea	2001
United Kingdom	EGN	ESRC Genomic Network (Cesagen, Innogen, Egenis, Genomics Forum	ESRC	2002
Netherlands	CSG, MCG	Centre for Society and Genomics; Societal component of Genomics Research	Netherlands Genomic Initiative	2002
Germany, Austria, Finland	ELSAGEN	Transnational Research Programme	GEN-AU, FFG. DFG, Academy of Finland	2008
Africa	H3Africa	Human Heredity and Health in Africa	Wellcome Trust (UK), NIH (US)	2013
Europe	RRI	Responsible Research and Innovation	Co-funded by EC	2014

Modified from Chadwick, R., & Zwart, H. (2013) From ELSA to responsible research and Promisomics. Life Sciences, Society and Policy 2013 **9**:*3 **DOI**: 10.1186/2195-7819-9-3.*

National and regional initiatives that anticipate and address **ethical, legal, and social implications (ELSI) or aspects (ELSA, in Europe)** of applications arising from genomics have focused on challenges with respect to protection of privacy, data protection and availability of sequence data. Table 7.2 summarizes a few examples of such initiatives.

7.6 MECHANISMS FOR PREVENTING GENETIC DISCRIMINATION

One of the most significant dilemmas facing policy makers has been the question of whether genetic information can be considered sufficiently different from other individual health information as to warrant special legislative attention. As a result, in many nations and in the European Union, protection of genetic information has not been laid down in general laws, but in more specific statutes that apply to health care information (Gerards, Heringa, & Janssen, 2005).

7.6.1 RIGHTS-BASED ADVOCACY

The framework of rights-based advocacy entails **respect** (acknowledgment of the rights) and **identification of a "gap" or "disadvantage"** in a person's ability to lead a **"dignified human existence"** in the absence of a particular right.

7.6.1.1 Human Rights

As defined by the **United Nations**:

> Human rights are rights inherent to all human beings whatever our nationality, place of residence, sex, national or ethnic origin, colour, religion, language, or any other status. We are all equally entitled to our human rights without discrimination. These rights are all interrelated, interdependent and indivisible.
>
> **United Nations Human Rights, n.d.**

The principle of universality of human rights was established in the *Universal Declaration on Human Rights* in 1948 and integrated into subsequent international conventions, declarations, resolutions and law. It mandates that governments act in ways that promote and protect human rights and fundamental freedom of individuals and groups. The human rights thus entail both rights and obligations — member states are obligated under international law to respect, to protect, and to fulfill human rights.

7.6.1.2 Universal Declaration on the Human Genome and Human Rights, 1997; and International Declaration on Human Genetic Data, 2003

The **United Nations Educational, Scientific and Cultural Organization** (UNESCO) **Bioethics Programme** was created in 1993 as part of the UNESCO's Social and Human Sciences Sector. The aim of the program is to define and promote a common ethical standard setting framework that States can use for developing their own bioethics policies. The **Universal Declaration on the Human Genome and Human Rights** was proposed under this program. **It prohibits discrimination on genetic grounds and protects the confidentiality of individual genetic information**. Adopted unanimously by the UNESCO's General Conference on November 11, 1997, it was endorsed the following year by the UN General Assembly. As a sequel to this Declaration, the **International Declaration on Human Genetic Data** was adopted in 2003. This declaration recognized

> *"that human genetic data have a special status on account of their sensitive nature since they can be predictive of genetic predispositions concerning individuals and that the power of predictability can be stronger than assessed at the time of deriving the data; they may have a significant impact on the family, including offspring, extending over generations, and in some instances on the whole group; they may contain information the significance of which is not necessarily known at the time of the collection of biological samples; and they may have cultural significance for persons or groups"*
>
> **United Nations Educational, Scientific, & Cultural Organization, 2003.**

Article 7 of the Declaration sought to prevent discrimination and stigmatization while **Article 14** advocated privacy and confidentiality

Although ratified by the United Nations, these Declarations are **not legally binding** on any country.

7.6.1.3 Universal Declaration on Bioethics and Human Rights, 2005

The **General Conference of UNESCO** adopted the *Universal Declaration on Bioethics and Human Rights* in October 2005 to deal with ethical issues raised by medicine, life sciences and associated technologies. While **Article 3** advocates respect for human dignity, human rights, and fundamental freedoms, **Article 9** states that privacy of the persons concerned and confidentiality of their personal information should be respected, and **Article 11** prohibits discrimination and stigmatization.

7.6.1.4 EU Charter of Fundamental Rights, 2000

Drafted by the European Convention, the *Charter of Fundamental Rights* was proclaimed by the European Parliament, the Council of Ministers and the European Commission on December 7, 2000 (The Charter of Fundamental Rights of the European Union, 2000). The document establishes certain political, social and economic rights for citizens of the European Union into law. The Charter became legally binding on EU institutions and on national governments on December 1, 2009, with the entry into force of the Treaty of Lisbon. The Charter has 54 articles divided into seven titles:

1. *Dignity:* guarantees right to life and prohibits slavery, torture, eugenic practices and cloning
2. *Freedoms:* guarantees privacy (Article 7), protection of personal data (Article 8) among others
3. *Equality*: prohibits all discrimination (Article 21) including on the basis of disability (Article 26), age, sexual orientation, culture, language, among others
4. *Solidarity:* ensures social and workers' rights including right to fair working conditions
5. *Citizen's Rights:* covers the rights of EU citizens to vote and to move freely within the EU
6. *Justice:* ensures the right to a fair trial, and effective remedy
7. *General provisions:* deals with the interpretation and application of the charter.

7.6.2 RIGHTS-BASED POLICY AND TOOLS

A rights-based approach aims at **strengthening the capacity of duty bearers** and to **empower the rights holders**.

7.6.2.1 Council of Europe Convention on Human Rights and Biomedicine, 1997

This constitutes the first and so far, only attempt to establish a legally binding instrument in Europe that covers core areas of medicine. It builds on the *European Convention on Human Rights* to create more precise international minimum standards for protection against misuse of new technologies in biology and medicine. The convention was tabled on 4 April 1997 at Oviedo, Spain (therefore known as the **Oviedo Convention**) and entered into force under international law on December 1, 1999. It has been signed by most of the member States of the Council of Europe (Council of Europe Portal, CETS No.164). The aims of the Oviedo Convention are based on relevant principles of **protecting the dignity, identity and integrity of human beings, safeguarding the interests of the individual over that of the society or science, and providing equal access to health care services**. Also included are protection of the patients' autonomy ("**informed consent**"), access to medical data, predictive genetic tests, interventions on human genome, tissue and organ removal from living donors for transplantation, use of body substances and medical research on persons. The Convention is further developed through Additional Protocols in specific fields:

1. prohibition of cloning in human beings (ETS No.168) in Paris, 1998
2. transplantation of organs and tissues of human origin (ETS No.186) in Strasburg, 2002

3. biomedical research (CETS No.195), in Strasburg, 2005
4. genetic testing for health purposes (CETS No, 203) in Strasburg, 2008

The **Protocol on Genetic Testing for Health Purposes** (Council of Europe Portal, CETS No.203) proposes to achieve the objective of protecting the dignity and identity of all humans and prevent improper use of genetic testing by restricting the use of genetic testing to health purposes only. It also **sets down rules for conduct of genetic tests including prior information and consent, and genetic counseling**.

7.6.2.2 United States: Title VII of the Civil Rights Act of 1964

This federal law protects employees from being discriminated against on the basis of sex, race, color, religion or national origin by their employers. The law has been extended to include genetic discrimination in cases of genetic disorders linked to race or ethnicity. It is however of limited use because:

- a strong correlation between a genetic trait and race or national origin has been established only for a few diseases;
- applicable only in cases where the employer engages in discrimination based on a genetic trait.

7.6.3 NATIONAL LEGISLATION

The United States of America has been the most proactive among nations in terms of enacting genetic non-discrimination legislation with a law that specifically prohibits discrimination on the basis of genetic information. Several other nations have modified existing laws to also address issues of genetic discrimination.

7.6.3.1 United States

In the United States, prior to 2008, existing non-discrimination laws such as the Americans with Disabilities Act of 1990, the Health Insurance Portability and Accountability Act (HIPAA) of 1996, and the HIPAA National Standards to Protect Patients' Personal Medical Records of 2002, were being interpreted to include genetic discrimination. In 2008, the Genetic Information Non-discrimination Act (GINA) was enacted.

7.6.3.1.1 Americans with Disabilities Act (ADA) of 1990

The disability based anti-discrimination laws such as the *Rehabilitation Act* of 1973 (Sections 501 and 505) and Title I of the *Americans with Disabilities Act* (http://www.ada.gov/), enforced by the **Equal Employment Opportunity Commission** (EEOC), do not explicitly address genetic information. However, they have **been interpreted to afford some degree of protection** against genetic discrimination as it prohibits discrimination against a person regarded as having a disability including **symptomatic genetic disabilities**. The ADA is however unable to protect against unexpressed genetic conditions. Thus unaffected carriers or individuals with late onset disorders are not covered by ADA. It also cannot protect workers from having to provide medical information deemed to be job related and consistent with business necessity. Another major concern is that it does not protect prospective workers who have been given job offers from having to provide genetic samples or information prior to joining work.

7.6.3.1.2 Health Insurance Portability and Accountability Act (HIPAA) of 1996

While this act does directly address the issue of genetic discrimination, **it applies to employer-based and commercially issued group health insurance only** (HHS.gov, n.d.). It **does not apply to private individuals seeking health insurance** in the individual market. The HIPAA prohibits group health plans from using genetic information or any health status related factor as a basis for denying, limiting eligibility, or enhancing premium of individuals for coverage. It also explicitly states that genetic information in the absence of a current diagnosis of illness shall not be considered a pre-existing condition, limits exclusion for pre-existing conditions to 12 months and prohibits such exclusion if the individual has been covered previously for that condition for 12 months or more. HIPAA however does not prohibit employers from refusing to offer health coverage as part of their benefit package.

7.6.3.1.3 HIPAA National Standards to Protect Patients' Personal Medical Records, 2002

This regulation is not specific for genetic information, but **protects all personal health information maintained by health care providers, hospitals, health care clearinghouses, health insurers**. It serves to give patients new rights to access their medical records and restrict non-consensual use and release of private health information. It helps to restrict disclosure of health information to the minimum required, establishes new criminal and civil sanctions, and permissions for access of records by researchers and others.

7.6.3.1.4 Genetic Information Non-discrimination Act (GINA), 2008

Since existing State and Federal legislation was insufficient to protect Americans from genetic discrimination, the GINA was proposed to **establish a national and basic standard of protection against discrimination** and allay concerns about the potential for discrimination so that the public could benefit from genetic testing, technology, research and new therapies (U.S. Equal Employment Opportunity Commission, n.d.). This legislation **specifically prohibits discrimination on the basis of information derived from genetic tests by US insurance companies and employers**. Under this law, insurers and employers are not permitted to request or demand a genetic test. Results of a genetic test cannot be used by insurance companies to reduce coverage or alter pricing, or be used by employers to make adverse employment decisions. It is also **unlawful to disclose genetic information about applicants, employees or members**. Limited exceptions to this non-disclosure rule include disclosure of relevant genetic information to government officials investigating compliance with GINA and for disclosures made pursuant to a court order. This law has been in effect since November 21, 2009.

7.6.3.2 *Rest of the world*

Most countries have provisions applicable for genetic testing in the general legal framework covering the health field such as in vitro fertilization (pre-implantation genetic testing), data protection and patient rights. Some examples of national laws are summarized in Table 7.3.

Table 7.3 National laws for governing the use of genetic testing/ prevention of genetic discrimination

Country	Name of Act	Year
Australia	Disability Discrimination Act	1992
Austria	Gene Technology Act	1995 (revised 2005)
Canada	Canadian Human Rights Act	1977
France	Bioethical law 2004-800 that modified the civil code and public health code.	2004
Germany	Human Genetic Examination Act	2010
Norway	Act on human medical use of biotechnology etc. 2003-12-05	2004
Portugal	Act on Personal Genetic Information and health information 12/2005.	2005
Spain	Act on biomedical investigations	2007
Sweden	Act on genetic integrity etc.	2006
Switzerland	Federal Act on Human Genetic Analysis	2007
United Kingdom	Data Protection Act	1998

KEY TAKEAWAYS

Mechanisms for preventing genetic discrimination:

I. Rights-based advocacy
   1. Universal Declaration of Human Rights, 1948
   2. UNESCO Universal Declaration on the Human Genome and Human Rights, 1997; International Declaration on Human Genetic Data, 2003
   3. UNESCO Universal Declaration on Bioethics and Human Rights, 2005
   4. EU Charter of Fundamental Rights, 2000
II. Rights-based policy and tools
   1. Council of Europe Convention on Human Rights and Biomedicine, 1997
   2. United States: Title VII of the Civil Rights Act of 1964
III. National Legislation
   1. United States
      i. Americans with Disabilities Act (ADA), 1990
      ii. Health Insurance Portability and Accountability Act (HIPAA), 1996
      iii. HIPAA National Standards to Protect Patients' Personal Medical Records, 2002
      iv. Genetic Information Non-discrimination Act (GINA),2008
   2. Rest of the world (examples)
      i. United Kingdom: Data Protection Act, 1998
      ii. Canada: Canadian Human Rights Act, 1977
      iii Australia: Disability Discrimination Act, 1992
      iv Germany: Human Genetic Examination Act, 2010
      v France: Bioethical Law 2004-800, 2004

7.7 SUMMARY

This chapter highlights the unique nature of the information obtained by genetic testing and the need to maintain confidentiality and privacy, and ensure fair use of the genetic information of an individual. While genetic testing can be useful for early diagnosis of genetic diseases and is of immense value to drug developers, it can potentially be misused by employers and insurance companies to deny services. Understanding issues of genetic discrimination has been the objective of studies on the ethical, legal and social implication of genomic information. Mechanisms to prevent discrimination on the basis of genetic information include international advocacy and policies on human rights as well as national laws, including several that specifically address the issue.

REFERENCES

Council for Responsible Genetics (2001, January) *Genetic discrimination*. Retrieved from http://www.council-forresponsiblegenetics.org/ViewPage.aspx?pageId = 85.

Council of Europe Portal (CETS No.164) Convention for the protection of Human Rights and Dignity of the Human Being with regard to the Application of Biology and Medicine: Convention on Human Rights and Biomedicine. CETS No 164. Retrieved from http://www.coe.int/en/web/conventions/full-list/-/conventions/treaty/164.

Council of Europe Portal (CETS No.203) *Additional Protocol to the Convention on Human Rights and Biomedicine concerning Genetic Testing for Health Purposes*. CETS No 203. Retrieved from http://www.coe.int/en/web/conventions/full-list/-/conventions/treaty/203.

Gene Veritas (2015) Human data 'precious' for Huntington's Disease drug hunters: An interview with Dr. Robert Pacifici [Video file] Retrieved from https://vimeo.com/120920753.

Gerards, J.H., Heringa A-W., & Janssen, H.L. (2005). Genetic Discrimination and Genetic Privacy in a Comparative Perspective. Intersentia. p.241.

HHS.gov (n.d.) *Health information privacy*. Retrieved from http://www.hhs.gov/hipaa/index.html.

Jewish Genetic Disease Consortium (n.d.) *Jewish Genetic Diseases*. Retrieved from http://www.jewishgenetic-diseases.org/jewish-genetic-diseases/

Markel, H. (1992). The stigma of disease: implications of genetic screening. *Am J Med*, *93*(2), 209–215.

Maya, C. (2015, July 4) (Reporter) *State's newborn screening programme wins laurels*. [The Hindu] Retrieved from http://www.thehindu.com/news/cities/Thiruvananthapuram/states-newborn-screening-programme-wins-laurels/article7385355.ece.

Morse, R. J., Leeds, J. M., Macdonald, D., Park, L., Toledo-Sherman, L., & Pacifici, R. (2011). Pharmaceutical Development for Huntington's Disease. In D. C. Lo, & R. E. Hughes (Eds.), *Neurobiology of Huntington's Disease: Applications to Drug Discovery* (p. 2011). Boca Raton (FL): CRC Press/Taylor & Francis. Chapter 8. Retrieved from: http://www.ncbi.nlm.nih.gov/books/NBK56003/.

National Human Genome Research Institute (n.d.) *The Ethical, Legal and Social Implications (ELSI) Research Program*. Retrieved from https://www.genome.gov/elsi/.

The Charter of Fundamental Rights of the European Union (2000) Official Journal of European Communities C364/1. Retrieved from http://www.europarl.europa.eu/charter/pdf/text_en.pdf.

United Nations Human Rights (n.d.) *What are human rights*. Retrieved from http://www.ohchr.org/EN/Issues/Pages/WhatareHumanRights.aspx.

United Nations Educational, Scientific and Cultural Organization (2003) *International Declaration on Human Genetic Data.* Retrieved from http://portal.unesco.org/en/ev.php-URL_ID = 17720&URL_DO = DO_TOPIC&URL_SECTION = 201.html.

U.S. Congress, Office of Technology Assessment (1990). *Genetic Monitoring and Screening in the Workplace: Results of a Survey-Background Paper, OTA- BA-455.* Washington, DC: U.S. Government Printing Office. Retrieved from http://ota.fas.org/reports/9020.pdf.

U.S. Equal Employment Opportunity Commission (n.d) *Genetic Information Discrimination.* Retrieved from https://www.eeoc.gov/laws/types/genetic.cfm.

World Health Organization (2016) *HGN activities in ELSI of human genomics.* Retrieved from http://www.who.int/genomics/elsi/elsiatwho/en/.

FURTHER READING

Tavani, H. T. (Ed.), (2006). *Ethics, Computing and Genomics* USA: Jones and Bartlett.

U.S. Department of Energy, *Genomes to Life Black Bag.* Retrieved from http://web.ornl.gov/sci/techresources/Human_Genome/publicat/jmmbbag.pdf.

To know more about Huntington's Disease and the efforts of the CHDI watch videos: Robert Pacifici //CHDI Part I available at https://vimeo.com/73003665, and Robert Pacifici //CHDI Part II available at https://vimeo.com/73007928.

BIODIVERSITY AND SHARING OF BIOLOGICAL RESOURCES

Sharing is sometimes more demanding than giving.
-Mary Catherine Bateson, writer, anthropologist

CHAPTER OUTLINE

8.1 Introduction 190
8.2 Concept of Biodiversity.. 191
    8.2.1 The Functions of Biodiversity.. 191
    8.2.2 Bioprospecting 194
    8.2.3 Concerns Regarding Bioprospecting 194
8.3 The United Nations Convention on Biological Diversity 195
    8.3.1 Conference of the Parties 196
    8.3.2 Bonn Guidelines on Access to Genetic Resources and Fair and Equitable Sharing
         of the Benefits Arising Out of Their Utilization (Bonn Guidelines on ABS, 2001) 198
    8.3.3 The Nagoya Protocol on Access and Benefit-sharing 199
8.4 The United Nations Convention on the Law of the Sea 201
8.5 FAO International Treaty on Plant Genetic Resources for Food and Agriculture 202
    8.5.1 Access and Benefit Sharing (ABS) Obligations 203
    8.5.2 Intellectual Property and Farmer's Rights 203
    8.5.3 Ex Situ Conservation Centers 203
8.6 Regional and National Significance and ABS Frameworks 204
    8.6.1 Category 1: Countries With No National Laws Specifically Devoted to ABS 205
    8.6.2 Category 2: Countries With a Biodiversity or Environmental Law
         With Provisions on ABS... 206
    8.6.3 Category 3: Countries With National Laws Specifically Devoted to ABS 206
8.7 Summary 209
References 209
Further Reading 209

An Introduction to Ethical, Safety and Intellectual Property Rights Issues in Biotechnology.
DOI: http://dx.doi.org/10.1016/B978-0-12-809231-6.00008-9

Plant samples in the gene bank at CIAT's Genetic Resources Unit, at the institution's headquarters in Colombia. [(*Pic by Neil Palmer (CIAT). By CIAT—genebank4Uploaded by mrjohncummings, CC BY-SA 2.0,* https://commons.wikimedia.org/w/index.php?curid=30332302)]

8.1 INTRODUCTION

From the dawn of civilization, man has depended on biological resources for food, feed, fiber, fuel, and pharmaceuticals. Plants and animals have been regarded as *nature's gift to humanity* and treated as a common resource, freely available to all. This has resulted in a free movement of plant and animal species across the globe. Early adventurers visited far off lands and returned bearing exotic flora and fauna, which they tried to establish in new habitats. The outcome of this is that many new crops were grown, and animals reared, in places far away from the centers of origin of the species. For example, maize was first domesticated in Central America and carried by Columbus's men to Spain in 1492, where it was initially grown as a garden curiosity, but now is a major feed, food, and fuel crop. With the development of modern biotechnology and the recognition that genetic resources (GR) are essential for the development of commercially important products, biological resources can no longer be treated as freely accessible commodities. Issues of ownership and access to GR have gained prominence in the face of biodiversity present in and around the centers of origin of plant, animal, and microbial species being mostly confined to the developing/third world nations. The majority of these nations have limited technological capabilities to convert the resource into commercially viable products. **A strong ethical framework that establishes fair and equitable sharing of the benefits arising from the use of biological resources alone would ensure that mankind continues to derive benefit from nature's bounty.** This chapter discusses international and select national strategies aimed at sustainable use of biodiversity and regulatory frameworks for ensuring access and benefit sharing (ABS). The content of this chapter is based on information from the website of the Convention of Biological Diversity (https://www.cbd.int/) and has drawn extensively from documents developed by the International Union for Conservation of Nature (IUCN) Environmental Law Programme [IUCN Environmental Policy Law Paper No. 54 (Carrizosa, Brush, Wright, & McGuire, 2004) and No. 83 (Greiber et al., 2012)] and Centre for International Sustainable Development Law Biodiversity & Biosafety Law Research Programme (Medaglia, Perron-Welch, & Phillips, 2014).

8.2 CONCEPT OF BIODIVERSITY

Biodiversity is the variability in living organisms existing in various ecosystems and ecological complexes on earth. It is now fairly well understood that this diversity is a result of evolution—changes in the genetic makeup of species occur, and having a use or evolutionary advantage persist, giving rise to new forms of the species. There appears to be a clear link between genetic diversity and its geographic distribution pattern since species show the greatest diversity in biodiversity "hot-spots" (see Fig. 8.1) situated around their **centers of origin**. For **cultivated plants**, it is generally accepted that **there are eight centers of origin as proposed by Vavilov in 1926**, most of them being in the tropical and subtropical regions of the world (see Box 8.1). Scientists estimate that there are at least 250,000 different species of plants, 30 million species of insects, and possibly equal numbers of bacteria, algae, and fungi on earth, only a fraction of which have been discovered and documented.

8.2.1 THE FUNCTIONS OF BIODIVERSITY

The centers of origin and biodiversity including the domesticated species and wild or weedy relatives are of **crucial importance to current and future breeding programs** as they serve as **reservoirs of useful genes**. Traits like drought tolerance, salinity tolerance, resistance to pests and diseases are usually seen in wild and weedy relatives and through a process of introgression, bred into high yielding crop plants to improve agronomic performance. With the development of genetic engineering, the process is far more efficient since select genes from unrelated plant species, or even microbial and animal sources, can be used for crop improvement.

Biological diversity is an important **source of products of use in several industries**. For example, enzymes produced by bacteria and fungi growing on leaf litter in tropical forests are used to pretreat lignocellulosic feed stock such as crop residues and fast growing grasses before fermentation to yield ethanol used as fuel. Other microbial enzymes are used for removal of pollutants from industrial effluents, for separation of metals from ores, for degradation of plastics, and several other diverse applications.

Perhaps, the most significant products derived from biological resources are **substances used in medicine**. These include enzymes and other proteins, alkaloids, tannins, and other phytochemicals. To cite a few examples, vincristine and vinblastine, compounds derived from periwinkle originating from Madagascar, are used in the treatment of cancer. Toxins produced by the cone snail are used to produce a pain medicine that is 1000 times more powerful than morphine. Artemisinin from sweet wormwood is used for treatment of malaria. The widely prescribed cancer drug Taxol is

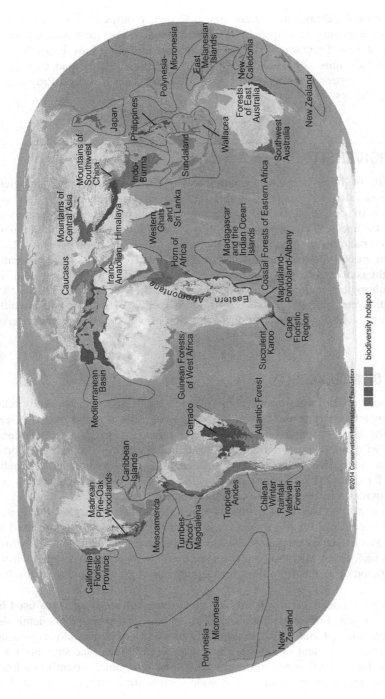

Conservation International (conservation.org) defines 35 biodiversity hotspots—extraordinary places that harbor vast numbers of plant and animal species found nowhere else. All are heavily threatened by habitat loss and degradation, making their conservation crucial to protecting nature for the benefit of all life on Earth.

biodiversity hotspot

©2014 Conservation International Foundation

FIGURE 8.1 Biodiversity Hotspots.

Conservation International (conservation.org) defines 35 biodiversity hotspots—extraordinary places that harbor vast numbers of plant and animal species found nowhere else. All are heavily threatened by habitat loss and degradation, making their conservation crucial to protecting nature for the benefit of all life on Earth. (*By Conservation International—Own work, CC BY-SA 4.0, https://commons.wikimedia.org/w/index.php?curid=34669870.*)

BOX 8.1 CENTERS OF ORIGIN OF CULTIVATED PLANTS

Based on information concerning about 250,000 samples gathered by a large group of Soviet botanists, including himself, Russian geneticist N. I. Vavilov identified the following eight basic geographic centers of origin of cultivated plants:

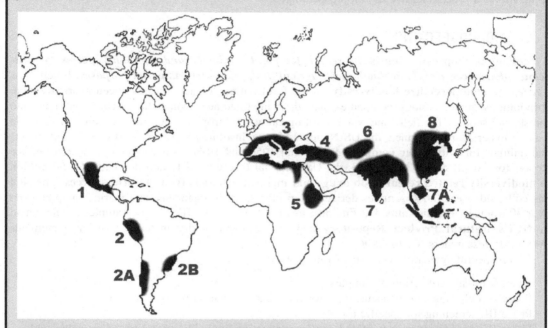

Vavilov's Centers of Origin of Cultivated Plants. (*By Redwoodseed (en-wiki [1]) [CC BY 3.0 (http://creativecommons.org/licenses/by/3.0)], via Wikimedia Commons*)

1. The **South Mexican** and **Central American center**: origin of about 90 food crops, industrial, and medicinal species, including corn, long-fiber cotton species, cacao, several species of beans and squash, and many fruit species.
2. The **South American Andes center**, the **Chilean Center**, and **Brazilian-Paraguayan Center**: home of many species of tuberous plants, above all potatoes, oca, ullucu, nasturtiums, cinchona, and coca.
3. The **Mediterranean center**: origin of about 11% of the cultivated plant species, including the olive, the carob, and many feed and vegetable crops.
4. The **Asia Minor center**: accounts for 4% of all cultivated plants, is the most important area of origin of crops raised in Europe, including bread grains, legumes, fruit crops, and grapes.
5. The **Ethiopian center**: accounts for about 4% of the cultivated plants. It is characterized by a number of endemic species and genera: for example, teff, *Guizotia*, a unique species of banana, and the coffee tree. It also has original cultivated endemic species and subspecies of wheat and barley.
6. The **Inner Asiatic center**: with wheats, rye, herbaceous legumes as well as seed sown root crops and fruits.
7. The **Indian** and **Indo Malayan Center**: rice, millets, legumes, sugarcane, spices, and root crops
8. The **East Asian center**: accounts for 20% of cultivated plants, including soybeans and various millet, vegetable, and fruit species.

Source: Centers of Origin of Cultivated Plants. (n.d.) *The Great Soviet encyclopedia, 3rd Edition.* (1970–1979). Retrieved from http://encyclopedia2.thefreedictionary.com/Centers+of+Origin+of+Cultivated+Plants.

derived from the bark of the Pacific yew, whereas aspirin is from the bark of the willow tree. More than half of the world's population relies on nature for medicines, and plants supply the active ingredients of most traditional medical products. Pharmaceutical chemists over the past few decades have made concerted efforts to seek new "leads" from biological sources for prospective drug applications, a process known as "bioprospecting."

8.2.2 BIOPROSPECTING

By definition, **bioprospecting is** *the search for plants, animals, and microbial species for academic, pharmaceutical, biotechnological, agricultural, and other industrial purposes*. It serves as a means to **commercialize biodiversity**. In the forefront of this activity have been pharmaceutical corporations, biotechnology companies, and their intermediates, which have led expeditions into forests and scoured the fields and waters in tropical and subtropical countries in search of valuable genetic material. For instance, the US National Cancer Institute (NCI) screened over 35,000 plants and animals for **anticancer compounds** between 1956 and 1976 (Congressional Research Service Report for Congress, 1993). In 1993, the US National Institutes of Health devoted US$ 60 million to **biodiversity related research on drugs and medical products from natural sources**. The NCI has collected over 50,000 samples derived from plants, microorganisms, and marine samples from over 30 tropical countries aimed at finding cures for cancer and AIDS. These samples are now held in **NCI's Natural Product Repository** and are available under material transfer agreements (MTAs) to researchers (Grifo, 1996).

Bioprospecting generally consists of four phases:

- Phase I: on-site collection of samples
- Phase II: culturing of organisms, isolation and characterization of specific compounds
- Phase III: screening for specific uses
- Phase IV: product development and commercialization, including patenting, trials, sales, and marketing

If well managed, bioprospecting can be advantageous for the developing countries since in addition to generating new products, it can generate income and provide an incentive for the conservation of biodiversity. However, if not managed properly, it can create environmental problems due to over-exploitation/habitat destruction, and socioeconomic issues related to unfair sharing of benefits.

Over the past couple of decades, large pharmaceutical companies have scaled back bioprospecting activities primarily due to the expenses involved and complex negotiations over intellectual property rights (IPR). The majority of natural products research is being done in laboratories in academic and government funded research institutes.

8.2.3 CONCERNS REGARDING BIOPROSPECTING

Biodiversity prospecting has been found to be **more efficient when organisms are collected with indigenous knowledge (IK)**—valuable chemical compounds are more easily identified by studying plants, microorganisms, and animals used in folk medicine than by random screening. The Rural Advancement Foundation International (since 2001 known as the Erosion, Technology, and Concentration Group) cites several examples of this at http://www.etcgroup.org/content/bioprospectingbiopiracy-and-indigenous-peoples, some of which are reproduced here:

- *86% of plants used by Samoan healers displayed significant biological activity when tested in the laboratory.*
- *Four times as many positive results in lab-tests for HIV activity was seen in crude extracts of plants used by a healer in Belize.*
- *Shaman Pharmaceuticals found that screening tests on a plant used by three different communities was 5000 times more effective than random screening for developing potentially profitable drugs.*

Since the overarching aim of bioprospecting is to derive a commercial benefit from biological resources, it is only fair that the knowledge of the indigenous people [know as traditional knowledge (TK) or IK, see Chapter 16: Protection of Traditional Knowledge Associated With Genetic Resources] is acknowledged and they receive a share of the profit arising from the exploitation of this knowledge. Failure to do so constitutes what has been dubbed as "**biopiracy**."

The politics of bioprospecting is complex due to differences of opinion regarding the extent of sharing and (with many companies avoiding the technical difficulties of using TK) has hindered natural product development. A need for a regulatory framework that is consistent, ethically sound and legally strong, that is not overly burdensome so as to impede scientific progress, was addressed at the Rio "Earth Summit" and resulted in the signing of the Convention on Biological Diversity (CBD) by 150 countries in June 1992.

KEY TAKEAWAYS
- Diversity of plants, animals, and microbes, known as "biodiversity," is an important source of products (especially medicines) and genes for future breeding programs.
- The search for useful organisms or their genes is known as "bioprospecting."
- Much of the biodiversity resides in third world countries.
- Exploitation of bioresources (including TK) without the consent of the owners of the resource constitutes "biopiracy."

8.3 THE UNITED NATIONS CONVENTION ON BIOLOGICAL DIVERSITY

In recognition of the need to protect the fast depleting biodiversity for the present and future generations, the **United Nations Environment Programme** in November 1988 convened the Ad Hoc Working Group of Experts to consider an international CBD. In 1989, the **Ad Hoc Working Group of Technical and Legal experts** was formed to prepare an **international legal instrument for conservation and sustainable use biodiversity**. Work done by this group which soon came to be known as the Intergovernmental Negotiating Committee resulted in the Adoption of the Agreed Text of the CBD at the Nairobi Conference on May 22, 1992. The **CBD** was opened for signature at the **United Nations Conference on Environment and Development** (the **"Earth Summit"** of

Rio de Janeiro) on **June 5, 1992**. The convention entered into force on December 29, 1993, 90 days after the 30th ratification (https://www.cbd.int/convention/default.shtml).

The three main objectives of the CBD are as follows:

1. *The conservation of biological diversity*
2. *The sustainable use of the components of biological diversity*
3. *The fair and equitable sharing of the benefits arising out of the utilization of GR*

(https://www.cbd.int/intro/default.shtml)

While on the one hand, the CBD stressed the **sovereign rights of nations over their biological resources**; on the other, the convention (also called the "biotrade convention") **encouraged members to facilitate access to GR** and to enact **measures to ensure fair and equitable sharing of benefits** accruing from them. This has mainly been implemented through bilateral agreements on **mutually agreed terms (MAT)** involving **prior informed consent (PIC)** of the resource provider. The CBD is basically a **legal framework** outlining in broad and general terms goals and policies to be pursued by member nations **through their own legislation and policies relevant within the national context**. One important feature of the convention is that it considers conservation to be inseparable from sustainable use. Also, it recognizes the unique contribution of communities in biodiversity conservation and in the creation and sustenance of ecosystems.

The Convention has 42 Articles and Annexes of which the first 18 Articles outline recommendations for interaction between national governments and prescribe codes of conduct regarding utilization of biological resources in research or industry (see Box 8.2). Article 19 deals with biotechnological applications and distribution of benefits, whereas Articles 20−42 address administrative, financial, and legal issues arising from CBD.

In order to be enforced, the Convention required ratification by 30 signatories. The CBD became legally binding on **December 29, 1993**. At present, 196 countries (see Fig. 8.2) are Parties to the convention; the United States of America is a signatory but not a Party to CBD (for list of countries see https://www.cbd.int/information/parties.shtml).

8.3.1 CONFERENCE OF THE PARTIES

The governing body of the CBD is the Conference of the Parties (COP) which meets every 2 years (or as required) to review progress of implementation of the Convention and to adopt programs of work. It also provides policy guidance. Assistance to the COP in the form of recommendations on technical aspects is provided by the Subsidiary Body on Scientific, Technical, and Technological Advice (SBSTTA) comprised of government representatives, nonparty observers, and subject experts from scientific communities and other relevant organizations. Ad hoc open-ended Working Groups are established from time to time to deal with specific issues. Current working Groups listed on the CBD web-page (https://www.cbd.int/convention/bodies/intro.shtml) are as follows:

- **The Working Group on ABS** is currently the forum for negotiating an international regime on ABS;
- **The Working Group on Article 8(j)** addresses issues related to protection of TK;
- **The Working Group on Protected Areas** is guiding and monitoring implementation of the program of work on protected areas;

BOX 8.2 CONVENTION ON BIOLOGICAL DIVERSITY (CBD)

The text of the Convention of Biological Diversity has 42 Articles and 2 Annexes. In the preamble to the document, the contracting parties aver the importance of biological diversity, the need to conserve it and to use it in a sustainable and responsible manner in order to prevent further erosion of biological diversity at source. The parties also recognize that "*economic and social development and poverty eradication are the first and over-riding priorities of developing nations,*" and that, special provisions are required to meet the needs of developing nations including access to technology for conservation activities and sustainable use of biological diversity. The first five Articles define the objective and scope of the convention, whereas Article 6 prescribes that each Contracting Party develop national strategies, plans or programs for meeting the objectives of the Convention. Codes of conduct with respect to ex-situ and in-situ conservation, sustainable use, research and training, and public education and awareness contributing to conservation and sustained use are specified in Articles 7 to 14. Article 8(j) of the CBD recognizes contributions of local and indigenous communities to the conservation and sustainable utilization of biological resources through traditional knowledge, practices, and innovations and provides for equitable sharing of benefits with such people arising from the utilization of their knowledge, practices and innovations. Article 15 recognizes "*the sovereign rights of States over their natural resources,*" but recommends that each Contracting Party "*endeavor to create conditions to facilitate access to genetic resources*" for environmentally sound uses and "*not impose restrictions that run counter to the objectives of the Convention.*" While Articles 16 and 19 discuss the legislative, administrative, and policy measures for access and transfer of technology, Articles 20 and 21 specifically address financial mechanisms for affair and equitable sharing, on mutually agreed terms, of commercial benefits arising from the utilization of the genetic resources. The Conference of Parties is established by Article 23 which would by consensus agree upon and adopt rules for the implementation of the Convention. Articles 24 to 42 deal with the administrative, financial, and legal issues that may arise from the Convention. At its first meeting, the Parties were to determine how to establish a clearing-house to promote and facilitate technical and scientific cooperation.

While the Convention is legally binding on member States, no legislative structure is imposed on unwilling nations.

Implementation of the CBD has been difficult. Direct compensation of indigenous populations has been problematic due to the often-one-sided nature of negotiations. For successful ABS contracts, the involvement of knowledgeable local institutions to facilitate negotiations between the indigenous community and the bioprospecting company is essential. Another problem encountered in the implementation of the CBD has been to determine royalty percentages for access to genetic resources used to create a new plant variety. This is because modern crop varieties are the result of decades of breeding and putting a value to each trait is impossible.

In reality, implementation of the CBD will not, by itself, prevent the erosion of biodiversity. Several factors contribute to the degradation of the environment and the loss of the world's biological diversity such as population growth, poverty, and unsustainable harvesting of biological resources. What the treaty has achieved however is to sensitize communities to the need for sustainable use of biological resources.

Source: Convention on Biological Diversity, Retrieved from https://www.cbd.int/doc/legal/cbd-en.pdf.

- **The Working Group on the Review of Implementation of the Convention** examines the implementation of the Convention, including national biodiversity strategies and action plans.
- **Open-ended Ad Hoc Intergovernmental Committee for the Nagoya Protocol on ABS** was established as an interim governing body for the Nagoya Protocol until the first meeting of the Parties to the Protocol at which time it will cease to exist.

KEY TAKEAWAYS

- The CBD, 1992, is an international legal instrument for conservation and sustainable use of biodiversity

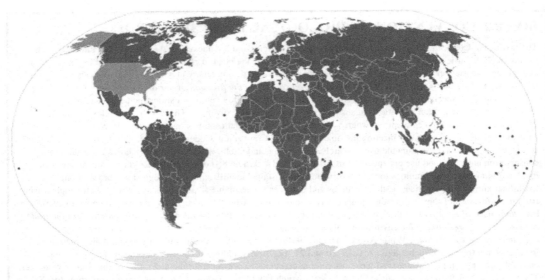

FIGURE 8.2 States parties to the Convention on Biological Diversity.

⎯Parties, ⎯Only signed, but not ratified, and ⎯nonsignatory.
[*By L.tak (Own work) [CC BY-SA 3.0* (http://creativecommons.org/licenses/by-sa/3.0)], *via Wikimedia Commons* (https://commons.wikimedia.org/wiki/File:Convention_on_Biological_Diversity2.svg)]

- The convention has three major objectives: conservation of biological diversity; sustainable use of biodiversity; and fair and equitable sharing of benefits from utilization of biodiversity
- Sharing of benefits is implemented through bilateral agreements on MAT involving PIC of the resource provider
- The governing body of the CBD is the COP
- The SBSTTA assists the COP
- Ad hoc open-ended Working Groups are established from time to time to deal with specific issues

8.3.2 BONN GUIDELINES ON ACCESS TO GENETIC RESOURCES AND FAIR AND EQUITABLE SHARING OF THE BENEFITS ARISING OUT OF THEIR UTILIZATION (BONN GUIDELINES ON ABS, 2001)

Adopted by the sixth Conference of Parties to the CBD in April 2002, these guidelines apply to all GR covered by the CBD with the exception of those covered by International Treaty on Plant Genetic Resources for Food and Agriculture (ITPGRFA). The primary aim of these guidelines is to **assist countries to develop their own access regulations.**

8.3.3 **THE NAGOYA PROTOCOL ON ACCESS AND BENEFIT-SHARING**

The *Nagoya Protocol on Access to GR and the Fair and Equitable Sharing of Benefits Arising from their Utilization to the* **CBD [Nagoya Protocol]** is an international agreement adopted in 2010 in Nagoya, Japan, which entered into force on October 12, 2014, having been ratified by 15 nations. With Germany having ratified the Protocol in April 2016, a total of 73 countries and the European Union (see Fig. 8.3) have ratified the Nagoya Protocol (https://www.cbd.int/abs/). It is a supplementary agreement to the CBD which provides a **transparent legal framework** for the **third objective of the CBD, that of fair and equitable sharing of benefits** arising from the utilization of GR. The Nagoya Protocol applies to both the **GR**, as well as **TK associated with GR**. It specifies **core obligations at the domestic level** for its contacting Parties with respect to ABS and compliance:

8.3.3.1 *Access obligations*

Contracting Parties are obligated to establish a legal framework that has clarity and transparency, provides fair and nonarbitrary rules and procedures, clear rules and procedures for PIC and MAT, and provides for issuance of permits when access is granted. The Parties are to promote and encourage research contributing to the conservation and sustainable use of biodiversity and the importance of GR for food security, while paying due regard to present or imminent threats to human, animal, or plant health.

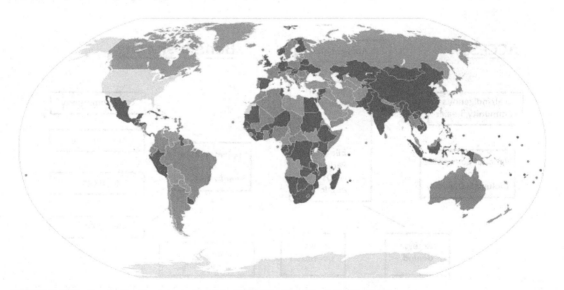

FIGURE 8.3 Parties to the Nagoya protocol to the Biological Diversity Convention.

▎—Parties, ▎—Only signed, but not ratified ▎—nonsignatory, but Biological Diversity Convention party, and ▎ nonsignatory, non-Biological Diversity Convention party.
[*By L.tak (Own work) [CC BY-SA 4.0* (http://creativecommons.org/licenses/by-sa/4.0)], *via Wikimedia Commons* (https://upload.wikimedia.org/wikipedia/commons/c/c6/NagoyaProtocol.svg)]

8.3.3.2 Benefit sharing obligations

Contracting Parties are to provide for fair and equitable sharing, subject to MAT, of benefits arising from the utilization (research as well as subsequent applications including commercialization) of GR. The benefits need not necessarily be monetary, but could be nonmonetary such as sharing of research results. Fig. 8.4 summarizes elements of a national policy on ABS.

8.3.3.3 Compliance obligations

A significant innovation of the Nagoya Protocol is the incorporation of specific obligations to **support compliance with domestic legislation** of contracting Parties providing the GR and the contractual obligations in the MAT. Contracting Parties are to ensure that GR in their jurisdiction have been accessed with PIC and that terms regarding the sharing of benefit have been mutually agreed upon. They are to monitor the utilization of the GR through all stages of development: research, development, innovation, pre-commercialization, and commercialization, even after it has left the country. In case of dispute, the contracting Parties are to **facilitate access to justice**, to seek recourse to legal systems and to cooperate and encourage resolution of disputes and alleged violations of MAT.

The Nagoya Protocol also addresses **Article 8(j) of the CBD on the role of TK** in the access and utilization of GR—to ensure PIC is obtained from communities with the right to consent to access GR and to ensure fair and equitable benefit-sharing with the community. This aspect is dealt

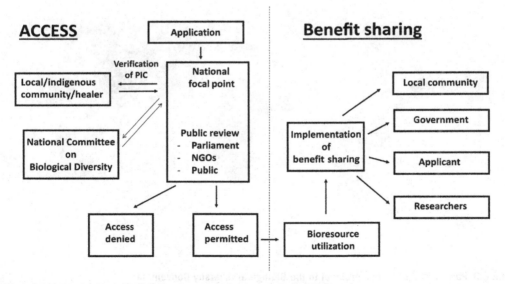

FIGURE 8.4 Access and Benefit Sharing from Valorization of Biological Resources.

Reprinted with permission from Timmermans, K. (2001). Trips, CBD and traditional medicines, concepts and questions. Report of the ASEAN workshop on the TRIPS agreement and traditional medicine, Jakarta, February 2001. 88p. 1 WHO 2001.

with in detail in Chapter 16, Protection of Traditional Knowledge Associated With Genetic Resources in this book.

8.3.3.4 Access and benefit-sharing clearing-house

The Access and Benefit-sharing Clearing-House (ABSCH) is a **key tool for facilitating the implementation of the Nagoya Protocol.** Established under Article 14 of the Nagoya Protocol, the ABSCH will facilitate information exchange and include an obligation to Parties to provide and update certain types of information.

KEY TAKEAWAYS

- The Nagoya Protocol is a supplementary agreement to the CBD which provides a transparent legal framework for the third objective of the CBD, that of fair and equitable sharing of benefits arising from the utilization of GR and applies both to GR, as well as TK associated with GR
- *Access obligations*: Contracting Parties are to establish a transparent legal framework for PIC and MAT
- *Benefit sharing obligations*: Contracting Parties are to provide for fair and equitable sharing, subject to MAT, of benefits arising from the utilization of GR
- *Compliance obligations*: In case of dispute, the contracting Parties are to facilitate access to justice, to seek recourse to legal systems, and to cooperate and encourage resolution of disputes and alleged violations of MAT
- *ABSCH*: a key tool for facilitating the implementation of the Nagoya Protocol, the ABSCH will facilitate information exchange.

8.4 THE UNITED NATIONS CONVENTION ON THE LAW OF THE SEA

The United Nations Convention on the Law of the Sea (UNCLOS), adopted in 1982 and entered into force in 1994, "*sets out the legal frameworks within which all activities in the oceans and seas must be carried out and is of strategic importance as the basis for national, regional and global action and cooperation in the marine sector*" (*UN doc A/RES/65/37, of 7 December 2010, Preambular para 4, at* www.un.org/Depts/los/general_assembly/general_assembly_resolutions.htm#2010).

The **UNCLOS recognizes different maritime zones** such as the internal waters and territorial sea, the exclusive economic zone (including the continental shelf up to 200 nautical miles), and the extended continental shelf (up to 350 nautical miles), over which the **coastal states can exercise sovereign rights over the exploitation, conservation, and management of living and nonliving natural resources.** The ABS for marine GR found in these areas are **subject to the laws of the coastal nation** and are **within the scope of the Nagoya Protocol.** Marine GR in the high seas and deep sea beds, in areas beyond national jurisdiction, are beyond the scope of the Nagoya Protocol. The treaty currently has 157 signatories and 167 Parties (Fig. 8.5).

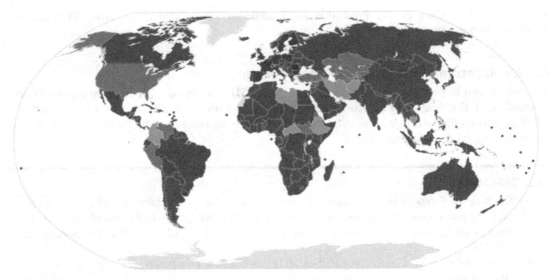

FIGURE 8.5 Map of parties to the United Nations Convention on the Law of the Sea.

—Parties, —Parties, dually represented by the European Union, —Signatories, and —nonmembers. *(By User: Canuckguy, User: Danlaycock et al.—File: International Court of Justice parties.svg, Public Domain, https://commons.wikimedia.org/w/index.php?curid=38655905)*

8.5 FAO INTERNATIONAL TREATY ON PLANT GENETIC RESOURCES FOR FOOD AND AGRICULTURE

The objectives if the **ITPGRFA** (http://www.planttreaty.org/) are similar to those of the CBD: **conservation, sustainable use, and fair and equitable sharing of the benefits derived from plant GR for food and agriculture**. It is a binding legal document adopted by the **United Nations Food and Agriculture Organization** (FAO) Conference on November 3, 2001. It entered into force on June 29, 2004, after being ratified, accepted, approved or acceded to by 40 countries. At present, the treaty has 139 Contracting Parties.

This treaty builds on the *International Undertaking on Plant GR* (henceforth referred to as the "Undertaking"), a nonbinding legal instrument approved by the FAO in 1983 to address the concerns of countries with regard to access and fair use of GR and the role of IPR as applied to biological materials. This Undertaking was especially important in resolving the issues of ownership of biological resources in view of the mounting tensions between the countries in the South (biodiversity-rich) and the countries in the North (technologically advanced but biodiversity-poor). The Commission on GR for Food and Agriculture of the FAO began the task of revision of the Undertaking in 1993 and adopted the revised version in July 2001. Subsequently, at the Conference in November 2001, **the revised version was adopted by consensus as the ITPGRFA**. With its adoption, existing regional/national laws and policies with respect to ABS are to be amended to avoid conflicts with overall ITPGRFA obligations and mandates.

The Treaty aims at:

> » *recognizing the enormous contribution of farmers to the diversity of crops that feed the world;*
> » *establishing a global system to provide farmers, plant breeders and scientists with access to plant genetic materials;*
> » *ensuring that recipients share benefits they derive from the use of these genetic materials with the countries where they have been originated.* (http://www.planttreaty.org/content/overview)

8.5.1 ACCESS AND BENEFIT SHARING (ABS) OBLIGATIONS

Under a **Multilateral System of ABS**, standardized *MTAs* approved by the ITPGRFA Governing Body allow access to 64 food and forage crops. Sharing is crucial because of the interdependence between countries for plant GR for breeding, conservation, and meeting food production goals. Under the ITPGRFA, fair and equitable sharing of benefits from facilitated access is through:

- exchange of information,
- access to and transfer of technology,
- capacity building, and
- sharing of monetary benefits from commercial use of the resources.

8.5.2 INTELLECTUAL PROPERTY AND FARMER'S RIGHTS

Article 12.2.d of the ITPGRFA **states that intellectual property (in the form of patents or plant breeder's rights) shall not be applied to plant GR in the form received from the Multilateral system**, thereby preventing recipients of GR from claiming IPR over those resources. However, Article 13.2.b.iii resolves to ensure that IPRs are respected and protected in access and transfer of plant GR so that access to GRs already protected by IPRs is consistent with national and international laws. The concept of *Farmer's Rights*, which had been developed in the Undertaking to compensate small farmers for conservation and development of plant GR efforts, is implemented at the national level through individual government action under Article 9 of the ITPGRFA.

8.5.3 EX SITU CONSERVATION CENTERS

A considerable portion of the world's collection of plant GR for food and agriculture existing in international agricultural research centers are held "in trust" for the benefit of humankind. Access to and use of these materials will also be facilitated by specific MTAs under ITPGRFA.

KEY TAKEAWAYS

UNCLOS, 1982:

- Coastal states can exercise sovereign rights over the exploitation, conservation, and management of living and nonliving natural resources over maritime zones of 200 nautical miles and extended continental shelf up to 350 nautical miles
- ABS for marine GR subject to laws of the coastal nation are within the scope of the Nagoya Protocol

FAO ITPGRFA:

- Objectives are similar to those of the CBD
- ABS obligations—access to 64 food and forage crops ensured under a multilateral system with standardized MTAs
- IPR (in the form of patents or plant breeder's rights) shall not be applied to plant GR in the form received from the Multilateral system, but IPRs are respected and protected in the MTAs
- Plant GR shall be held in ex situ conservation centers, access to which shall be facilitated by specific MTAs

8.6 REGIONAL AND NATIONAL SIGNIFICANCE AND ABS FRAMEWORKS

Articles 3 and 15 of CBD recognize the sovereign rights of states over the GR in their territories, but expects Parties to facilitate access to GR on mutually agreed upon terms by enacting national legislation. Since the CBD came into force in 1993, designing ABS frameworks has involved experts from many disciplines such as genetics, molecular biology, biochemistry, environmental science, ecology, as well as economists, sociologists, and political scientists in debates on contractual arrangements. In designing and implementing ABS policies at the national level, countries have encountered a wide variety of technical, political, social, and legal difficulties. Problems associated with the implementation have ranged from difficulty in deciding ownership, allocating specific royalty percentages and direct compensation to indigenous populations, to deciding on IPR.

The **main provisions of ABS laws and policies** in various nations include:

1. *Ownership:* The ownership of GR is a novel concept and in most countries common law principles regarding sovereign rights on natural resources have been applied to GR. Ownership of GR would determine access conditions, rules, procedures, and rights over these resources.
2. *Scope:* Implementation of ABS policies are facilitated by a precise definition of the scope, such as whether the policies are applicable only to the organism or also to derivatives or parts (e.g., proteins or gene sequences derived from the biological resources); whether it includes only resources in the natural habitat (in-situ) or from collections (ex-situ); or whether use and exchange of resources by indigenous communities and TK should be addressed, and so on.
3. *Access procedure:* Includes the access permits to be obtained from various agencies that administer biological resources.
4. *PIC:* The mandate of the CBD states that access to GR should be granted on MAT, but most nations require a PIC to be obtained from indigenous people in addition to that of the government.
5. *Benefit sharing and compensation mechanism:* Contracts specifying details are usually negotiated so that bioprospecting companies share benefits with researchers, collectors, and collaborators. Compensation could be through monetary benefits or nonmonetary benefits such as transfer of technology, training opportunities, scientific cooperation in conservation, public awareness, and exchange of information.
6. IPR *and protection of TK:* Under the CBD, Parties are not obliged to have a patent system for protection of inventions derived from biodiversity or TK, but under the Trade-Related Aspects of IPR Agreement, member countries must provide IPR protection.

7. *In-situ biodiversity conservation and sustainable use:* The CBD under Article 8 listed several activities aimed at conservation and sustainable use of biodiversity including establishing a system of protected areas and promoting the protection of ecosystems and habitats while simultaneously adopting measures to reduce the impact of use of biological resources.

8. *Enforcement and monitoring:* This is essential for a meaningful implementation of ABS laws and policies and to minimize the effects of use of biological resources.

Fair and equitable sharing of biological resources is a major objective of the CBD; however, it has been the adoption of the Nagoya Protocol in 2010 that has provided the impetus for implementation of ABS measures. Based on the implementation of ABS measures, Medaglia et al. (2014) have categorized countries into three groups:

1. *countries that refer to ABS in their national biodiversity strategy and action plan (NBSAP) or their environmental or biodiversity legislation, but have not yet regulated ABS in detail (these measures generally provide for the development of ABS regulations and include some general elements to be addressed);*

2. *countries that have a biodiversity or environmental law with some general provisions on ABS or access to biological resources, which may include a provision for the establishment of a regulation on ABS; and*

3. *countries that have addressed ABS in greater detail. This latter group of countries have established competent national authorities, procedures for prior informed consent, procedures for the development of mutually agreed terms, including benefit-sharing, and compliance measures.*

In the following sections, a few examples of national and regional implementation of ABS in each of the three categories is described.

8.6.1 CATEGORY 1: COUNTRIES WITH NO NATIONAL LAWS SPECIFICALLY DEVOTED TO ABS

This category includes countries such as United States of America, Canada, and New Zealand. In the United States, jurisdiction governing ABS is divided among different federal and state departments and agencies, as well as private landowners. A system of ABS has been created for the **national parks system—the US Code of Federal Regulations prohibits the "sale or commercial use of natural products"** collected from national parks and requires permission from the National Park Service for proposed research activities under the *National Parks Omnibus Management Act* of 1998. Similarly, in Canada, there are no national laws specifically devoted to ABS, and jurisdiction of ABS is shared with provincial and territorial governments and Aboriginal communities.

New Zealand is a biodiversity "hotspot" with three natural World Heritage Sites, committed to preserving its biodiversity and the TK of its indigenous communities. The country is not a signatory to the Nagoya Protocol, and laws on ABS in New Zealand are uncoordinated. Many biological species are protected under the Wildlife Act of 1953.

8.6.2 CATEGORY 2: COUNTRIES WITH A BIODIVERSITY OR ENVIRONMENTAL LAW WITH PROVISIONS ON ABS

There are several countries in this category, many of which, subsequent to ratification of the Nagoya Protocol are in the process of drafting or formalizing ABS laws. For example, **Mexico**: rules regulating access to GR are found in several federal laws such as the *Ecological Equilibrium and Environmental Protection General Act*, the *Sustainable Forestry Development Act*, and the *Wildlife General Act*, although the country lacks a regulatory framework for GR. Similarly, in **South Africa**, ABS is governed by sections of South Africa's *Biodiversity Act* of 2004, amendments to the *Patents Act* made in 2005, the *Bioprospecting and Access and Benefit-Sharing Regulations* of 2008, and the *Bioprospecting Guidelines* issued in 2012.

In **Australia**, one of the objectives of the *National Strategy for the Conservation of Australia's Biological Diversity* is to ensure that benefits arising from use of the country's biological resources accrue to Australia. Access to GR is regulated under the *Environment Protection and Biodiversity Conservation Act* of 1999 and the *Environment Protection and Biodiversity Conservation Regulations 2000*. Under this Regulation, permits to access biological resources are issued by the Department of the Environment, Water, Heritage, and the Arts (DEWHA). Model contracts to assist parties develop benefit sharing agreements with access providers have been developed by DEWHA. Regional organizations such as the Great Barrier Reef Marine Park Authority and the Australian Government Antarctic Division have also been empowered for overseeing ABS by laws such as the *Great Barrier Reef Marine Park Act 1975*, and the *Great Barrier Reef Marine Park Regulations 1983*.

8.6.3 CATEGORY 3: COUNTRIES WITH NATIONAL LAWS SPECIFICALLY DEVOTED TO ABS

The Executive Order 247 of 1995, which established the framework for access to GR in the Philippines, was the world's first ABS law to be enacted. Rules and regulations for implementing this order were by the Administrative Order No. 20 issued by the Department of Environment and Natural Resources in 1996. Underlying features of ABS frameworks include:

1. *Competent National Authority (CNA):* This could be a new organization created by ABS measure or an already existing organization
2. *PIC:* Application for access usually require details of whether the access to be granted is for commercial or noncommercial uses, as well as consent from the access provider in the geographical area from where the GR is to be accessed
3. *MAT:* Contracts which specify whether the resource is being accessed for research or commercial uses and how the benefits are to be shared with the competent authority or indigenous people/local community/resource providers are encouraged by the majority of national ABS frameworks
4. *Compliance measures:* Most countries specify penalties or sanctions for offenses or infractions to the provisions of the ABS law/regulation/guideline. These could be in the form of fines, imprisonment, bans on bioprospecting, or denial of access to GR.

Table 8.1 summarizes examples of national and regional ABS laws. It is apparent that the **biodiversity rich nations have been more proactive in formalizing laws to regulate access, ensure that the resources and associated TK is adequately protected,** and have mechanisms to

Table 8.1 Examples of National and Regional ABS Laws

Region/ Country	Legislation	Competent National Authority	PIC & MAT	Compliance
Costa Rica	Biodiversity Law, 1998 (BL)	Technical Office (TO) of National Biodiversity Administration Committee (CONAGEBIO), Ministry of Environment and Energy	Specified in Chapter 5, Relevance of Ethics in Biotechnology of BL	Article 112 establishes a system of fines for illegal access
Nicaragua	Biodiversity Law, 2012	Natural Heritage Direction General, Ministry of Environment and Natural Resources	National System of Licenses and Permits, PIC is required from Indigenous peoples, ethnic groups, and local or municipal authorities	Permits can be revoked by the Directorate for Biodiversity and Natural Resources for not fulfilling obligations, infringement could result in fines, revocation or suspension of the permit, or the requirement to pay compensatory damages
Panama	General Environment Law No. 41, 1998, Executive Decree No.25, 2009 (ED 25)	National Environmental Authority-created Unit for Access to Genetic Resources (UNARGEN) under the Directorate of Protected Areas and Wildlife	ED 25 establishes that all petitions shall be presented to the single processing window for UNARGEN's evaluation. Also establishes the parties to the Access Contract	Infringements have sanctions ranging from written warnings to fines and cancelation of contract
Peru	Law on the Conservation and Sustainable Utilization of Genetic Resources, 1997; Law for the Protection of Access to Peruvian Biological Diversity and to Collective Indigenous Knowledge (the Anti-Biopiracy Law); National Decree No. 003-2009-MINAM, 2009	Ministry of Environment	Permitting process and the signature of access contracts is the responsibility of the Administrative and Execution Authorities 173	National system for the monitoring of genetic resources, noncompliance attracts sanctions, including cancelation, suspension of the permit, fines, seizure of the material, and cancelation of the register

(Continued)

Table 8.1 Examples of National and Regional ABS Laws *Continued*

Region/ Country	Legislation	Competent National Authority	PIC & MAT	Compliance
India	Biological Diversity Act of India 2002 and its 2004 Rules	National Biodiversity Authority (NBA), the State Biodiversity Board (SBB) and the Biodiversity Management Committees (BMC)	2004 Rules establish procedures for (i) access to GR or TK, (ii) the transfer of research results internationally, (iii) receiving prior approval for patent applications, and (iv) third party transfer	Noncompliance attracts sanctions, including cancelation, suspension of the permit, fines, seizure of the material, and cancelation of the register
Syria	Act for the Protection and Exchange of Plant Genetic Resources No. 20, 2009	National Plant Genetic Resources Authority, part of the General Commission for Scientific Agricultural Research	No reference has been made in the Act to the concepts of prior informed consent and mutually agreed terms, and the only requirement for access is permission from the designated national authority	
Ethiopia	Proclamation in 2006 and Regulations in 2009	Ethiopian Institute of Biodiversity Conservation (IBC)	IBC has issued a code of conduct establishing the basic principles for access and utilization of GR and TK, including integrity and good faith, confidentiality, conservation and sustainable use, PIC, MAT, and benefit sharing	Monitoring to ensure that use of the GR is in compliance with the standard MAT
European Union	Regulation (EU) No 511/2014 of the European Parliament and of the Council, 2014	Competent authorities of Member States	Access permits, where applicable; mutually agreed terms, including benefit-sharing arrangements, where applicable	Member States should ensure that infringements of the rules implementing the Nagoya Protocol are sanctioned by means of effective, proportionate and dissuasive penalties

Medaglia J.C., Perron-Welch, F. & Phillips, F.-K. (2014). Overview of national and regional measures on access and benefit sharing. *Challenges and opportunities in implementing the Nagoya Protocol. Third Edn. CISDL Biodiversity and Biosafety Law Research Programme. Retrieved from http://www.cisdl.org/aichilex/files/Global%20Overview%20of%20ABS% 20Measures_FINAL_SBSTTA18.pdf.*

compensate or share with communities/owners of the resource the benefits arising from the use of the biological resource. Countries such as Costa Rica, Panama, Ethiopia, and India are centers of origin of several useful plant and animal species (see Box 8.1) and have elaborate biodiversity laws and compliance strategies.

8.7 SUMMARY

Mankind has a moral obligation to preserve nature for the present and future generations. Biological resources are a precious source of food, feed, fiber, fuel, and pharmaceuticals. The United Nations CBD signed in 1992 represents a comprehensive international effort to promote conservation, sustainable use, and equitable sharing of biological resources. A supplementary agreement to the CBD, the Nagoya Protocol, provides a transparent legal framework for the third objective of the CBD, that of fair and equitable sharing of benefits arising from the utilization of GR and associated TK. ABS is made transparent to prevent biopiracy by PIC from the government or CNA, and from holders of TK where applicable, and through MAT or bilateral contracts between bioprospecting companies and the government.

REFERENCES

Congressional Research Service Report for Congress (1993). *Biotechnology, indigenous peoples, and intellectual property rights*. The Library of Congress. Retrieved from https://www.hsdl.org/?view&did=717508.

Grifo, F. T. (1996). *Chemical bioprospecting: an overview of the International Cooperative Biodiversity Groups Program. Biodiversity, biotechnology, and sustainable development in health and agriculture: Emerging connections. Pan American Health Organization Scientific Publication No. 560.* Washington, D.C.: EUA.

FURTHER READING

Carrizosa, S., Brush, S. B., Wright, B. D., & McGuire, P. E. (Eds.), (2004). *Accessing biodiversity and sharing the benefits: Lessons from implementation of the convention on biological diversity* Gland, Switzerland and Cambridge, UK: IUCN. xiv + 316 pp. Retrieved from https://www.cbd.int/financial/bensharing/g-abs-iucn.pdf.

Greiber, T., Peña Moreno, S., Åhrén M., Carrasco, J.N., Kamau E.C., Medaglia J.C., Olivia, M.J., & Perron-Welch, F. in cooperation with Natasha Ali and China Williams (2012) *An explanatory guide to the Nagoya Protocol on access and benefit-sharing.* IUCN, Gland, Switzerland. xviii + 372 pp. Retrieved from https://cmsdata.iucn.org/downloads/an_explanatory_guide_to_the_nagoya_protocol.pdf.

Medaglia J.C., Perron-Welch, F. & Phillips, F.-K. (2014) *Overview of national and regional measures on access and benefit sharing.* Challenges and opportunities in implementing the Nagoya Protocol. Third Edn. CISDL Biodiversity and Biosafety Law Research Programme. Retrieved from http://www.cisdl.org/aichilex/files/Global%20Overview%20of%20ABS%20Measures_FINAL_SBSTTA18.pdf.

Web Resource
Website of Convention on Biological Diversity: https://www.cbd.int/.

ENSURING SAFETY
IN BIOTECHNOLOGY

*Hindsight is a wonderful thing but foresight is better, especially when it
comes to saving life, or some pain!*
-**William Blake, English poet**

CHAPTER OUTLINE

9.1 Introduction 212
9.2 Safety Issues in Recombinant DNA Technology 213
    9.2.1 Effect of Genetically Modified Organisms on the Environment 213
    9.2.2 Foods Derived from Genetically Modified Organisms 214
    9.2.3 Safety Issues in the Use of Recombinant DNA Technology in Medicine.......... 214
9.3 International Organizations, Treaties, and Conventions Addressing Biosafety 215
    9.3.1 Cartagena Protocol on Biosafety to the Convention on Biological Diversity 217
    9.3.2 United Nations Food and Agricultural Organization (FAO) Instruments That Deal
        With Issues Pertaining to Biosafety 221
    9.3.3 Codex Alimentarius Instruments That Deal With Issues Pertaining to Food Safety 222
9.4 Regulatory Oversight for Handling of Genetically Modified Organisms 223
    9.4.1 Organization for Economic Cooperation and Development (OECD) Guidelines 223
    9.4.2 European Union Regulatory Approach.. 224
    9.4.3 National Institutes of Health (NIH) Guidelines and Regulation of GMOs
        in the United States 225
    9.4.4 Other National Frameworks for Regulation of GMOs...................................... 228
9.5 Summary 231
References 231
Further Reading 232

An Introduction to Ethical, Safety and Intellectual Property Rights Issues in Biotechnology.
DOI: http://dx.doi.org/10.1016/B978-0-12-809231-6.00009-0

(Left to right): Maxine Singer, Norton Zinder, Sydney Brenner, Paul Berg at the **Asilomar Conference**, February 1975. At this time their affiliations were as follows:

Singer: National Cancer Institute, National Institutes of Health, Bethesda, MD

Zinder: Rockefeller University, New York, NY

Brenner: Medical Research Council, Cambridge, England

Berg: Stanford University School of Medicine, Stanford, CA

(Photo courtesy: National Academy of Science)

9.1 INTRODUCTION

The need for regulating modern biotechnology that could potentially give rise to new organisms was first voiced by the scientists in the forefront of recombinant DNA technology at the Gordon research conference on nucleic acids held in 1973. The major concerns were those regarding safety and the uncertainties in predicting possible adverse effects on laboratory workers, other humans and the environment. The United States National Academy of Sciences asked the then Director of National Institutes of Health (NIH), Paul Berg, to constitute a committee to examine the pros and cons of the new technology. The "Berg Committee" recommended that the NIH constitute a **Recombinant DNA Advisory Committee (RDAC)** to (1) oversee an experimental program to evaluate potential biological and ecological hazards of certain types of recombinant DNA (rDNA) molecules, (2) develop procedures to minimize spread of such molecules within human and other populations, and (3) devise guidelines for laboratory workers while handling potentially dangerous rDNA molecules. The committee also requested a **voluntary moratorium** on rDNA research involving antibiotic resistance genes, bacterial toxins, cancers, and tumor development genes.

To invite discussion on the new technology, the Berg Committee organized the **International Congress on Recombinant DNA Molecules** in February 1975 at the **Asilomar Conference Centre in California** (Berg, Baltimore, Brenner, Roblin, & Singer, 1975). The (Asilomar) conference was attended by around 140 participants from various walks of life including molecular biologists, lawyers, leaders of various religions, journalists, and the lay public. The conference helped to formulate recommendations for experimental designs, and conditions to ensure safe handling of rDNA molecules. This included containment conditions commensurate with the risk associated with handling microorganisms and potential pathogenic microbes. This chapter reviews the different mechanisms that has since been put into place in order to ensure safety in rDNA research and products derived through use of the technology.

9.2 SAFETY ISSUES IN RECOMBINANT DNA TECHNOLOGY

In the early days of the development of recombinant DNA technology, scientists were mainly concerned with the accidental creation of pathogenic strains of microbes and their effect on laboratory staff and, in the event of it escaping from the laboratory, on human and animal health. The public on the other hand, possibly as a result of science fiction in the print and visual media, were concerned about the creation of "monsters' and "engineered humans." In the aftermath of the Asilomar Conference of 1975, **genetic engineering experiments were confined to the K12 strain of the common gut bacterium *Escherichia coli***, which had been cultured since 1920, and were unable to grow outside of the laboratory. Also recommended were four levels of physical containment and three levels of biological containment (see Chapter 11: Laboratory Biosafety and Good Laboratory Practices). These recommendations have since been elaborated into guidelines for researchers and for release of products (genetically modified organisms as well as substances derived from them) into the environment, and in some regions and countries, as specific legally binding rules, discussed in further sections. Some of the safety concerns that warrant the need to regulate rDNA technology are discussed here.

9.2.1 EFFECT OF GENETICALLY MODIFIED ORGANISMS ON THE ENVIRONMENT

The effect of genetically modified organisms (GMOs) on the environment has been hotly debated with opponents of the technology claiming that GMOs poison the soil and cause **"genetic" pollution** resulting in the **spread of antibiotic resistant genes** that could compromise health care. The root of the issue lies in vector constructs used in rDNA technology having genes conferring antibiotic resistance. These genes serve as **selectable markers** to distinguish and separate the transformed organisms from untransformed ones. The argument is that detritus of the GMOs left in the field could result in soil bacteria acquiring antibiotic resistance genes (by a process referred to as **"horizontal gene transfer"**). Since antibiotics are the primary mode of treatment for most human and veterinary diseases, development of antibiotic resistant bacterial strains is of concern. Modern day transformation vectors have mechanisms to remove the antibiotic resistance marker gene shortly after the transformation (referred to as **"clean gene technology"**).

A large proportion of GM crops being commercially cultivated have genes from bacteria which give herbicide tolerance and/or insect resistance. Pollen transmission of herbicide tolerance genes could result in "**super-weeds**" that are difficult to remove from the fields. Insect resistance is mostly conferred by genes for toxins produced by strains of *Bacillus thuringiensis* (Bt). **Bt toxins are considered safe as they are host specific.** The toxin is released **only by proteolytic cleavage of a crystal protein in the gut of lepidopteran insects** and is digested similar to any other protein by other insects and animals. However, there have been reports of the direct effect of Bt toxins on nontarget organisms. For instance, in 1999, it was reported that the larvae of the Monarch butterfly were killed when fed on pollen from Bt corn (see Chapter 4: Recombinant DNA Technology and Genetically Modified Organisms). The issue drew considerable media attention as the Monarch is a familiar and beloved butterfly in North America and not a pest as its larvae grow on milkweed. The report raised concerns whether GMOs could result in a loss of biodiversity or affect other organisms in the same ecosystem.

In 1999, two scientists, William Muir and Richard Howard of Purdue University, Indiana, United States, proposed a hypothesis, which they dubbed the **"Trojan gene" hypothesis.** The hypothesis was based on studies on a common aquarium fish called the Japanese medaka (*Oryzias latipes*) that had been genetically modified with a human Growth Hormone gene. The GM fish were larger, became sexually mature faster, produced more eggs, and the larger males attracted four times as many mates, but only about two-thirds of the hybrids survived to maturity. By **computer modeling**, their hypothesis stated that **60 transgenic fish could lead to extinction of a 60,000-fish healthy wild stock in 40 generations** (Muir & Howard, 1999).

9.2.2 FOODS DERIVED FROM GENETICALLY MODIFIED ORGANISMS

Whether food produced by GMOs is **safe to eat** has been a contentious issue with highly polarized viewpoints. Doubts regarding the safety of GM potatoes were cast by the study conducted by Arpad Pusztai of Rowett Institute in 1998 (see Chapter 4: Recombinant DNA Technology and Genetically Modified Organisms). Also, in the absence of adequate monitoring mechanisms and labeling laws, mixing of products from GM crops engineered for production of feed (e.g., Starlink corn), or pharmaceuticals, on grocery shelves is worrisome.

9.2.3 SAFETY ISSUES IN THE USE OF RECOMBINANT DNA TECHNOLOGY IN MEDICINE

Recombinant DNA technology has contributed to health care in two important ways: production of pharmaceutically important proteins (biopharmaceuticals) and gene therapy for replacement of defective genes.

9.2.3.1 Biopharmaceuticals

Recombinant DNA technology has been effectively used to produce various human proteins in microorganisms, such as insulin and growth hormone, used in the treatment of diseases (see Chapter 4: Recombinant DNA Technology and Genetically Modified Organisms). Unlike chemically synthesized drugs, these are biomacromolecules—primarily endogenous proteins, and present a variety of special considerations and concerns:

- whether the molecule produced through rDNA technology is biologically equivalent to the naturally occurring one
- as these are mostly proteins, will they result in immunogenic reactions that would limit their usefulness.

Testing of these compounds presents unique problems. For instance, since they are endogenously produced, assessing pharmacokinetics and metabolism is difficult. Also, since they are available only in small quantities, traditional testing protocols that involve progressively increasing dosages until adverse effects occur, may not be possible. For conventional pharmaceutical safety assessment, the compound is to be tested separately in at least two mammalian species of which one must be a nonrodent. With substances with specific activity in humans, the evaluation in rodent and other model species may not be appropriate. Differences in immunological sensitivities in animal and human systems can have disastrous effects as was seen in the TeGenero trial for testing an antibody TGN1412 intended to treat rheumatoid arthritis and B-cell chronic lymphocyte leukemia (see Chapter 5: Relevance of Ethics in Biotechnology).

9.2.3.2 Gene therapy

Gene therapy aims to treat/cure/prevent disease by replacing a defective gene with a normal one using recombinant DNA technology. Most human clinical trials in gene therapy are still in the research stage with only over 400 trials conducted in about 3000 patients in order to treat for single gene disorders, cancers, and AIDS. Scientists agree that this is the most powerful application of rDNA technology, but have been cautious in its application due to associated risks as exemplified by the case of Jesse Gelsinger (see Chapter 4: Recombinant DNA Technology and Genetically Modified Organisms). Ensuring safety of patients in clinical trials has led to the development of better risk assessment in clinical trials (see Chapter 10: Risk Analysis).

KEY TAKEAWAYS

Safety issues in recombinant DNA technology include:

- "Gene pollution" of the environment resulting in "superweeds," antibiotic-resistant microbes
- Health effects of foods from GMOs
- Allergenicity/adverse immune reactions/effectiveness of pharmaceutical compounds produced using rDNA technology
- Risks in gene therapy

9.3 INTERNATIONAL ORGANIZATIONS, TREATIES, AND CONVENTIONS ADDRESSING BIOSAFETY

Given the immense potential applications of biotechnology and the need to minimize potential adverse effects of its products on the environment and on human health, several international organizations have focused on drafting agreements and guidelines addressing biosafety issues. Although there is no single comprehensive legal instrument that addresses all aspects of GMOs and their products, Glowka (2003) has compiled a list of 15 instruments, some binding on all nations, while others are nonbinding, listed in Table 9.1.

In 1985, an **Informal Working Group on Biosafety** was formed by members of United Nations Industrial Development Organization (UNIDO), United Nations Environment Programme (UNEP), World Health Organization (WHO), and later Food and Agricultural Organization (FAO). This group was instrumental in issuing the *UNIDO Voluntary Code of Conduct for the Release of Organisms into the Environment (1992)*, the *OECD Safety Considerations for Biotechnology (1992)*, the *FAO Draft Code of Conduct on Biotechnology (1993)* and later, the *UNEP International Technical Guidelines for Safety in Biotechnology (1995)*. Parallel discussions on conservation, and access and benefit sharing, of biological diversity resulted in the adoption of the *Convention on Biological Diversity* (CBD) and the *Agenda 21* at the United Nations Conference on Environment and Development (UNCED) in 1992. Both the CBD (in Article 19) as well as the

Table 9.1 Binding and Nonbinding Instruments Addressing Biosafety Issues Associated With the Use, Handling, Transfer, and Transboundary Movement of GMOs and Products

Binding Biosafety Instruments	Nonbinding Biosafety Instruments
United Nations Convention on the Law of the Sea (1982)	World Conservation Union (IUCN) Position Statement on Translocation of Living Organisms (1987)
Convention on Biological Diversity (CBD) (1992)	Agenda 21, Chapter 16, Protection of Traditional Knowledge Associated with Genetic Resources (Environmentally Sound Management of Biotechnology) (1992)
World Trade Organization (WTO) Agreement on the Application of Sanitary and Phytosanitary Measures (1994)	Organization for Economic Cooperation and Development (OECD) Safety Considerations for Biotechnology (1992)
WTO Agreement on Technical Barriers to Trade (1994)	FAO preliminary draft International Code of Conduct on Plant Biotechnology as it Affects the Conservation and Utilization of Plant Genetic Resources (1992)
UN Food and Agriculture Organization (FAO) International Plant Protection Convention (IPPC) (1997)	United Nations Industrial Development Organization (UNIDO) Voluntary Code of Conduct for the Release of Organisms into the Environment (1992)
United Nations Economic Commission for Europe (UNECE) Convention on Access to Information, Public Participation in Decision-making and Access to Justice in Environmental Matters (1998)—Aarhus Convention	FAO Code of Conduct on Responsible Fisheries (1995)
CBD Cartagena Protocol on Biosafety (2000)	United Nations Environment Programme (UNEP) Technical Guidelines for safety in Biotechnology (1995)
FAO International Treaty for Plant Genetic Resources for Food and Agriculture (2001)	FAO Code of Conduct for the Import and Release of Exotic Biological Control Agents (1996)

Reference: Glowka, L. (2003) Law and modern biotechnology. FAO Legislative Study 78. Food and Agriculture Organization of the United Nations, Rome.

Agenda 21 (in chapter 16, section 4) called for the development of an international agreement on safety procedures in adoption of biotechnology.

9.3.1 CARTAGENA PROTOCOL ON BIOSAFETY TO THE CONVENTION ON BIOLOGICAL DIVERSITY

The United Nations CBD (see Chapter 8: Biodiversity and Sharing of Biological Resources) Article 19, paragraphs 1 and 2, dealt with participation in research and sharing of benefits of biotechnology. In paragraph 3, it recommended that the Parties consider a protocol on biosafety. Subsequent to the signing of the CBD, a panel of experts was established by UNEP to recommend the contents of a protocol. At the second meeting of the Conference of the Parties (COP2) in November 1995 an *Open-ended Ad-Hoc Working Group on Biosafety* (the BioSafety Working Group, BSWG) was formed to elaborate a protocol focusing on the **transboundary movement of Living Modified Organisms (LMOs)** resulting from modern biotechnology. After several rounds of meetings and negotiations between different Groups of countries (see Box 9.1), the draft protocol was presented at an extraordinary meeting (ExCOP) held in Cartagena, Columbia, on February 22, 1999.

BOX 9.1 INTERNATIONAL NEGOTIATIONS AND THE CARTAGENA PROTOCOL ON BIOSAFETY

The objective of the Cartagena Protocol on Biosafety, which is a supplementary to the United Nations *Convention on Biodiversity* (CBD), was to cobble a legally binding instrument that would sustain trade in products of modern biotechnology while simultaneously protecting the environment and biodiversity. Although negotiations on the protocol were initiated with the view to extend the environmental biodiversity agenda of the CBD to biotechnology, it soon evolved into a transatlantic trade dispute, one that pitted trade interests against environmental and health concerns. The need for such a treaty arose because of the burgeoning trade in genetically modified organisms (GMOs) being met with increasing consumer and regulatory resistance in several countries (Falkner, 2000). The United States (the world's largest exporter of biotechnology products) alleged unfair trade restrictions and violation of the World Trade Organization principles by the European Commission due to stringent rules governing the release of GMOs and reliance on the precautionary principle in regulating GMOs.

Negotiations on the protocol were contentious due to different positions maintained by

1. countries which were producers of GMOs and therefore interested in trade;
2. the biodiversity rich nations (such as the third world countries) who were worried that import of these organisms or their products could have adverse effects on their genetic resources;
3. and regions (such as the European Union) concerned about the safety of foods/feeds from organisms produced by modern biotechnology.

There were mainly five negotiating groups of countries:

1. the Miami group which included USA, Canada, Australia, Argentina, Chile, and Uruguay, which were major producers and exporters of GMOs;
2. the Like-Minded Group, which included the majority of the G-77 countries (mostly developing countries including China);
3. the European Union (EU);
4. the Central and Eastern European group; and
5. the Compromise group (Japan, Korea, Singapore, New Zealand, Mexico, Norway, and Switzerland).

(Continued)

BOX 9.1 (CONTINUED)

Major contentious issues related to the Protocol were:

1. *Scope:* The Miami Group sought to limit the scope of the Protocol in order to avoid the infinite expansion of trade protectionism. The debate centered around whether the scope should be extended to LMOs used as pharmaceuticals for humans, LMOs destined for contained use, and LMOs in transit.
2. *Advance informed agreement (AIA) Procedure:* The Miami group insisted that the AIA procedure apply only to LMOs and not "commodities" (GMOs intended for food/feed/processing). However, the developing nations wanted the Cartagena Protocol to apply to all GMOs and their products. The argument that the Miami group had against the inclusion of commodities was that the products did not pose any real environmental threat. Also, the requirement for notification and informed consent would severely hamper trade and distribution, as products from GM crops would have to be segregated and handled separately from those from non-GM crops.
3. *Application of the precautionary principle:* The precautionary principle advocates taking preventive measures in the face of uncertainty regarding the environmental or health impacts of genetically modified organisms. The EU wanted the precautionary principle to be emphasized by inclusion in the Protocol. However, the Miami Group insisted on the "science-based" methods of risk assessment prevalent in the United States which had been approved by the World Trade Organization (WTO).
4. *Relationship with other international agreements:* The compatibility of the Cartagena Protocol with other agreements especially the WTO agreements such as the Agreement on the Application of Sanitary and Phytosanitary Measures (the SPS Agreement) and the Agreement on Technical Barriers to Trade (the TBT Agreement) was also extensively discussed. Although the Miami group wanted a clause to be inserted in the Protocol to the effect that provisions of the Protocol would not affect the rights and obligations under any existing international agreements such as the WTO agreements, this was not acceptable to most of the developing nations as they wanted to retain the right to restrict or prohibit imports of LMOs. Many developing nations did not have laws to regulate import of LMOs and were looking to strengthen their regulatory powers with regard to trade in GMOs through a global biosafety standard.
5. *Socioeconomic considerations:* Most of the developing countries wanted that socioeconomic considerations should be made part of the risk assessment and risk management procedures for LMOs and should factor into decision-making on the import of LMOs. However, most of the developed nations felt that socioeconomic considerations are issues of national domestic concern and should remain beyond the scope of the protocol.
6. *Liability and redressal for damage:* The developing countries wanted that exporting parties be held liable for damage to the environment or to human health caused by LMOs exported from within its jurisdiction. A consensus on this could not be achieved, although Article 27 of the Protocol states an endeavor toward this would be made.

Negotiations on the Biosafety Protocol began in 1995 with the formation of the open-ended ad hoc working Group on Biosafety (BSWG). The BSWG met five times between 1996 and 1998 to try and finalized a draft treaty by 1999, but failed to create sufficient common ground. The sixth meeting in Cartagena held in February 1999, characterized the final bargaining phase and debate on the draft. Juan Mayr, Chair of the ExCOP meeting, established a group of ten negotiators (five from the Like-Minded group, two from the Miami Group, and one each from the other three groups), to reach an agreement on the text (Mayr, n.d.). However, despite their best efforts, consensus could not be reached, and two more informal meetings were required, in Montreal in July 1999 and Vienna in September 1999, before the draft could be signed in Montreal in January 2000.

References

Falkner, R. (2000). Regulating biotech trade: the Cartagena Protocol on Biosafety. *International Affairs*, 76(2), 299−313. Retrieved from https://www.cbd.int/doc/articles/2002-/A-00143.pdf.

Mayr, J. (n.d.) *Doing the impossible: The final negotiations of the Cartagena Protocol*. In: Cartagena Protocol on Biosafety: From Negotiations to Implementation. CBD News Special Edition, The Secretariat of the Convention on Biological Diversity, Montreal, Canada. pp. 10−12. Retrieved from https://www.cbd.int/doc/publications/bs-brochure-02-en.pdf.

See also

Moore, P. & Yang, W. (2004) *Trade, Biodiversity and Sustainable Development: Proceedings of the Training Workshop, October 29−31, 2003*, Beijing China, IUCN, Bangkok, Thailand and Gland, Switzerland and IISD, Winnipeg, Canada. viii + 178 pp.

However, as the Conference of the Parties could not finalize the draft, the *Cartagena Protocol on Biosafety* was finally adopted at Montreal, Canada, on January 29, **2000** (http://bch.cbd.int/protocol). It established an open-ended *ad-hoc* **Intergovernmental Committee for the Cartagena Protocol** on Biosafety (**ICCP**) to undertake the preparations for the first **meeting of the Parties to the protocol (MOP or COP-MOP)**. The Protocol entered into force on September 11, 2003, and currently has 170 parties including 167 United Nations members, State of Palestine, and the European Union (EU).

The Cartagena Protocol does not use the term "genetically modified organism," but instead uses an apparently equivalent term *"living modified organism"* (LMO). By definition, **LMOs are biological entities capable of transferring or replicating genetic material (including sterile organisms, viruses, and viroids) that possess novel combinations of genetic material obtained by the use of modern biotechnology**. Modern biotechnology refers to techniques not used in traditional breeding and selection. It includes application of in vitro nucleic acid techniques, or fusion of cells, that overcome natural physiological, reproductive, or recombination barriers. **"LMO Products" are, by definition, processed material derived from LMOs** that contain detectable novel combinations of replicable genetic material got through the use of modern biotechnology (e.g., puree or paste from GM tomatoes, or flour from GM corn).

9.3.1.1 Advance informed agreement

While establishing rules and procedures for safe handling and use of LMOs, the Cartagena Protocol is especially focused on the **safe transfer and transboundary movement of LMOs**. For ensuring protection of biological diversity against potential risks when LMOs are to be intentionally introduced into the environment or shipped across boundaries, the Protocol establishes the **advance informed agreement (AIA)** procedure set out in Article 7–10 and 12. This procedure allows Parties intending to import LMOs to make **informed choices on whether to import, and for handling LMOs in a safe manner**. Parties to the protocol must ensure safety in the handling, packaging, and shipment of LMOs, and this includes proper documentation specifying the identity and details of the LMOs being shipped, and adequate mechanisms to manage risks. Decisions on import of LMOs are to be based on scientific risk assessment established by principles and methodologies suggested by the protocol. The process is facilitated by the **Biosafety Clearing-House** established by Article 20 of the Protocol. **The AIA procedure applies only to the first transboundary movement of LMOs** by the Party of import. The AIA procedure **does not apply to**:

1. LMOs in transit
2. LMOs intended for contained use
3. LMOs to be used directly as food, feed, or for processing.

However, Parties have the right to regulate importation or to exempt certain LMOs from application of AIA procedures on the basis of **national legislation**. Importing Parties retain the right to decide against importing LMOs on **precautionary principles** in cases when risk analysis is inconclusive in establishing safety.

9.3.1.2 Biosafety Clearing House

The Biosafety Clearing-House (BCH) (http://bch.cbd.int/) aims to facilitate exchange of information on LMOs and to *"assist the Parties to better comply with their obligations under the Protocol. Global access to a variety of scientific, technical, environmental, legal and capacity building information is provided in the six official languages of the UN."*

9.3.1.3 Assessment and review

The Conference of Parties serving as the meeting of the Parties (COP-MOP) is required to undertake an evaluation of the effectiveness of the Protocol 5 years after the entry into force of the Protocol and at periodic intervals thereafter (Article 35).

9.3.1.4 Capacity building

During the negotiations, the developing countries had raised the issue of not having the capability to assess and implement biosafety measures in respect of LMOs. Article 22 of the Protocol specifically addresses this issue. It requires Parties to cooperate in the development and/or strengthening of human resources and institutional capacities in biosafety taking into account the needs of the developing countries for financial resources and access to and transfer of technology and know-how.

9.3.1.5 Compliance

Article 34 of the Protocol requires the COP-MOP to adopt procedures and mechanisms to ensure compliance. A Compliance Committee has been formed to provide advice and assistance and to address cases of noncompliance.

9.3.1.6 Liability and redressal

During the negotiations, the Parties were unable to come to a consensus on the issue of liability and redressal in the event of LMOs causing damage to the environment, but Article 27 resolved to address the issue in subsequent meetings. Consequently, the first meeting of the COP-MOP established an *Open-ended Ad hoc Working Group of Legal and Technical Experts on Liability and Redressal* to address this. In October 2010, a supplementary treaty to the Cartagena Protocol on Biosafety known as the **Nagoya-Kuala Lumpur Supplementary Protocol on Liability and Redress** was adopted by governments, which specifies response measures that need to be taken in the event of damage to biodiversity by LMOs.

KEY TAKEAWAYS

Cartagena Protocol on Biosafety:

- Supplementary to the CBD focuses on the transboundary movement of LMOs adopted in 2000
- LMOs are defined as biological entities capable of transferring or replicating genetic material (including sterile organisms, viruses, and viroids) that possess novel combinations of genetic material obtained by the use of modern biotechnology
- "LMO Products" are by definition, processed material derived from LMOs that contain detectable novel combinations of replicable genetic material got through the use of modern biotechnology (e.g., puree or paste from GM tomatoes or flour from GM corn).

- Safe transfer and transboundary movement of LMOs are ensured by the **AIA** procedure facilitated by the **Biosafety Clearing-House**
- Importing Parties retain the right to decide against importing LMOs on precautionary principles in cases when risk analysis is inconclusive in establishing safety.
- The Protocol has provisions for capacity building in developing nations, ensuring compliance and addressing issues of liability and redressal.

9.3.2 UNITED NATIONS FOOD AND AGRICULTURAL ORGANIZATION (FAO) INSTRUMENTS THAT DEAL WITH ISSUES PERTAINING TO BIOSAFETY

FAO's mandate on food and agriculture includes crops, livestock, forestry, and fisheries and has been addressing biosafety and related aspects since the late 1990s, before the Cartagena Protocol came into force. Several instruments of the FAO deal directly or indirectly with the application of biosafety protocols:

9.3.2.1 The International Plant Protection Convention

The International Plant Protection Convention (https://www.ippc.int/en/) is an international plant health agreement, established in 1952, to which 182 contracting parties currently adhere. The purpose of the IPPC is to *"to protect cultivated and wild plants by preventing the introduction and spread of pests."* **The governing body of the IPPC is the Commission on Phytosanitary Measures (CPM).** The IPPC aids in harmonization of phytosanitary matters affecting trade and establishes standards to ensure that phytosanitary measures are not used as unjustified barriers to trade. It thus has a prominent role in the World Trade Organization's Agreement on the Application of Sanitary and Phytosanitary Measures (the **SPS Agreement**, see also Chapter 10: Risk Analysis). **Any LMO that can be considered a plant pest** falls within the scope of the IPPC.

9.3.2.2 The Commission on Genetic Resources for Food and Agriculture

The FAO Commission on Plant Genetic Resources was established in 1983 and renamed as the *Commission on Genetic Resources for Food and Agriculture* in 1995. The main objective of the CGRFA is to ensure conservation and sustainable use of genetic resources for food and agriculture as well as fair and equitable sharing of benefits derived from them. The Commission develops and monitors the *Global System on Plant Genetic Resources* (started in 1983) and the *Global Strategy for the Management of Farm Animal Genetic Resources* (started in 1993). In 2001, the FAO adopted the *International Treaty on Plant Genetic Resources for Food and Agriculture* (ITPGRFA, see Chapter 8: Biodiversity and Sharing of Biological Resources). Since 1992, the CGRFA has been working on a *Draft Code of Conduct on Biotechnology as it relates to genetic resources for food and agriculture*. The draft code is a holistic document that addresses a wide range of issues involving plant genetic resources including technical, legal, economic, social, ecological, ethical, in addition to biosafety issues. It has eight listed objectives, three of which emphasize biosafety practices.

KEY TAKEAWAYS

FAO and biosafety:

- *International Plant Protection Convention*—aids in harmonization of phytosanitary matters affecting trade. LMOs that can be considered a plant pest falls within the scope of the IPPC.
- *Commission on Genetic Resources for Food and Agriculture*—ensures conservation and sustainable use of genetic resources for food and agriculture as well as fair and equitable sharing of benefits derived from them. *Draft Code of Conduct on Biotechnology as it relates to genetic resources for food and agriculture* emphasizes biosafety practices.

9.3.3 CODEX ALIMENTARIUS INSTRUMENTS THAT DEAL WITH ISSUES PERTAINING TO FOOD SAFETY

The Codex Alimentarius Commission (http://www.fao.org/fao-who-codexalimentarius/en/) is an **intergovernmental body** with more than 180 members **established by the FAO and the World Health Organization (WHO)** in 1963 to *"develop harmonized international food standards, which protect consumer health and promote fair practices in food trade."* The Codex Commission is the **primary international forum addressing the food safety aspects of GMOs and their products**.

At its 26th session in 2003, the Codex Alimentarius Commission adopted Principles and Guidelines on foods derived from biotechnology. *"These are overarching principles on the risk analysis of foods derived from modern biotechnology and guidelines for food safety assessment of foods derived from recombinant-DNA plants and microorganisms"* (and animals in 2008 Amendment) (Codex Alimentarius Commission, 2009). These include:

1. *Principles for the risk analysis of foods derived from modern biotechnology* (CAC/GL 44-2003)
2. *Guideline for the conduct of food safety assessment of foods derived from recombinant-DNA plants* (CAC/GL 45-2003)
3. *Guideline for the conduct of food safety assessment of foods produced using recombinant-DNA microorganisms* (CAC/GL 46-2003)
4. *Guideline for the conduct of food safety assessment of foods derived from recombinant-DNA animals* (CAC/GL 68-2008)

The principles of risk assessment to identify hazards include a safety assessment that compares the food derived from modern biotechnology with its conventional counterpart (see more in Chapter 10: Risk Analysis).

The **Codex Committee on Food Labeling (CCFL)** began work in 1993 on developing guidelines for labeling of GM food products. The guidelines for mandatory GM labeling were however **strongly opposed by the United States**, supported by several others such as Canada, Mexico, Argentina, Costa Rica, and Australia. A *Proposed Draft Compilation of Codex Texts Relevant to Labeling of Foods Derived from Modern Biotechnology* (the "GM Guidelines") was finally adopted by the 39th session of the CCFL held in Quebec City, Canada, in May 2011 (Codex Alimentarius Commission, 2011). It permitted the **voluntary labeling of GM food**.

9.4 REGULATORY OVERSIGHT FOR HANDLING OF GENETICALLY MODIFIED ORGANISMS

In the 1980s in addition to international organizations (e.g., OECD) and regional groups (e.g., EU), several nations interested in the potential applications afforded by modern biotechnology were involved in discussions on safety and the need for regulatory oversight. **Denmark,** for instance, was the first European country to adopt **legislation specifically on rDNA** with the **Gene Technology Act of 1986**. Most other nations prefer the model of **voluntary guidelines as in the United States** and extend existing laws for managing risks associated with research and deliberate release of GMOs into the environment.

9.4.1 ORGANIZATION FOR ECONOMIC COOPERATION AND DEVELOPMENT (OECD) GUIDELINES

In 1983, the OECD with a view toward **harmonizing regulatory processes regarding safety issues in modern biotechnology** established an *ad hoc* group of governmental experts. This *Ad Hoc* **Committee** in 1986 published a report titled *"Recombinant DNA Safety Considerations"* (dubbed the "Blue Book") and introduced the safety considerations and risk assessment procedures which in many respects became the basis for regulation of GMOs and gene technology in many parts of the world. Two concepts related to risk assessment and safety considerations linked to rDNA organisms were introduced in the Blue Book: the review of potential risks on a "**case-by-case basis**," and the "**step-by-step procedure**" of progressively decreasing physical containment. The "case-by-case" regulatory principle is connected to the risk assessment procedures and is used to separate management of specific GMO applications from other GMO applications that the regulatory authority receives. The rationale for this principle is that each GMO transformation event may differ and should have a separate peculiar evaluation in order to evaluate all possible hazards and risks of that specific GMO. The "step-by-step procedure" is used during the research and development stages and involves careful characterization of the GMO and assessment of the safety and performance data collected at each research stage (e.g., the laboratory, contained enclosures like an animal house or green house), before small- and large-scale field testing. The detection of hazards or negative potential at any stage would require the organism to be brought back to a higher containment level, or the termination of the experiment (see Chapter 10: Risk Analysis).

Discussions of safety in biotechnology by the OECD's Group of National Experts (GNE) ensured that the majority of the OECD countries developed and enacted regulations regarding oversight of GMOs in the late 1980s and early 1990s before GMOs entered the market in 1995. Since 1995, the **OECD's Working Group on Harmonization of Regulatory Oversight in Biotechnology** and the **OECD's Task Force for the Safety of Novel Food and Feed** (the "Task Force," established in 1999) continue to monitor biosafety regulations and risk assessment guidelines. The Task Force also maintains a **database on GMOs, BioTrack** (accessible at http://www.oecd.org/biotrack). Current information that is important in food and safety assessment is brought out by the OECD in the form of "**consensus documents**." Agreed upon by consensus among the Task Force participants, these documents **provide a technical tool for regulatory officials, industry, and interested parties**. Although **not legally binding**, the documents are **mutually acceptable** to member countries.

9.4.2 EUROPEAN UNION REGULATORY APPROACH

In the EU, regulation of research on GMOs in contained research environments, as well as release of GMOs or their products into the environment, is by means of **Directives (which are implemented by national laws in the member states), and Regulations (which are directly applicable)**. The earliest EU Directives, **90/219/EEC on contained use of genetically modified microorganisms (GMMs) and 90/220/EEC on deliberate release of GMOs (both experimental as well as market releases)**, were adopted in 1990 and entered into force in 1991. The containment Directive was implemented at the national level and allowed member states to specify stricter regulations than what was outlined in the articles. The Directive on GMO release depended on cooperation between competent authorities in member countries in decision-making. Through a **summary notification information format system**, it was possible for member countries to comment on experimental releases and take them into consideration while reviewing applications for national releases. However, due to disagreements between different EU authorities on the implementation of the directive and risk assessment procedures, and also differences in opinion on safety issues between scientists, the biotech industry and nongovernmental organizations (NGOs) which resulted in a "de facto moratorium" on GMO approvals in 1998, Directive 90/220/EEC was replaced by **Directive 2001/18/EEC**. This new directive which came into force in April 2001 mandates a prerelease authorization procedure. This requires a case-by-case risk assessment of direct and indirect, immediate and delayed effects of GMOs on human health and the environment. (The term "risk to human health" in this directive refers to risks that arise from environmental exposure; food safety is not part of this assessment.) **Public registers of releases** including cultivation sites are to be established. **Authorization is for a period of 10 years**, but can be renewed after reviewing monitoring reports carried out during the period of marketing and use. **Public participation** and **transparency** are key principles in the EU policy.

In 1993, regulation of the marketing of GMOs for medical products for use in humans and veterinary was removed from the EU Directive 90/220/EEC into a **separate Regulation (EC) No. 2309/93**. Subsequent to the constitution of the European Agency for the Evaluation of Medicinal Products (EMEA) in 1995, this regulation of pharmaceuticals from GMOs has been replaced by a new **Regulation (EC) No. 726/2004**, which entered into force on May 20, 2004.

In 2009, Directive 90/219/EEC of April 23, 1990, regulation on the contained use of genetically modified microorganisms, which had been substantially amended several times, was recast as **Directive 2009/41/EC on May 6, 2009**. Paragraph 5 of the directive requires that the "*contained use of GMMs should be such as to limit their possible negative consequences for human health and the environment, due attention being given to the prevention of accidents and the control of waste*." The containment and other protective measures applied to a contained use must correspond to the classification of the contained use. Further, "*People employed in contained uses should be consulted in accordance with the requirements of relevant Community legislation, in particular **Directive 2000/54/EC** of the European Parliament and of the Council of September 18, 2000 on the protection of workers from risks related to exposure to biological agents at work (seventh individual Directive within the meaning of Article 16(1) of Directive 89/391/EEC).*"

Currently, the EU regulatory framework for GMOs includes regulations not only for contained use (**Directive 2009/41/EC**) and deliberate release (**Directive 2001/18/EEC**), but also for placing

GMOs on the market as food, feed or for processing (**GM Food/Feed Regulation 1829/2003**), for labeling and traceability (**Traceability Regulation 1830/2003**) and for transboundary movement (**Transboundary Movements Regulation 1946/2003**).

As of 2015, Directive 2001/18/EC has been amended "*as regards the possibility for the Member States to restrict or prohibit the cultivation of genetically modified organisms (GMOs) in their territory*" by a new directive: **Directive (EU) 2015/412 of March 11, 2015**. Under this new directive, governments can ask biotech companies whose GM crops have already been authorized for cultivation in the EU, or are pending approval, not to market their crops on their territory. The Commission website lists notifications for national bans (http://ec.europa.eu/food/plant/gmo/authorization/cultivation/geographical_scope_en.htm).

KEY TAKEAWAYS

EU Regulation	Old Regulation	Purpose
Directive 2009/41/EC (2009)	Directive 90/219/EEC (1990)	Contained use of GMMs
Directive 2001/18/EEC (2001)	Directive 90/220/EEC (1990)	Deliberate release of GMOs into the environment
Directive (EU) 2015/412 (2015)	Amendment to Directive 2001/18/EC	Member States can restrict or prohibit the cultivation of genetically modified organisms (GMOs) in their territory
Directive 2000/54/EC (2000)		Protection of workers from risks related to exposure to biological agents at work
Regulation (EC) No. 726/2004 (2004)	Regulation (EC) No. 2309/93 (1993)	GMOs for medical products for use in humans and veterinary
GM Food/Feed Regulation 1829/2003 (2003)		Placing GMOs on the market as food, feed or for processing
Traceability Regulation 1830/2003 (2003)		Labeling and traceability
Transboundary Movements Regulation 1946/2003 (2003)		Transboundary movement

---

9.4.3 NATIONAL INSTITUTES OF HEALTH (NIH) GUIDELINES AND REGULATION OF GMOs IN THE UNITED STATES

Based on the discussions and the recommendations of the Gordon conference, the Berg Committee, and the Asilomar Conference, **the first set of guidelines for the safe handling of rDNA molecules in contained research and production systems was developed by the National Institutes of Health (NIH) and published in 1976**. Although **mandatory** for research conducted in laboratories receiving federal funds in the United States, the guideline is **voluntary for privately funded research institutions and industry** (see Box 9.2 for a brief overview of the NIH guidelines). The NIH guidelines thus represent the first step in formulating a regulatory approach for handling GMOs. In keeping with scientific developments, the guideline has been revised several times over the past decades, and over the years, the focus of safety has shifted from contained research and production systems to deliberate release of GMOs for various applications into the environment.

BOX 9.2 NIH GUIDELINES FOR RESEARCH INVOLVING RECOMBINANT DNA MOLECULES (NIH GUIDELINES)

Now (April 2016): **NIH Guidelines for Research Involving Recombinant or Synthetic Nucleic Acid Molecules (NIH Guidelines)**

The NIH Guidelines were first published in the Federal Register on June 23, 1976, and has in the past four decades been revised several times in keeping with developments in recombinant DNA technology. Crafted by the RDAC of the Office of Science Policy, National Institutes of Health, with local review, the guidelines have helped to preserve public trust and permitted the science to progress in a manner that did not compromise public safety. The first major revision of the guidelines was in 1978 which relaxed certain restrictions deemed no longer necessary. Further revisions in 1984, 1986, 1989, and 1990 incorporated guidelines on human gene therapy and focused on ethical and social implications of human research with recombinant DNA. The 1994 revision of the guidelines adopted Appendices P (plants) and Q (animals) that had originally been developed by USDA. The 1997 and 2000 revisions amended the requirements for submission of gene transfer protocols, while the 2002 revision harmonized federal requirements for reporting safety information. The Guidelines amended in 2013, and again in 2016, has explicit inclusion of certain basic and clinical research with synthetic nucleic acid molecules. The NIH guidelines now covers recombinant and synthetic nucleic acids, defined as:

1. molecules that (1) are constructed by joining nucleic acid molecules and (2) can replicate in a living cell (i.e., recombinant nucleic acids);
2. nucleic acid molecules that are chemically or by other means synthesized or amplified, including those that are chemically or otherwise modified but can base pair with naturally occurring nucleic acid molecules (i.e., synthetic nucleic acids); or
3. molecules that result from the replication of those described in (1) or (2) above

The Guidelines comprise of five Sections and 15 Appendices.

Section I covers the scope and general applicability of the guidelines in addition to definitions relevant to the guidelines. The Guidelines are applicable to all research involving recombinant or synthetic nucleic acids within the United States or its territories, and all research supported with NIH funds.

Section II elaborates the safety considerations. It recognizes four classes (*Risk Groups 1 to 4*) of human etiological agents on the basis of hazard, listed in **Appendix B** of the Guidelines. Section II-B specifies the containment requirements for each Risk Group. **Appendix G** describes four levels of *Physical containment* (Biosafety Level 1 to 4) to be practiced for each corresponding Risk Group of etiological agents. **Appendix I** describes *Biological containment* for microorganisms. For research involving plants, **Appendix P** describes four biosafety levels (BL1-P to BL4-P) for *Physical and Biological Containment for Recombinant or Synthetic Nucleic Acid Molecule Research Involving Plants*. A similar guideline for research involving animals (BL1-N to BL4-N) for *Physical and Biological Containment for Recombinant or Synthetic Nucleic Acid Molecule Research Involving Animals* is described in **Appendix Q**.

Section III describes six categories of experiments involving recombinant or synthetic nucleic acid molecules:

1. experiments that require **Institutional Biosafety Committee (IBC) approval, RAC review, and NIH Director approval before initiation** (Section III-A),
2. experiments that require **NIH OSP and Institutional Biosafety Committee approval before initiation** (Section III-B),
3. experiments that require **Institutional Biosafety Committee and Institutional Review Board approvals and RAC review** before research participant enrollment (Section III-C),
4. experiments that require **Institutional Biosafety Committee approval before initiation** (Section III-D),
5. experiments that require **Institutional Biosafety Committee notification** simultaneous with initiation (Section III-E), and
6. experiments that are **exempt from the NIH Guidelines** (Section III-F).

Section IV assigns roles and responsibilities in ensuring that research is conducted in full conformity with the provisions of the NIH Guidelines. Responsibilities are assigned to:

1. The Institution (Section IV-B)
2. Institutional Biosafety Committee (IBC) (Section IV-B-2)

(Continued)

BOX 9.2 (CONTINUED)

   3. Biological Safety Officer (BSO) (Section IV-B-3)
   4. Plant, Plant Pathogen, or Plant Pest Containment Expert (Section IV-B-4)
   5. Animal Containment Expert (Section IV-B-5)
   6. Human Gene Therapy Expertise (Section IV-B-6)
   7. Principal Investigator (Section IV-B-7)
   8. NIH Director (Section IV-C-1)
   9. Recombinant DNA Advisory Committee (Section IV-C-2)
 10. Office of Science Policy (OSP) (Section IV-C-3)

  Section V comprises of footnotes and references of Sections I to IV
  Although the NIH guidelines is not legally binding, it has served as a comprehensive document guiding practitioners of modern biotechnology, not only in the United States, but in all other countries and regions around the world. It has also formed the basis of several legal instruments providing oversight to research and application of genetic manipulations.

Source: NIH Guidelines: http://osp.od.nih.gov/sites/default/files/NIH_Guidelines.html.

The NIH guidelines categorizes **human etiological agents into four groups based on hazard analysis based on their relative pathogenicity**:

1. *Risk Group 1 (RG1):* agents not associated with diseases in humans, includes microbes which are generally regarded as safe (GRAS)
2. *Risk Group 2 (RG2):* agents associated with diseases that are rarely serious and for which adequate prophylaxis is available
3. *Risk Group 3 (RG3):* agents which are associated with serious or lethal diseases for which preventive or therapeutic interventions *may be* available
4. *Risk Group 4 (RG4):* agents likely to cause serious or lethal diseases for which preventive or therapeutic interventions *are usually not* available.

The Guidelines recognize **six categories of experiments** involving recombinant DNA based on risk involved, each of which require permissions from different regulatory bodies (see Box 9.2). For more details on recommendations made in the guidelines for ensuring safety in laboratories engaged in research in recombinant DNA technology see Chapter 11: Laboratory Biosafety and Good Laboratory Practices, and for large-scale applications see Chapter 12: Recombinant DNA Safety Considerations in Large-Scale Applications and Good Manufacturing Practice.

In 1976, the US White House established the **Office of Science and Technology Policy (OSTP)** with a committee to oversee the regulatory framework for rDNA technology. In 1986, the OSTP concluded that rDNA technology was not inherently dangerous and that GMOs were not inherently riskier than non-GMOs and therefore did not require any special regulatory attention. Instead, the current legislations and regulations could be adapted to deal with the products of this technology. Under a **coordinated framework** established by the OSTP, the **United States Department of Agriculture (USDA)** is the lead organization overseeing the introduction of GM crops, the **Federal Food and Drug Agency (FDA)** reviews GMO applications for food/feed and in the pharmaceutical sector and the **Environment Protection Agency (EPA)** monitors and evaluates

GMOs with pesticidal properties (McHughen & Smyth, 2007). Most GMO applications require reviews from two or even three agencies. Legislative authority in each agency is rendered by federal acts such as the *Plant Protection Act (PPA) of 2000*, the *Federal Food, Drug and Cosmetic Act (FD&C Act) of 1938*, and the *National Environment Protection Act (NEPA) of 1970*. The EPA is given the authority to regulate pesticidal properties in GM plants under the *Federal Insecticide, Fungicide and Rodenticide Act (FIFRA)* (7 U.S.C. s/s 135 et seq. 1972) and *FD&C Act*. Clinical applications of gene therapy are regulated by the FDA under the *FD&C Act* and the *Public Health Service Act (PHS Act)*.

9.4.4 OTHER NATIONAL FRAMEWORKS FOR REGULATION OF GMOs

While most countries have felt the need to regulate GMOs, there is little consistency between countries in the manner in which it is achieved. At least some of these differences in approach stem from the regulation end-points—whether the GMO is meant for food/feed or other purposes, and its environmental impact. In most countries, multiple ministries/departments are involved in providing regulatory oversight.

- *Australia*
  Australia regulates genetically modified crops and organisms through a structured system that includes both Federal and State laws. Use of GM crops is regulated by the Commonwealth *Gene Technology Act 2000*, and the *Gene Technology Regulations*. The Act also establishes the **Office of the Gene Technology Regulator (OGTR)** to implement decisions. Western Australian components of the national scheme are the *Western Australian Gene Technology Act 2006*, and the *GM Crops Free Areas Act 2003*. There had been a moratorium on the commercial cultivation of GM crops in Western Australia since 2004, but the new *Genetically Modified Crops Free Areas Repeal Act 2015* repealed the 2003 Act and amended the **Biosecurity and Agricultural Management Act of 2007** (Parliament of Western Australia, 2016).

  GM foods are regulated by the **Food Standards Australia New Zealand (FSANZ)** under *Standard 1.5.2—Food produced using Gene Technology* in the Food Standards Code and includes mandatory premarket approval (including a food safety assessment) and mandatory labeling requirement.

- *Canada*
  Evaluating the safety and nutritional quality of foods from plants with novel traits (PNTs) including transgenic plants and their viable plant parts is the responsibility of **Health Canada**, under the *Food and Drugs Act*, and the **Canadian Food Inspection Agency (CFIA)**. Import of PNTs is regulated by the *Directive D-96-13: Import Requirements for Plants with Novel Traits, including Transgenic Plants and their Viable Plant Parts*. Oversight for research trials and confined environment release is provided by *Directive 2000-07: Conducting Confined Research Field Trials of Plants with Novel Traits in Canada;* and for unconfined environmental releases by *Directive 94-08: Assessment Criteria for Determining Environmental Safety of Plants with Novel Traits*.

  Canada is one of the largest producers of GM canola and raises several other GM crops. According to Canadian laws, details regarding the GMOs to be imported or to be grown are to be submitted by importers/manufacturers to Health Canada for permits. Labeling of foods from GMOs is voluntary.

- *China*

  China does not have a law specifically regulating GMOs. Oversight is primarily provided by the *Regulations on Administration of Agricultural Genetically Modifies Organisms Safety (GMO Regulations)* enacted by the State Council in 2001 (revised in January 2011) that regulates not only crops, but also animals, microorganisms and their products (Zhang, 2014). The **Ministry of Agriculture (MOA)** is responsible for biosafety management in China assisted by the **Office of Agricultural Genetic Engineering Biosafety Administration (OGEBA)**. The GMO Regulations establishes a national agriculture **Biosafety Committee** to evaluate applications for GMO Safety Certificates. Marketing of GMOs and products (including foods and feeds) from GMOs requires a permit from the MOA, and labeling is required.

- *India*

  In India, the **Department of Biotechnology (DBT)**, Ministry of Science and Technology, monitors the safety aspects of experiments in recombinant DNA technology. In 1990, the DBT formulated the *Guidelines for rDNA research in India*. These guidelines were further revised in 1994 to cover research and development activities on GMOs; transgenic crops; large-scale production and deliberate release of GMOs, plants, animals, and products into the environment; and shipment and importation of GMOs for laboratory research. In 1998, DBT brought out separate guidelines for carrying out research in transgenic plants called the *Revised Guidelines for Research in Transgenic Plants*. These also include the guidelines for risk analysis of toxicity and allergenicity of transgenic seeds, plants, and plant parts.

  The legal foundations of the Indian biotechnology system are contained in the *Environment (Protection) Act of 1986 (EPA, 1986)*. Three provisions of the EPA elaborated in Sections 6, 8, and 25 form the basis of the biosafety regulations. Section 6 enables the Indian government to enact rules on procedures, safe-guards, prohibitions and restrictions for the handling of hazardous substances; Section 8 prohibits the handling of hazardous substances except in accordance to proper safe-guards and procedures; and Section 25 empowers the government to continue this task and adopt specific rules and guidelines in the field of biosafety. These provisions of the Environment (Protection) Act provide the legal background to the *Rules for Manufacturing, Use, Import, Export, and Storage of Hazardous Microorganisms, Genetically Engineered Organisms or Cells*. This is also known as *Biosafety Rules* or, simply, the Rules of 1989 and is the key piece of the Indian legislation on. The biosafety rules are driven by six multi-layered decision-making structures: (1) the Genetic Engineering Approval Committee (**GEAC**), (2) the Review Committee on Genetic Manipulation (**RCGM**), whose activities would be assisted by the Monitoring and Evaluation Committee, (3) the Recombinant DNA Advisory Committee (**RDAC**), (4) the Institutional Biosafety Committee (**IBSC**), (5) the State Biotechnology Coordination Committee, and (6) District Level Biotechnology Committee. For activities involving large-scale commercial use and release of hazardous microorganisms, import of GMOs and recombinants in research and industrial production, approval from the GEAC, based in the **Ministry of Environment and Forests (MoEF)**, is mandatory.

  The multiplicity of regulatory agencies has been perceived as one of the factors that negatively affects the functioning of the Indian biotech sector. As these agencies are often placed under the control of different ministries, and operate at very different administrative levels, there is often a lack of coordination that causes significant confusion and delays in commercialization, and a lack of confidence in the Indian regulatory system. This has led to controversies related to biomedicines and new transgenic crops. In 2003, the union government appointed a task force on agricultural biotechnology headed by M.S. Swaminathan. In its report submitted to the Government, the task force raised serious questions on the competence of the GEAC to give final clearance for commercial cultivation and large-scale release of GMOs. It recommended constitution of a National Biotechnology Regulatory Authority (NBRA) with two arms, one for agricultural and food technology and the other for medical and pharmaceutical technology. The authority would replace the GEAC under the MoEF. In November 2007, after 2 years of consultations with several stakeholders, the Indian government approved the National Biotechnology Development Strategy (NBDS) which devised a comprehensive 10-year road map for the Indian biotech sector. The NBDS defined three general goals: development of human resources, strengthening of the infrastructure, and promotion of trade and industry. The NBDS also proposed the creation of the NBRA as an independent statutory body with wide-encompassing functions relating to the biosafety approval of genetically modified products and processes. Some months later, in July 2008, the National Biotechnology Regulatory Act was drafted to establish the NBRA under the DBT. As per the government directive, the NBRA would be set up as an independent, autonomous and professionally led body to provide a single window mechanism for biosafety clearance of genetically modified products and processes. The draft is awaiting approval by the government.

• *South Africa*

The *Genetically Modified Organisms Act of 1997 (GMO Act)* is the primary legislation dealing with contained use, trial or commercial release, import, and export of GMOs and their products in South Africa. The Act establishes three regulatory authorities for implementation of its objectives, namely: an **Executive Council**, **Registrar**, and an **Advisory Committee**. The Act requires permits and registration for research on GMOs. Risk assessments for demonstrating safety to environment and notification to public are required for production and marketing of GMOs (Goitom, 2014).

KEY TAKEAWAYS

National frameworks for regulation of GMOs:

- **United States:**
  - **Office of Science and Technology Policy (OSTP)**
  - OSTP established the **coordinated framework** involving the
    - **United States Department of Agriculture (USDA)** (introduction of GM crops)
    - **Federal Food and Drug Agency (FDA)** (reviews GMO applications for food/feed and pharmaceuticals)
    - **Environment Protection Agency (EPA)** (monitors and evaluates GMOs with pesticidal properties)
- **Australia:**
  - **Office of the Gene Technology Regulator**
  - GM foods are regulated by the **Food Standards Australia New Zealand (FSANZ)**

- **Canada:**
  - **Health Canada**, and **CFIA**
- **China:**
  - **Office of Agricultural Genetic Engineering Biosafety Administration (OGEBA)** under **MOA**
    - **Biosafety Committee**
- **India:**
  - **DBT** under Ministry of Science and Technology
    - RCGM regulates rDNA research and applications
    - GEAC under MoEF oversees marketing and release of GMOs
- **South Africa:**
  - Three regulatory bodies: **Executive Council**, **Registrar**, and **Advisory Committee**.

9.5 SUMMARY

One of the first questions that scientists dealt with in developing applications using recombinant DNA technology was the safety of the research itself. In a remarkable example of self-governance, at an international meeting held in 1975, scientists working in rDNA technology agreed on strict biological and physical safe-guards, and a moratorium on the research till guidelines could be formulated. The first set of guidelines for rDNA research brought out by the National Institutes of Health in 1976 was created with inputs from scientists, lawyers, policy-makers, and the public. Since then, several legally binding and nonbinding international agreements have focused on ensuring safety to human health and to the environment in research and in environmental releases of genetically modified organisms and products derived from them. Of especial significance is the Cartagena Protocol on Biosafety, a supplementary to the United Nations CBD. The protocol seeks to protect biodiversity by an AIA which helps importing countries to make informed decisions regarding the safety of genetically modified organisms, thus ensuring safety in transboundary movement of organisms produced by modern biotechnology. Several regions (e.g., EU) and nations (e.g., Denmark, Australia, and South Africa) have specific legislation to provide oversight, while others prefer the model of voluntary guidelines, as in the United States, and extend existing laws for managing risks associated with research and deliberate release of GMOs into the environment.

REFERENCES

Berg, P., Baltimore, D., Brenner, S., Roblin, R. O., III, & Singer, M. F. (1975). Summary Statement of the Asilomar Conference on Recombinant DNA Molecules. *Proceedings of the National Academy of Sciences*, 72(6), 1981–1984.

Codex Alimentarius Commission (2009). *Foods derived from modern biotechnology*, (2nd Edition). Rome: World Health Organization, Food and Agriculture Organization of the United Nations. Retrieved from ftp://ftp.fao.org/codex/Publications/Booklets/Biotech/Biotech_2009e.pdf.

Codex Alimentarius Commission. (2011). *Proposed Draft Compilation of Codex Texts Relevant to Labelling of Foods Derived from Modern Biotechnology.* In Report of the Thirty Ninth Session of the Codex Committee on Food Labelling Appendix III. Retrieved from http://www.codexalimentarius.net/download/report/765/REP11_FLe.pdf.

Goitom, H. (2014). Restrictions on genetically modified organisms: South Africa. *Library of Congress.* Retrieved from https://www.loc.gov/law/help/restrictions-on-gmos/south-africa.php.

McHughen, A., & Smyth, S. (2007). US regulatory system for genetically modified [genetically modified organism (GMO), rDNA or transgenic] crop cultivars. *Plant Biotechnology Journal*, 6(1), 2—12. Retrieved from http://onlinelibrary.wiley.com/doi/10.1111/j.1467-7652.2007.00300.x/full.

Muir, W. M., & Howard, R. D. (1999). Possible ecological risks of transgenic organism release when transgenes affect mating success: Sexual selection and the Trojan gene hypothesis. *Proceedings of the National Academy of Sciences*, 96(24), 13853—13856.

Zhang, L. (2014). Restrictions on genetically modified organisms: China. *Library of Congress.* Retrieved from https://www.loc.gov/law/help/restrictions-on-gmos/china.php.

FURTHER READING

Glowka, L. (2003). *Law and modern biotechnology.* Rome: FAO Legislative Study 78. Food and Agriculture Organization of the United Nations. Retrieved from ftp://ftp.fao.org/docrep/fao/006/y4839E/y4839E00.pdf.

Lynch, D., & Vogel, D. (2001). *The Regulation of GMOs in Europe and the United States: A Case-Study of Contemporary European Regulatory Politics.* Council on Foreign Relations Press. Retrieved from http://www.cfr.org/agricultural-policy/regulation-gmos-europe-united-states-case-study-contemporary-european-regulatory-politics/p8688.

Parliament of Western Australia (2016) Genetically Modified Crops Free Areas Repeal Bill 2015. Retrieved from http://www.parliament.wa.gov.au/482565B60082E1C5/0/9195C3B0AA930D4548257F0100182661?Open&Highlight=2,sitesearchyes,genetically%20modified%20crops%20free%20areas%20act.

Pew Initiative on Food and Biotechnology. (n.d.). *Guide to regulation of genetically modified food and agricultural biotechnology products.* Retrieved from http://www.pewtrusts.org/~/media/legacy/uploadedfiles/wwwpewtrustsorg/reports/food_and_biotechnology/hhsbiotech0901pdf.pdf.

Websites

Canadian Food Inspection Agency, Flow chart—Regulation of plants with novel traits in Canada: http://www.inspection.gc.ca/plants/plants-with-novel-traits/general-public/eng/1337380923340/1337384231869.

Directive (EU) 2015/412: http://eur-lex.europa.eu/legal-content/EN/TXT/?uri=OJ:JOL_2015_068_R_0001.

Directive 2009/41/EC: http://eur-lex.europa.eu/legal-content/EN/TXT/?uri=CELEX:32009L0041.

GMO legislation in the European Union: http://ec.europa.eu/food/plant/gmo/legislation/index_en.htm.

Library of Congress: Restrictions on Genetically Modified Organisms: https://www.loc.gov/law/help/restrictions-on-gmos/index.php.

Office of the Gene Technology Regulator, Australian Government: http://www.ogtr.gov.au/.

The Cartagena Protocol on Biosafety: http://bch.cbd.int/protocol.

RISK ANALYSIS

10

*Risk management is a more realistic term than safety. It implies that hazards are ever-present,
that they must be identified, analyzed, evaluated and controlled or rationally accepted.*
-Jerome F. Lederer, American aviation-safety pioneer.

CHAPTER OUTLINE

10.1 Introduction 234
10.2 Risk Assessment 234
10.3 Risk Management 237
10.4 Risk Communication 238
10.5 Risk Assessment for Genetically Modified Microorganisms 238
10.6 Risk Assessment for Genetically Modified Crops 240
    10.6.1 Stacked Events 241
    10.6.2 Pest-Risk Analysis for Quarantine Pests Including Analysis of Environmental Risks and
           Living Modified Organisms 241
10.7 Risk Assessment for Transgenic Animals 243
    10.7.1 World Organization for Animal Health 243
10.8 Safety Assessment of Foods Derived from Genetically Modified Organisms 247
    10.8.1 Codex Alimentarius Commission 247
    10.8.2 Postrelease Monitoring 249
10.9 Safety Assessment in Clinical Trials 249
10.10 Precautionary Principle in GMO Regulation 250
10.11 Summary 251
References 251
Further Reading 252

An Introduction to Ethical, Safety and Intellectual Property Rights Issues in Biotechnology.
DOI: http://dx.doi.org/10.1016/B978-0-12-809231-6.00010-7

Advanced Genetic Sciences scientist Julie Lindemann sprays Frostban on strawberry plants in the **first approved release of genetically engineered organisms**.

(Photo courtesy: The Alicia Patterson Foundation)

10.1 INTRODUCTION

Some of the key concerns regarding genetically modified organisms (GMOs) and products derived from GMOs are whether it could cause harm to human or animal health due to pathogenicity, toxicity, or allergenicity, and whether it could cause harm to the environment due to persistence, invasiveness, or adverse effects on other species occupying the same ecological niche. An integral part of ensuring safety in the development and release of GMOs and their products is an assessment of the risks involved. A scientific estimation of the effect on human and animal health and on the environment caused by the use or presence of GMOs or their products is mandatory for decision-making by regulatory authorities. The earliest **methods of risk assessment** advocated a **"case-by-case"** basis as well as a **"step-by-step"** procedure of progressively decreasing physical containment in risk analysis. This chapter discusses the three aspects of risk analysis: risk assessment, risk management, and risk communication in the context of GMOs and products derived through biotechnology.

10.2 RISK ASSESSMENT

Environmental risk assessment has been defined as the *evaluation of risks to human health and to the environment, whether direct or indirect, immediate or delayed, which experimental deliberate release or deliberate release by placing GMOs on the market may pose* (Directive 2001/18/EC of the European Parliament & of the Council, 2001). Risk assessment can be *retrospective*—for reasons of determining the cause so as to prevent a recurrence of an adverse event, or *prospective*—to predict possible hazardous outcomes. The latter is advocated for the release of GMOs since, in the event of a problem, it may be impossible to reverse the process or recall/eliminate all GMOs from the receiving environment. Prospective risk assessment suffers from the fundamental problem of **induction**—being able to accurately predict outcomes based on prior knowledge of the organism, the genetic characteristic involved, and the ecology of the receiving environment.

For risk assessment to be meaningful, it should be **science-based** and of a **high scientific standard**. It is usually done in a stepwise manner with the first level of testing being done under contained laboratory conditions and subsequent testing being done in progressively larger scale—an initial small-scale field trial followed by a larger scale field trial, the results of which are submitted for consideration of the statutory regulatory authority for market release (see Fig. 10.1). **Risk assessment takes into account the intrinsic potential harmful characteristic or "hazard" and the likelihood and magnitude of the hazard being realized. "Risk" is thus a product of "magnitude" and the "likelihood of consequences" of a specific hazard.** In the context of biotechnology that involves the use of organisms, the United Nations Environment Program Technical Guidelines for Safety in Biotechnology (UNEP, n.d.) considers that risk assessment involves the following components and interaction between them:

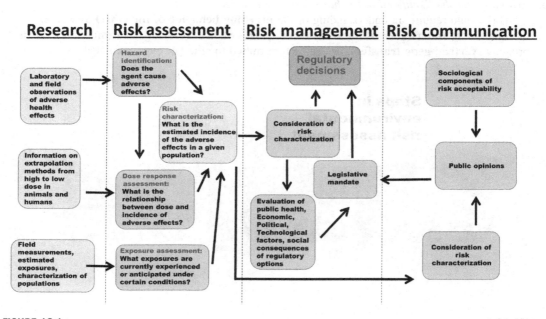

FIGURE 10.1

Risk analysis.

1. The characteristics of the organisms involved, including any newly introduced traits
2. The manner in which the organisms are to be used
3. The characteristics of the area and other organisms that might be affected.

Both **direct effects due to the GMO itself** (e.g., toxicity or allergenicity) as well as **indirect effects** (e.g., effects due to interaction with other organisms in the environment) are taken into account during risk assessment. Typically, the steps as summarized in Fig. 10.2 include the following:

1. *Identification of hazards*

   From a **complete description** of the organism taking into account the **"foreign" gene integrated** into the GMO, the **vector** used, and the **method** used in the creation of the transgenic organism, an **identification of characteristics** that may cause adverse effects is made.

2. *Evaluation of the potential consequences of adverse effect if it occurs*

   This includes **estimating the potential for production of substances toxic or allergenic** to human beings or another species. Also, **environmental effects** such as changes in soil/water chemistry or changes in nitrogen and carbon recycling brought about by decomposition processes of the GMO. Of equal concern is the **persistence of the trait in the environment—** whether the introduced gene would confer any ecological fitness advantage to the GMO, thus altering the ecological balance of the receiving environment.

3. *Evaluation of the likelihood of the hazards being realized*

   This would require an understanding of the **breeding behavior of the GMO** as well as the **manner of inheritance of the introduced gene**—whether it is only transmitted to its own progeny (**vertical gene transfer**), or can be transmitted to allied or even unrelated species

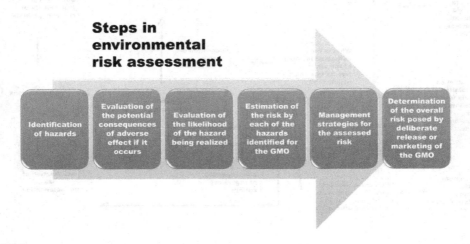

FIGURE 10.2

Six-step process of risk assessment.

(**horizontal gene transfer**); also, whether the gene expression is **stable** over consecutive generations. The inheritance as well as the stability of the introduced gene is **used in estimations of the persistence and possible invasiveness** of the introduced GMO in the receiving environment.

4. *Estimation of the risk by each of the hazards identified for the GMO*

The **actual risk is a combination of the magnitude of the hazard and the likelihood of the hazard occurring**. The magnitude can be described as **negligible, low, moderate, or severe**, as also the probability of the hazard being realized can be estimated as being **negligible, low, moderate, or high**. An estimation of this **helps to decide on a risk-management strategy** to minimize the consequences of release of the GMO in the market. For example, even a severe adverse effect may have a negligible likelihood of occurring, so the overall risk involved may be low/moderate, and release of the GMO may not warrant any special management practices. Hence, risk assessment would have to be done on a **case-by-case basis** for most GMOs.

5. *Management strategies for the assessed risk*

Physical and biological containment strategies have been devised in order to ensure safety while working with GMOs (see Chapter 11: Laboratory Biosafety and Good Laboratory Practices). A **step-by-step procedure** is usually used to manage assessed risk in which the GMO is assessed for safety and performance at different stages of development from laboratory stage, to contained enclosures (e.g., animal houses or green houses), to small-scale field testing, and finally large-scale field trials before release into the market. Detection of adverse effects at any of these stages would require the GMO to be brought back to a higher containment level, or the experiment to be terminated.

6. *Determination of the overall risk posed by deliberate release or marketing of the GMO*

The **overall risk is a function of the assessed risk and the effective management of the risk**. The overall risk due to deliberate release of GMOs or their products is dependent on the **immediate as well as long-term effects** due to the interaction of the GMO with other organisms in the receiving environment. The estimation of overall risk is therefore dependent on a careful monitoring and documentation of the effects of the GMO over time.

10.3 RISK MANAGEMENT

The objective of risk management is to **minimize or mitigate the assessed risk** so as to release the GMO or the product into the market. Even if the risk assessed is high and environmental release is not recommended, it may still be possible to derive benefit from the GMO by:

- **Release in a restricted (geographical) area** to limit transfer of genes to related species, or in ecosystems which may not be affected by the introduction of the GMOs
- **Physical confinement**—growing the GMO in suitable sequestered facilities.

An important part of risk management is also the **post-release monitoring and documentation** (record keeping) in order to identify and control unforeseen effects and/or long-term effects of the GMO in the receiving environment.

10.4 RISK COMMUNICATION

Unlike nuclear technology that also has issues of safety in use but was developed in high security secret facilities, scientists in the forefront of rDNA technology have invited public participation in determining the course of development and the applications of the technology. Public perceptions of safety have played a significant role in regulation of GMOs and release of GMOs and their products into the market. Risk communication to all stakeholders is crucial to risk perception and perforce should be relevant, transparent, and unambiguous to be effective. In order to be comprehensible, and in order to put the assessed risk in perspective, comparisons are made with the risks associated with different but related sources of risk [e.g., genetically modified (GM) crops with genes conferring resistance to insect pests may be compared with the risks associated with pesticide use in non-GM crops].

10.5 RISK ASSESSMENT FOR GENETICALLY MODIFIED MICROORGANISMS

Applications of genetically engineered microorganisms (GEMs, or genetically modified microorganisms, GMMs) that require release into the environment include use in mining, remediation of toxic pollutants from soil, use as biosensors, for nitrogen fixing, bacterial or viral pesticides, and bacteria which reduce frost damage. Typically, risk analysis of GEMs involves testing in a microcosm or greenhouse (Krimsky, 1996). The first field test for a GEM was carried out in 1987 in a strawberry field in California, United States, after 4 years of regulatory and legal deliberations. The test was for a strain of *Pseudomonas syringae*, called the "ice-minus" strain, with the putative ability to protect crops from frost damage due to the prevention of ice crystal formation (see Box 10.1). Risk assessment in the case of GEMs is especially challenging as the effects of the microorganisms on the environment are not as easily observed as those of plants or animals. The effects are also considerably harder to predict as microbes can multiply rapidly and frequently exchange genetic material with other species (horizontal gene transfer). As microbes show considerable diversity and the same strain may have different characteristics in different environments, results of risk analysis cannot be extrapolated to different strains or habitats.

Given the complexity of risk assessment for GEMs, two approaches have been advocated: a **biology-centric approach** that focuses on the host organism and the introduced gene, and an **ecology-centric approach** that seeks to model the ecological consequences of environmental release in a simulated environment, the microcosm. The rationale for the biology centric approach is that **microbes in nature already exchange genetic material** across species, and that **genetic engineering done in the laboratory is ecologically insignificant compared to what already occurs in nature**. Therefore, the risk assessment should only look at the **effect of the phenotype of the GEM in the receiving environment**. What is assessed is the specific properties of the microorganisms, for instance, the population genetics, epidemiology, and pathogenesis. For an example of risk assessment conducted for a GM *Escherichia coli* bacterial biosensor, see Kesselman and Ojeda (2014). The **second approach conducts the risk assessment in microcosms**

BOX 10.1 "ICE-MINUS" *PSEUDOMONAS SYRINGAE*

Rime ice on plant leaves [*By Emmanuel Boutet, via Wikimedia Commons*]

In the 1970s, scientists at the US Department of Agriculture found that plants infected with a strain of a bacterium *Pseudomonas syringae* were more vulnerable to frost damage than healthy plants. Plant pathologist Stephen Lindow of University of California, Berkeley, discovered that a protein on the surface of the bacterial cell wall, called the "ice nucleation active" (Ina) protein, facilitated ice formation by acting as nucleating centers for ice crystals. By mutating the gene responsible for the protein, it was possible to create "ice-minus" strains of the bacterium which when sprayed on crop plants conferred frost resistance by preventing ice crystallization on the plant surface. Ice-minus strains could also be produced by rDNA technology, and the strain was developed as a potential commercial product called "Frostban" by a biotech company, Advanced Genetic Sciences (Baskin, 1987). Field testing of this GMM for market approval however proved to be controversial. The Environmental Protection Agency (EPA) classified the ice-minus bacteria as a pesticide as the "ice-plus" unmodified *Pseudomonas syringae* was considered a "pest," and agents to displace or mitigate its effects would be considered pesticides (Miller, 2010). The company applied to the US government for permission to conduct field trials in fenced off fields of strawberry and potato in 1983, but the tests were delayed by 4 years. The delay was partly due to legal challenges by environmental activists foremost of whom was Jeremy Rifkin of the Foundation on Economic Trends. Rifkin sued the National Institutes of Health (NIH) alleging that Environmental Impact Assessment had not been conducted, and the effect of the bacteria on global weather patterns and ecosystems had not been considered. The field trials of the ice-minus bacteria proved to be safe and effective in preventing frost damage but were subjected to an extraordinarily long review by EPA and NIH. The difficulties in regulatory approvals and the anticipated downstream costs of pesticide registration discouraged further research, and Frostban was never commercialized.

References
Baskin, Y. (1987). *Testing the future.* The Alicia Patterson Foundation. Retrieved from http://aliciapatterson.org/stories/testing-future.
Miller, H. I. (2010). Feds freeze out frost fix. Retrieved from http://www.forbes.com/2010/01/29/frost-agriculture-epa-science-opinions-contributors-henry-i-miller.html.

or greenhouses under contained conditions and therefore **poses no risk to the environment**. It however suffers from the **difficulty of replicating field conditions in the simulated environment** in order to get a meaningful estimate of the actual risk in the receiving environment. What has been recommended is a case-by-case quantitative risk assessment for deliberate release and a scientific management of the risk.

KEY TAKEAWAYS

Risk assessment for GMMs or GEMs takes two approaches:

- Biology centric approach—focuses on the host microbe and the introduced gene (the specific properties of the GMM)
- Ecology-centric approach—focuses on modeling the ecological consequences of environmental release in a simulated environment (microcosm)

10.6 RISK ASSESSMENT FOR GENETICALLY MODIFIED CROPS

GM crops warrant special consideration and risk assessment prior to release into farmer's field. Reproduced below are the issues identified in Directive 2001/18/EC of the European Parliament and of the Council (2001) that are to be addressed for the cultivation of GM crops:

1. *"Likelihood of the GMO to become persistent and invasive in natural habitats under the conditions of the proposed release(s).*
2. *Any selective advantage or disadvantage conferred to the GMO and the likelihood of this becoming realized under the conditions of the proposed release(s).*
3. *Potential for gene transfer to other species under conditions of the proposed release of the GMO and any selective advantage or disadvantage conferred to those species.*
4. *Potential immediate and/or delayed environmental impact of the direct and indirect interactions between the GMO and target organisms (if applicable).*
5. *Potential immediate and/or delayed environmental impact of the direct and indirect interactions between the GMO with nontarget organisms, including impact on population levels of competitors, prey, hosts, symbionts, predators, parasites, and pathogens.*
6. *Possible immediate and/or delayed effects on human health resulting from potential direct and indirect interactions of the GMO and persons working with, coming into contact with, or in the vicinity of the GMO release(s).*
7. *Possible immediate and/or delayed effects on animal health and consequences for the feed/food chain resulting from consumption of the GMO and any product derived from it, if it is intended to be used as animal feed.*
8. *Possible immediate and/or delayed effects on biogeochemical processes resulting from potential direct and indirect interactions of the GMO and target and nontarget organisms in the vicinity of the GMO release(s).*
9. *Possible immediate and/or delayed, direct, and indirect environmental impacts of the specific techniques used for the management of the GMO where these are different from those used for non-GMOs".*

Risk assessment for GM crops is usually done by the **six-step process outlined earlier** (see Fig. 10.1) and the **risks are characterized by comparison between the GM crops with that of the non-GM parental line(s)** in an application of the principle of *substantial equivalence* (i.e., no difference exists between the GM and non-GM crop other than that due to the introduced gene, e.g., herbicide tolerance or insect resistance). *Post-market environmental monitoring* of GM crops is by *general surveillance* to identify adverse effects on human health or the environment not anticipated by the environment-risk assessment, and by *case-specific monitoring* to confirm the validity of assumptions made during the environment-risk assessment. The various steps are summarized in Fig. 10.3.

10.6.1 STACKED EVENTS

GMOs containing multiple GM events generated by crossing existing GMOs are referred to as "stacked events." When individual events have been separately assessed and authorized for release, only under exceptional conditions do regulatory bodies recommend the risk assessment to be conducted for stacked events. This is because it is generally understood that combining characters by crossing in plants does not pose a risk and has been practiced for centuries.

10.6.2 PEST-RISK ANALYSIS FOR QUARANTINE PESTS INCLUDING ANALYSIS OF ENVIRONMENTAL RISKS AND LIVING MODIFIED ORGANISMS

Hosted by the United Nations Food and Agriculture Organization is the **International Plant Protection Convention** (IPPC, https://www.ippc.int/en/) which is an international treaty (signed in

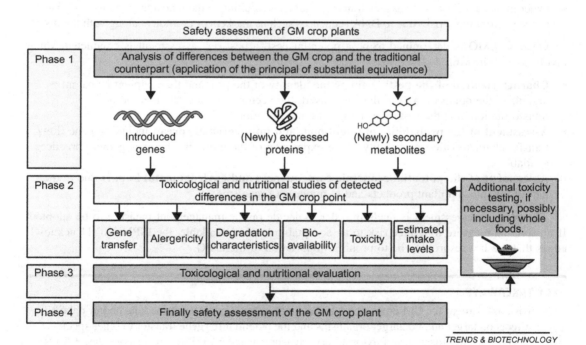

TRENDS & BIOTECHNOLOGY

FIGURE 10.3

Safety assessment of GM crop plants.

*Source: Reprinted by permission from Elsevier: [Trends in Biotechnology] Kok, E.J. & Kuiper, H.A. (2003) Comparative safety assessment for biotech crops. Trends Biotechnol. **21 (10)**, 439–444. doi:10.1016/j.tibtech2003.08.003), 1 (2003).*

1951) with currently 182 parties (as of September 2015) that sets phytosanitary standards for plants. The IPPC is the only international standard setting body for plant health recognized by the World Trade Organization's (WTO) Agreement on the *Application of Sanitary and Phytosanitary Measures* **(the SPS Agreement)**. In 2004, the governing body of the IPPC, the Commission on Phytosanitary Measures endorsed a supplement integrated standard *ISPM No. 11: Pest Risk Analysis for Quarantine Pests including Analysis of Environmental Risks and Living Modified Organisms (LMOs)*, which include guidance on evaluation of potential phytosanitary risks to plants and plant products due to LMOs. The standard does not apply only to GM plants, but also GM insects, fungi, and bacteria that may produce direct or indirect harm to plants. **In this risk assessment, LMOs are considered as phytosanitary risk/quarantine pest until decided otherwise.** The phytosanitary risks are associated with the following:

- Changes in adaptive characteristics that encourage spread of the LMO, such as, host range, pest resistance or pesticide resistance, tolerance to adverse environments, or growth rate and vigor
- Adverse effect on gene flow due to crossing species barriers, transfer of resistance genes, or increased pathogenicity
- Adverse effects on nontarget organisms due to changes in host range of the LMO, capacity to vector other pests, or effects on other organisms such as soil microflora
- Genotypic or phenotypic instability
- Other injurious effects such as new traits in organisms resulting in them posing phytosanitary risks, or the phytosanitary risks due to DNA sequences such as markers or promoters along with the insert.

Once an LMO is determined to be a potential pest, a pest-risk assessment is conducted which involves the following:

- **Characterization of the pest**: Defining the identity of the pest and the recipient or parent organism, the donor organism, the gene insert, the vector construct, the method of transformation, and the nature of the genetic modification
- **Assessment of the probability of intentional and unintentional spread** including gene flow/transfer characteristics, probability of the expression of the trait, and the management practices available
- **Assessment of the potential economic consequences and environmental impact** due to the harm to plants and plant products and to nontarget organisms.

The pest-risk assessment is made in order to decide on the management measure to be adopted. If no suitable measure to reduce risk to an acceptable level is available, the ISPM No. 11 acknowledges that the last resort may be to prohibit importation of the LMO.

KEY TAKEAWAYS
- Risk assessment for GM crops needs to take into account possible immediate and/or delayed effects on target and nontarget organisms and the potential for gene transfer to other species.
- The IPPC Commission on Phytosanitary Measures standard ISPM No. 11 considers LMOs as a phytosanitary risk/quarantine pest until decided otherwise. Management measures are adopted to reduce risks.

10.7 RISK ASSESSMENT FOR TRANSGENIC ANIMALS

Transgenic (or genetically engineered, GE) animals have been created for a number of diverse applications such as ornamental pets, for pharming (expression of pharmaceutically important substances) and for food (see Chapter 4: Recombinant DNA Technology and Genetically Modified Organisms). Although many ornamental pets such as the glofish and hypoallergenic dogs and cats have found their way into the market without apparent regulatory sanction, most countries regulate the release into the environment of animals for food and pharming. Concerns regarding environmental release of transgenic animals are essentially similar to those for GM crops and **risk assessment follows the six-step process**. Crucial to decisions of release of GE animals is an **estimation of the effect on other related organisms in the receiving environment**. Uncertainties with regard to this can delay the approval process for environmental release as in the case of transgenic salmon created by AquaBounty (see Box 4.1) and transgenic mosquitoes for control of diseases transmitted by *Aedes* mosquitoes (see Box 4.2). For an account of risk analysis and regulation of GE animals in the United States and an example of how it is done, see Box 10.2.

10.7.1 WORLD ORGANIZATION FOR ANIMAL HEALTH

The World Organization for Animal Health (*Office International des Epizooties*, OIE, www.oie.int) is an intergovernmental organization created in 1924 addressing a need to fight diseases at a global level and has a total of 180 member states (as of 2014). It is recognized as the reference organization responsible for setting standards related to animal health, and in trade in animals and animal products, by the SPS Agreement of the WTO. In 2005, OIE members adopted the *Resolution No. XXVIII: Applications of Genetic Engineering for Livestock and Biotechnology*, which requested the constitution of an **Ad Hoc Group on Biotechnology**. At a subsequent meeting in 2008, the **Ad Hoc Group on Vaccines** and the **Ad Hoc Group on Diagnostic Tests** have been

---

BOX 10.2 REGULATION OF GE ANIMALS IN THE UNITED STATES

In the United States, the Food and Drug Administration (FDA) regulates genetically engineered animals under the new animal drug provisions [Investigational New Animal Drug (INAD), or a New Animal Drug Application (NADA)] of the Federal Food, Drug and Cosmetic Act [FD&C Act, 21 USC321 et seq.] and the National Environmental Policy Act (NEPA). The FDA defines "drug" as *"articles intended for use in diagnosis, cure, mitigation, treatment or prevention of disease"* in man or animals, and/or articles *"intended to affect the structure or any function of the body of man or other animals"* [FD&C Act, sec.201(g)(1)]. New human drugs are administered by the Center for Drug Evaluation and Research (CDER) or the Centre for Biological Evaluation and Research (CBER), whereas new animal drugs are administered by the Centre for Veterinary Medicine (CVM).

The FDA defines GE animals as *"those animals modified by rDNA techniques, including the entire lineage of animals that contain the modification"* (CVM, 2009). The CVM guidance for GE animals, GFI #187 (CVM, 2009) mandates that all GE animals are subject to premarket approval requirements. In order to assess transgenic animals and their edible products intended for marketing, the FDA has developed a hierarchical risk-based approach (Veterinary Medicine Advisory Committee Briefing packet, 2010). The approach relies on the cumulative weight of evidence provided by all steps of the review rather than a single critical study. The different steps in the weight of evidence evaluation of GE animals include:

(Continued)

BOX 10.2 (CONTINUED)

Step 1: Product definition: includes a description of the animal such as the common name, genus species, ploidy, and zygosity of the GE animal; the rDNA construct and the number of copies of the same; the flanking genomic sequences.

Step 2: Molecular characterization of the construct: includes information for identifying and characterization of the rDNA construct; purpose of the modification; details of the assembly.

Step 3: Molecular characterization of the GE animal: includes information on the event that identifies and characterizes the GE animal, method of transformation of the initial GE animal, and nature of breeding strategy used to produce the lineage progenitor (the animal from which animals to be commercialized are derived); details of genomic location of the insert, number of copies of the rDNA construct at insertion site.

Step 4: Phenotypic characterization of the GE animal: includes an evaluation of the expression of the introduced trait. Examines whether the rDNA construct or its expression causes any direct effects (such as toxicity to the resulting GE animal) or indirect effects (such as affecting the expression of other proteins). This involves compiling data on the health and physiological status of the GE animals.

Step 5: Durability—genotypic and phenotypic plan: involves collecting information on whether the specific event defining the GE animal is stably inherited and the phenotype is consistent and in keeping with the predicted expression (durable). The durability plan is linked to the postapproval reporting requirements.

Step 6: Food/feed/environmental safety:

a. *Food/feed safety*: includes a comparison of whether consuming edible products from the GE animal pose any risk to humans or animals compared to that from nontransgenic animal. Both direct toxicity effects (including allergenicity associated with the expression product of the construct) and indirect effects (such as unintended food/feed consumption hazards).

b. *Environmental safety:* includes an environmental assessment (EA) report mandated by the National Environmental Policy Act (NEPA) and the FDA environmental impact regulation in 21CFR 25 for a new animal drug application (NADA). The EA requires details of the use and disposal of the GE animal including the types and extent of physical and biological containment that will be implemented, potential escape or release and spread of the GE animal, and the potentially accessible ecosystems. For approval, the EA report should indicate a **Finding of No Significant Impact (FONSI)**. An Environmental Impact Statement would have to be prepared in cases where the EA results in a finding that a significant environmental impact may result.

Step 7: Claim validation: includes substantial evidence to be provided by the sponsor to validate the claim being made regarding the GE animal.

FDA approval hinges on a team-based assessment of the application by an interdisciplinary team of subject experts. The decision-making process is transparent with the findings of safety and effectiveness review being made available to the public for comments. A statement of approval is issued by the FDA only after consideration of both the committee recommendation and the public comments.

Example: **Environmental Assessment (EA) for Bc6 rDNA construct in GTC 155-92 goats expressing recombinant human antithrombin III (rhAT or ATRYN)**

Applicant: GTC Biotherapeutics, Inc., NADA 141-294

The 22-page EA document was presented for public display in January 2009 by the Centre for Veterinary Medicine of the US FDA (Centre for Veterinary Medicine, 2009). The first introductory section gives the background and historical antecedents of animals as sources of human medicines and the use of rDNA-based technologies for the production of medicines. It also describes domestic and feral goats, and recombinant human antithrombin (rhAT or ATryn), and its uses. The second section establishes that the document was prepared in accordance with Guidance 187 in response to the request for NADA approval submitted by GTC biotherapeutics. The third section gives the details of the risk-based review leading to EA. The product is defined as a specific hemizygous diploid line of domestic goat (*Capra aegagrus hircus*) containing 5 copies of the Bc6 rDNA construct located at the GTC155-92 site, expressing the human gene for antithrombin, used for treatment of human disease, in the mammary gland of goats derived from lineage progenitor 155-92. The section also contains a diagram giving details of the construct and the characterization of the molecular construct in the GE animal lineage. In evaluating

(Continued)

BOX 10.2 (CONTINUED)

the consequences of the insertion of the rDNA sequence, the CVM reviewed the quality of the sequencing, number of insertion sites, the site itself and the possible disruption of other genes, and analysis of open reading frames within and around the insertion site, and came to the conclusion that the insertion site was well characterized in the lineage progenitor and subsequent generations. The review of submitted data did not identify any specific hazards that are intrinsic to the insertion of the Bc6 rDNA construct into the GE goats and did not find sequences arising from the insertion with a potential to pose hazards to animals, humans, or the environment. The phenotypic characterization of the GE animal was reviewed and the CVM concluded that there were no apparent differences in health, mastitis, nutrition, and reproductive status of the GE goats when compared to the non-GE goats in the same farm; hence, the insertion of the rDNA did not pose an increased risk. An analysis of the husbandry and containment of the GE goats documented the following:

1. *GE animal production facilities and practices*: The GE goat production herd is housed in a USDA registered farm owned by GTC in central Massachusetts, and goats on the farm are certified scrapie free by APHIS veterinary inspectors, and all activities are overseen by GTC's Institutional and Animal Care and Use Committee.

2. *GTC's Massachusetts farm, goat housing, and the surrounding environment*: GTC's acquisition of the GTC farm site in Massachusetts was governed by criteria designed to minimize risk of disease spread and to ensure containment: No evidence of occupation of bovine or sheep or goats for 5 years prior to purchase (to reduce risk of environmental pathogens and scrapie); no significant environmental risks on or close to the property or activity that could pose risk to the herd; suitability of the terrain for agricultural processes and availability of water that meets National Primary Drinking Water Standards. Animal housing built on the property consisted of state-of-the-art barns with internal penning, conforming to animal care and welfare regulations for animal spaces. Passive ventilation is provided, and goats are allowed access to outdoor paddocks via doors in each pen. No free-range pasturing is allowed, and on-site goats are contained by duplicate barrier systems. Access to site is highly restricted; the farm is surrounded on all sides by wooded, semirural areas.

3. *Animal identification, segregation, and husbandry*: GE animals are assigned a unique identification number at birth (recorded in a master list) which is associated with the animal by a permanent ear tattoo applied within 24 hours of birth, a subcutaneous transponder, and a physical tag attached to a neck chain or velcro collar. Monthly herd-wide inventory of animals is conducted to identify missing or illegible tags. Goats are segregated before sexual maturity into age and size cohorts.

4. *Animal health and biosecurity*: Strict adherence to currently accepted guidelines on animal health and welfare ensures animal health. Biosecurity is ensured by monitoring and restricting where necessary the incoming materials and flow of personnel/visitors.

5. *Physical security and animal containment*: A double barrier between all the 155-92 GE goats in the production herd and the outside environs serves as physical security and animal containment. The farm is surrounded by a heavy 6-foot high, chain-link perimeter fence with gated access. The fence extends 18 in. below ground to prevent animals from moving in or out. The farm is under video surveillance.

6. *Disposal of waste products and carcasses*: Farm wastes including liquids, solids, and by-products from processing are collected and disposed of in accordance with local, state, and federal regulations. During the CVM site visit, it was confirmed that no compost, manure, or GE animal carcasses were applied to the land or buried on the farm. Instead the solid and liquid waste generated on the farm is disposed off-site at a sewage treatment plant.

Section 4 of the report has an analysis and characterization of potential environmental impacts. The risk-related questions considered were as follows:

- *Risk associated with the GE goats while under confinement:*
  - Risk of gene flow via mobilization of the Bc6 rDNA construct—considered to be minimal as the construct was introduced by microinjection and did not involve viral vectors

(Continued)

BOX 10.2 (CONTINUED)

- Risk of direct toxicity resulting from increased environmental concentrations of rhAT—the product of the gene construct was similar to the naturally occurring forms of this protein. It presents no intrinsic hazard and is expected to rapidly degrade in the environment. It is expressed only in milk of lactating goats and no other tissue; hence, change in concentration in the environment is not expected
- Risk of disease spread from confined housing of 155-92 goats—possibility of disease transmission (and gene flow) to and from the GE goats is low due to the containment and biosecurity measures
- Risks that may be associated with the disposal of GE animal wastes or carcasses—highly controlled processes for waste disposal and incineration of carcasses make it unlikely for any hazard due to waste disposal
- *Likelihood that 155-92 goats will escape from confinement*: Escape of goats is unlikely due to two independent physical barriers, 24-hour security, daily staff checks, and video surveillance. The GE goats have identification systems which facilitate identification and recapture in the unlikely event of escape
- *Likelihood of harm in the event that GE goats escape from confinement*: Risk-related questions to evaluate potential environmental harms in the event of escape of GE goats include the following:
  - Likelihood of survival, reproduction, and establishment of 155-92 goats in an area outside the GTC farm
  - Likelihood of dispersion to new habitats
  - Likelihood of survival, reproduction, and establishment in the new habitats
  - Direct and indirect effects that may result in the receiving habitats
  These were all estimated to be very low as two or more animals would have to escape from multiple enclosures (as sexes are segregated) and evade recovery. The containment provided and low probability of escape, survival, and interbreeding with domestic or feral goats in the area make the likelihood of spread of the Bc6 rDNA negligible.
- *Risk analysis for the Pennsylvania GE goat facility:* GTC has a second facility for goat-holding with similar physical containment as the farm in Massachusetts; hence, the risk is no greater than that for the farm in Massachusetts.
- *Risk analysis for animal transport*: Occasionally, 8—12 goats would have to be transferred from the Pennsylvania farm to the one in Massachusetts. The transfer is done by a truck with at least two accompanying persons. The likelihood of escape of the GE goats into the environment during transit is estimated to be very low, and the chance of escapees evading identification and recapture, negligible.

The risk analysis therefore concluded that *the 155-92 GE goat herds at the GTC farm in Massachusetts and the holding facility in Pennsylvania are unlikely to result in significant effects to the quality of human environment. The Bc6 rDNA construct is not likely to mobilize and spread to other organism, and the gene product does not pose an intrinsic hazard; therefore, 155-92 GE goats in confinement are not believed to present any significant risk to the environment.*

On February 6, 2009, the FDA-approved ATryn for treatment of patients with hereditary antithrombin deficiency, and the CVM approved the goats that are used to manufacture ATryn. The drug had been approved earlier in 2006, for use in the European Union countries by the European Medicines Agency.

References

Centre for Veterinary Medicine (2009, January 29). *Environmental Assessment for Bc6 rDNA construct in GTC 155-92 goats expressing recombinant human antithrombin III (rhAT or ATRYN)*. Retrieved from http://www.fda.gov/downloads/AnimalVeterinary/DevelopmentApprovalProcess/GeneticEngineering/GeneticallyEngineeredAnimals/UCM163814.pdf.

CVM (2009). Guidance for Industry Regulation of Genetically Engineered Animals Containing Heritable Recombinant DNA Constructs (GFI #187). Retrieved from http://www.fda.gov/downloads/AnimalVeterinary/GuidanceComplianceEnforcement/GuidanceforIndustry/UCM113903.pdf.

Veterinary Medicine Advisory Committee Briefing packet (2010, September 20). *AquAdvantage Salmon*. Food and Drug Administration Centre for Veterinary Medicine. [Archived document] Retrieved from http://www.fda.gov/downloads/AdvisoryCommittees/CommitteesMeetingMaterials/VeterinaryMedicineAdvisoryCommittee/UCM224762.pdf.

Further Reading

FDA (2015). *An overview of Atlantic salmon, its natural history, aquaculture, and genetic engineering*. Retrieved from http://www.fda.gov/AdvisoryCommittees/CommitteesMeetingMaterials/VeterinaryMedicineAdvisoryCommittee/ucm222635.htm.

FDA (2016). *Oxitec mosquito*. Retrieved from http://www.fda.gov/AnimalVeterinary/DevelopmentApprovalProcess/GeneticEngineering/GeneticallyEngineeredAnimals/ucm446529.htm.

established. The major task of these groups is to develop standards, recommendations, and guidelines in order to enable countries to harmonize technical standards for regulation of biotechnology-derived animal health products and GM production animals.

10.8 SAFETY ASSESSMENT OF FOODS DERIVED FROM GENETICALLY MODIFIED ORGANISMS

The safety assessment of foods derived from GM plants and animals is based on guidelines devised by several international organizations, primarily the Codex Alimentarius Commission of the Food and Agriculture Organization (http://www.fao.org/fao-who-codexalimentarius/en/).

10.8.1 CODEX ALIMENTARIUS COMMISSION

Created in 1963 to develop food standards, guidelines, and codes of practice under the Joint Food and Agriculture Organization/World Health Organization Food Standards Programme, Codex Alimentarius Commission has currently (as on October 16, 2015) 187 members.

In 2003, the Task Force on Food Derived from Biotechnology of the Codex Alimentarius adopted the following:

- *Principles for the Risk Analysis of Foods Derived from Modern Biotechnology (amended in 2008, 2011),*
- *Guideline for the Conduct of Food-Safety Assessment of Foods Derived from Recombinant-DNA Plants (amended in 2008),*
- *Guideline for the Conduct of Food-Safety Assessment of Foods Produced using Recombinant-DNA microorganisms (amended in 2008),* and
- *Guideline for the Conduct of Food-Safety Assessment of Foods Derived from Recombinant-DNA Animals (amended in 2008).*

The approach for risk analysis is based on the principle that the safety of foods derived from new plant or animal varieties including those derived from recombinant-DNA plants and animals is assessed relative to the conventional counterpart which may exist, having a history of safe use. Both intended and unintended effects (which may be "predictable" or "unexpected") are taken into account. The guideline introduces the concept of *substantial equivalence* as a key step in the safety assessment process in order to overcome the difficulties in applying traditional toxicological and risk-assessment procedures to whole foods. The guideline emphasizes that the **concept of substantial equivalence is not a safety assessment in itself, but serves to identify similarities and differences between the food derived from a GMO and the conventional counterpart.** The following description of the recommended framework of food-safety assessment is paraphrased from the guideline available at www.fao.org/input/download/standards/10021/CXG_045e.pdf.

1. *Description of the recombinant-DNA plant (animal)*: The description should include the identity of the crop (animal species), the transformation event, and the purpose for which the modification was done.

2. *Description of the host plant (animal) and its use as food*: This includes the scientific and common names of the organism, taxonomic classification, a history of its use and development, an identification of its genotype and phenotype including any known toxicity, allergenicity, or adverse impact on human health, and its history of safe use as food.

3. *Description of the donor organism(s)*: This includes its scientific and common name, taxonomic classification, information with regard to food safety including antinutrients, allergens, and toxic substances, and for microorganisms, information on pathogenicity and relationship to known pathogens.

4. *Description of the genetic modification(s)*: This should include details of the gene construct used, the description of the vector and the transformation method. All genetic elements including marker genes, regulatory sequences, the location and orientation of the final construct should be described.

5. *Characterization of the genetic modification(s)*: A comprehensive characterization of the genetic modification with respect to the description of the DNA insert to the molecular and biochemical information of the expression of the insert is to be carried out. Also to be determined are changes if any in gene expression of host genes due to the process of transformation, and if any new fusion proteins are expressed.

6. *Safety assessment*:
   a. *Expressed substances (nonnucleic acid substances)*: The chemical nature of the newly expressed substances and whether new metabolites are toxic is to be determined. Proteins resulting from the inserted gene should be assessed for potential allergenicity.
   b. *Compositional analyses of key components*: A comparison is to be made between the concentration of key components in the transgenic organism with that of the conventional counterpart and the statistical significance of any changes are to be assessed with respect to naturally occurring variations for the parameter.
   c. *Evaluation of metabolites*: The genetic modification could in some cases result in changes in the level of various metabolites. Potential accumulation of metabolites could have adverse effects on human health. Conventional methods of establishing safety (as of chemicals in food) of such alterations in nutrient profile are to be made.
   d. *Food processing*: Heat stability of endogenous toxicants or bioavailability of important nutrients after processing, including home preparations, is to be established.
   e. *Nutritional modification*: Although an assessment of this is made under "compositional analyses of key components," in cases where the GMO has been created to intentionally alter nutritional quality or functionality, additional nutritional assessment is to be conducted. As there are geographical and cultural variations in the manner in which food is consumed, nutritional changes may have a greater impact in some cultural populations than others; hence, the assessment has to be made in the appropriate background and for specific use.

7. *Other considerations*:
   Use of antibiotic resistance marker (ARM) genes: Although the transfer of genes from food products to gut microbes or human cells is considered to be rare, as it is not possible to entirely rule out the occurrence of horizontal gene transfer, the use of alternate transformation technologies that do not rely on ARM genes is advocated in future development of transgenics for food. ARM genes that encode resistance to clinically used antibodies should not be present in foods.

Foods from GM crops have been commercially produced since 1994 when Calgene first marketed the **FlavrSavr** delayed ripening tomatoes in the United States, followed by Bt corn, potatoes, and soybean, and virus resistant squash by various companies such as Monsanto, Syngenta, and Ciba-Geigy. Spain began cultivating Bt corn in 1998 and for years has been the only EU country growing GM crops in any sizeable scale. **The GM food industry operates on the basic tenet that no credible evidence exists that suggests that GM foods damage the environment or is harmful to human or animal health.**

KEY TAKEAWAYS
- Guidelines for assessment of foods derived from GMOs developed by the Codex Alimentarius Commission of the Food and Agriculture Organization is based on the concept of *substantial equivalence*
- Foods from GMOs are assessed relative to the conventional counterpart which may exist, having a history of safe use

10.8.2 POSTRELEASE MONITORING

Postrelease/postcommercialization/postmarket monitoring that continues to observe and record health and environmental effects of GMOs subsequent to its release is essential for ensuring the safety of future generations. Such observations should logically start *before* the release of the GMO so as to detect changes in the receiving environment subsequent to the release. Practical problems associated with this activity include identifying who should implement the monitoring and the cost involved. The EU Directive 2001/18/EC (2001) holds the producer responsible for postmarket monitoring, although due to apprehensions that it may not be carried out independently and transparently, it has been suggested that monitoring should be the responsibility of national governments, despite the cost involved (Bardocz and Pusztai, 2007).

10.9 SAFETY ASSESSMENT IN CLINICAL TRIALS

Safety issues in testing of novel investigational drugs are of special concern in the area of biotechnology due to the rapid pace of development of biologics. In order to harmonize the scientific and technical aspects of pharmaceutical product registration, regulatory authorities and experts from the pharmaceutical industry of Europe, Japan, and the United States have formed the *International Conference on Harmonization (ICH) of Technical Requirements for Registration of Pharmaceuticals for Human Use* (http://www.ich.org). The primary objective of the ICH is to reduce or obviate the need to duplicate the testing required during the research and development of new medicines and to facilitate global development and availability of new medicines. **ICH guidelines-E6 Guidance** (ICH, 1996) (the ICH Good Clinical Practice guidelines, ICH GCP) serve to maintain quality, safety, efficacy, and regulatory obligations in clinical trials to protect public health and have been adopted as law in several countries. The guidelines were originally developed for the trials necessary for regulatory approval of drugs (new chemical entities) but now include biologics that undergo clinical testing. In keeping with the ICH GCP guidelines, a clinical investigation requires the development of a **clinical protocol** which describes how a clinical trial will be conducted and ensures the safety of trial subjects and integrity of the data collected. In addition to details of the background, rationale, methodology, statistical considerations, and organization of the clinical trial, **the protocol specifies the assessment of safety and the reporting of adverse events**. Four levels of **protocol risk** are recognized (TraCS Institute, 2008), number one representing no more than minimal risk, and numbers two, three, and four, representing greater than minimal risk:

1. *No greater than minimal risk*: The probability and magnitude of harm or discomfort is no more than in ordinary life or a routine medical or psychological examination (e.g., noninterventional observations of behavior, or physical examination, or blood draw). *Requires minimal intensity monitoring.*
2. *Minor increase over minimal risk*: A medium to high probability of the occurrence of a low-severity event that is completely reversible (e.g., headache from a lumbar puncture), or the likelihood of a serious harm occurring is low. *Requires low-intensity monitoring.*
3. *Moderate risk*: Medium to high probability of a moderate severity event occurring, but adequate surveillance and protection to identify adverse reactions promptly and to minimize their effects is available (e.g., pneumonia from a bronchoscopy). *Requires moderate intensity monitoring.*
4. *High risk*: Increased probability for generating serious adverse events, prolonged or permanent (e.g., a new drug for which there is limited or no available safety data in humans). *Requires high-intensity monitoring.*

KEY TAKEAWAYS

- The *International Conference on Harmonization (ICH) of Technical Requirements for Registration of Pharmaceuticals for Human Use* serves to harmonize the scientific and technical aspects of pharmaceutical product registration.
- **ICH guidelines-E6 Guidance** (the ICH GCP guidelines, ICH GCP) serve to maintain quality, safety, efficacy, and regulatory obligations in clinical trials to protect public health and have been adopted as law in several countries.
- Four levels of protocol risk are recognized, number one representing no more than minimal risk, and numbers two, three, and four, representing greater than minimal risk.

10.10 PRECAUTIONARY PRINCIPLE IN GMO REGULATION

Balancing the interests of different stakeholders such as the biotech industry and organizations and the public, while ensuring safety and protection of the environment, human, animal, and plant health, is a major task facing regulatory authorities. **Under conditions of scientific uncertainty, the Precautionary Principle has been advocated as a normative principle for making practical decisions** (Myhr, 2007). Although the Precautionary Principle is not defined in any treaty, it has nevertheless become an integral part of several international agreements for regulation of GMOs such as the Cartagena Protocol on Biosafety (see Chapter 8: Biodiversity and Sharing of Biological Resources). It is also integral to the EU Directive 2001/18/EC for deliberate release of GMOs into the environment. Guidelines regarding the implementation of the Precautionary Principle communicated by the Commission of the European Communities (2000) states that *"where actions were deemed necessary, measures based on precautionary principle should be inter alia*:

- *Proportional to the chosen level of protection*
- *Nondiscriminatory in their application*
- *Consistent with similar measure taken*
- *Based on an examination of the potential benefits and costs of action, or lack of action*
- *Subject to review in light of new scientific data*
- *Capable of assigning responsibility for producing the scientific evidence necessary for a more comprehensive risk assessment."*

10.11 SUMMARY

The three aspects of risk analysis are: risk assessment, risk management, and risk communication. GMOs and their products are perceived to be different from their nonmodified counterparts. Consequently, decisions regarding their release into an environment and marketing depends on the outcome of a rigorous science-based risk assessment. The actual risk posed by the introduction of a GMO is a function of the assessed risk and risk management. Major components of the risk assessment are: the characteristics (including the newly introduced traits) of the organism, the manner in which the organism is to be used, and the characteristics of the receiving environment including other organisms that may be affected. In instances where the perceived risk is high, hazards could be mitigated by confining the GMO to suitable containment facilities. When faced with scientific uncertainty, in order to protect the environment, human, animal, and plant health, the Precautionary Principle has been advocated for decisions regarding the release of GMOs and their products. Postrelease/postmarketing monitoring is an important part of risk assessment and serves to prevent harm due to delayed or indirect effects of GMO release. Risk communication is of tantamount importance for public acceptability of GMOs and their products.

REFERENCES

Bardocz, S., & Pusztai, A. (2007). Post-commercialization testing and monitoring (or post-release monitoring) for the effects of transgenic plants. In T. Traavik, & L. C. Lim (Eds.), *Biosafety First*. Tapir Academic Publishers. Retrieved from http://genok.no/wp-content/uploads/2013/04/Chapter-32.pdf.

Commission of the European Communities (2000) *Communication from the commission on the precautionary principle*. Retrieved from http://eur-lex.europa.eu/LexUriServ/LexUriServ.do?uri = COM:2000:0001:FIN: en:PDF.

Directive 2001/18/EC of the European Parliament and of the Council (2001) Directive 2001/18/EC of the European Parliament and of the Council of 12 March 2001 on the deliberate release into the environment of genetically modified organisms and repealing Council Directive 90/220/EEC - Commission Declaration. Retrieved from http://eur-lex.europa.eu/legal-content/EN/TXT/?uri = celex%3A32001L0018.

ICH (1996) *Guideline for good clinical practice E6(R1).* International Conference on Harmonisation of Technical Requirements for Registration of Pharmaceuticals for Human Use. Retrieved from http://www.ich.org/fileadmin/Public_Web_Site/ICH_Products/Guidelines/Efficacy/E6/E6_R1_Guideline.pdf.

Kesselman, V.S.G. and Ojeda, O.D.C. (2014) Risk assessment of a genetically modified *Escherichia coli* bacteria to a biosensor for the detection of copper in water for educational and research purposes. *Risk assessment dossier of the iGEM Zamorano Project.* Retrieved from http://2014.igem.org/wiki/images/d/d0/Zamorano_Dossier_Evaluacion_Riego_Ingles.pdf.

Krimsky, S. (1996). Risk assessment of genetically engineered microorganisms: from genetic reductionism to ecological modelling. In A. Dommelen (Ed.), *Coping with deliberate release—the limits of risk assessment, International Centre for Human and Public Affairs* (pp. 15–45). Tilburg: Buenos Aires. Retrieved from http://emerald.tufts.edu/~skrimsky/PDF/Risk%20Assessment%20GEMs.PDF.

Myhr, A. I. (2007). The precautionary principle in GMO regulations. In T. Traavik, & L. C. Lim (Eds.), *Biosafety First.* Tapir Academic Publishers. Retrieved from http://genok.no/wp-content/uploads/2013/04/Chapter-29.pdf.

TraCS Institute (2008) *Protocol risk assessment and monitoring guidelines.* Clinical and Translational Research Center. Retrieved from https://tracs.unc.edu/docs/regulatory/CTRC_Protocol_Risk_Assessment_Guidelines.pdf.

UNEP (n.d.) *International technical guidelines for safety in biotechnology.* United Nations Environmental Programme, Nairobi. Retrieved from http://www.unep.org/biosafety/Documents/Techguidelines.pdf.

FURTHER READING

Advisory Committee on Releases to the Environment (ACRE) Report 3 (2013) *Towards a more effective approach to environmental risk assessment of GM crops under current EU legislation.* Retrieved from https://www.gov.uk/government/uploads/system/uploads/attachment_data/file/239893/more-effective-approach-gmo-regulation.pdf.

Wilson Center Synthetic Biology Project (n.d.) *Navigating the Regulatory Landscape: Oxitec case study.* Retrieved from http://www.synbioproject.org/site/assets/files/1388/synbio_case_study_gmo_mosquito.pdf.

LABORATORY BIOSAFETY AND GOOD LABORATORY PRACTICES

11

Risk comes from not knowing what you're doing.
-Warren Buffett, American business magnate, investor, philanthropist.

CHAPTER OUTLINE

11.1 Introduction 254
11.2 Risk Categories of Microorganisms 255
11.3 Biosafety Levels 256
    11.3.1 Physical Containment 257
    11.3.2 Biological Containment 263
11.4 Physical and Biological Containment for Research Involving Plants 263
11.5 Physical and Biological Containment for Research Involving Animals 267
11.6 Good Laboratory Practice 270
11.7 Summary 271
References 271

Paul Berg opening a jar under a protective hood.

Photo courtesy: Stanford University Archives.

An Introduction to Ethical, Safety and Intellectual Property Rights Issues in Biotechnology.
DOI: http://dx.doi.org/10.1016/B978-0-12-809231-6.00011-9

11.1 INTRODUCTION

Handling organisms including microorganisms under laboratory conditions is an essential part of biotechnology research and applications. Responsible and safe handling of microorganisms (potentially pathogenic) is necessary to ensure the health of laboratory personnel, the community, and the environment. This chapter reviews practices and protocols that have been established at the international and national level to ensure biosafety in institutions involved in biotechnology research and development.

In the early days of the development of recombinant DNA technology, the consensus in the scientific community was that due to the potential to generate new (potentially harmful) forms of organisms, mere **good microbiological techniques (GMT)** alone would not suffice to ensure safety of workers in this area of research. Consequently, in June 1976, shortly after the Asilomar Conference (see Chapter 9: Ensuring Safety in Biotechnology), the US National Institutes of Health (NIH) brought out the *NIH Guidelines for Research Involving Recombinant Nucleic Acid Molecules* (henceforth referred to as "**NIH Guidelines**"). (The guidelines have undergone several revisions, the most recent one being in April 2016.) The guidelines recognize six classes of experiments based on the risk involved in the research, which require sanction from different regulatory bodies (see Chapter 9: Ensuring Safety in Biotechnology, Box 9.2).

Recognizing that biological safety is an important international issue, the United Nations World Health Organization (WHO) in 1983 published the *Laboratory Biosafety Manual* (henceforth referred to as "**WHO manual**") establishing basic concepts and practices in safe handling of pathogenic microorganisms. **The document encouraged countries to develop national codes of practice for implementation within their geographic boundaries**. The WHO manual has since been revised twice, in 1993 and again in 2004. The third edition incorporates biosecurity concepts in addition to making specific recommendations for biosafety in handling genetically modified organisms.

Adhering to international standards and incorporating biosafety practices in national policies is important not only to protect plant, animal or human life, and health of its citizens, but also in trade in biotech products. Exporting countries need to demonstrate that the measures it applies to its exports achieve the same level of health protection as in the importing countries in order to avoid barriers in trade. Providing oversight to global rules of trade between nations is the mandate of the World Trade Organization. This organization administers the General Agreement on Tariffs and Trade (GATT) (see Chapter 13: Relevance of Intellectual Property Rights in Biotechnology). Article 20 of the GATT allows governments to regulate trade to address biosafety of their citizens provided they do not discriminate or use this clause to disguise protectionism. Countries therefore have made efforts to develop guidelines and appropriate legal frameworks for biosafety regulation. Although the developed countries (e.g., the United States) and regions (e.g., European Union) as leaders in the development of modern biotechnology started to develop these frameworks in the mid-1970s and early 1980s, the developing nations generally started the development of national biosafety systems more recently. The WHO manual has served as an important resource document and is consequently reflected in national regulatory instruments for ensuring biosafety.

11.2 RISK CATEGORIES OF MICROORGANISMS

In both the NIH guidelines and the WHO manual, **four categories of microorganisms used in laboratory work are recognized**. The basis of the classification is the **risk of infection** to laboratory workers and, in the event of escape from the laboratory, to the community. Assigning a microorganism to a risk category is dependent on an initial risk assessment made by the investigator and is based on current knowledge of the:

1. **Pathogenicity** of the organism (all microorganisms do not cause diseases),
2. **Host range** and **mode of transmission** of the organism,
3. **Local availability of effective measures to prevent a disease outbreak**, and
4. **Local availability of effective treatment**.

   The Appendix B of the NIH guidelines, *Classification of Human Etiological Agents on the Basis of Hazard*, as does Table 1 of the WHO manual, recognizes four **Risk Groups** of microorganisms:

- *Risk Group 1*: Microorganisms **unlikely to cause human or animal diseases** and thus pose little or no risk to individuals and to the community; sometimes designated as **generally regarded as safe** (GRAS) organisms (e.g., asporogenic *Bacillus subtilis* or *Bacillus licheniformis*, the K-12 strain of *Escherichia coli*)
- *Risk Group 2*: Microorganisms that are pathogenic, but **unlikely to pose a serious hazard** to laboratory workers, livestock, the community, or the environment as effective treatment and preventive measures to limit spread of infection are available. These organisms thus are of moderate risk to the individual and low risk to the community (e.g., bacterial agents— *Aeromonas hydrophila*, *E. coli*, *Klebsiella* spp., *Salmonella* spp.; fungal agents—*Penicillium marneffei*, *Blastomyces dermatitidis*; parasitic agents—*Ascaris* spp., *Trypanosoma* spp.; viruses—adenoviruses, Coronaviruses, Papilloma viruses)
- *Risk Group 3*: Microorganisms that are **pathogenic and can cause serious human or animal diseases, but are not contagious** or have effective treatment and preventive measures. These organisms pose high risk to the individual, but low risk to the community (e.g., bacterial agents—*Brucella* spp., *Francisella tularensis*, *Rickettsia* spp.; fungal agents—*Coccidiodes immitis, Histoplasma* spp.; Viruses and prions—Togaviruses, Flaviviruses such as the Japanese encephalitis virus, West Nile virus, Pox viruses, prions such as the transmissible spongiform encephalopathies, retroviruses such as human immunodeficiency virus, rhabdovirus)
- *Risk Group 4*: Microorganisms that usually **cause serious diseases in humans** and animals and can be readily transmitted either directly or indirectly from one to the next individual. **Effective treatment and preventive measures are usually unavailable**. This class of organisms thus poses a **high risk** to individuals and to the community (e.g., viral agents such as the Lassa virus, Ebola virus, Marburg virus, Herpes virus simiae, Kayasanur Forest disease, Central European encephalitis, and as yet unidentified hemorrhagic fever agents).

   The NIH guidelines recognize that this classification is **dependent on current knowledge** of pathogenicity, and with the development of better therapeutic and preventive measures, pathogens may be assigned to a lower risk category. Different countries may assign the same organism to different risk groups, possibly because the same organism is more virulent in certain parts of the world than others depending on climatic conditions and other factors. Also, any strain more virulent than the wild-type parent strain should be assigned to a higher risk group.

11.3 BIOSAFETY LEVELS

Both the NIH guidelines and the WHO manual recommend **four Biosafety levels (BLs) 1 to 4** for handling organisms corresponding to the four risk groups. Implementation of safety procedures in each level relies on:

1. **Standard practices** of GMT
2. **Physical barriers** provided by special procedures, equipment, and laboratory installations commensurate with the estimated biohazard.

Appendix G of the NIH guidelines describes four BLs of *Physical Containment* summarized in Table 11.1 for standard laboratory experiments. For large-scale (over 10 L) research or production, physical containment requirements are defined in Appendix K (see Chapter 12: Recombinant DNA Safety Considerations in Large-Scale Applications and Good Manufacturing Practice) (WHO Laboratory Biosafety Manual, 2004). The **BL assigned for specific research work** depends on the **assessed risk group of the organisms** handled as well as **professional judgment** of risk associated with the activity.

Table 11.1 Relation of Risk Groups to Biosafety Levels, Practices, and Equipment

Risk Group	Biosafety Level	Laboratory Type	Laboratory Practices	Safety Equipment
1	Basic— Biosafety Level 1	Basic teaching, research	GMT	None; open bench work
2	Basic— Biosafety Level 2	Primary health services; diagnostic services, research	GMT plus protective clothing, biohazard sign	Open bench plus BSC for potential aerosols
3	Containment— Biosafety Level 3	Special diagnostic services, research	As Level 2 plus special clothing, controlled access, directional airflow	BSC and/or other primary devices for all activities
4	Maximum containment— Biosafety Level 4	Dangerous pathogen units	As Level 3 plus airlock entry, shower exit, special waste disposal	Class III BSC, or positive pressure suits in conjunction with Class II BSCs, double-ended autoclave (through the wall), filtered air

BSC, *biological safety cabinet*; GMT, *good microbiological techniques.*
Reprinted with permission from WHO Laboratory Biosafety Manual, 2004. Third edition, retrieved from http://www.who.int/csr/resources/publications/biosafety/en/Biosafety7.pdf.

11.3.1 PHYSICAL CONTAINMENT

The first principle of physical containment is **strict adherence to good microbial practices**, hence, **all personnel** directly or indirectly working with recombinant or synthetic nucleic acids **should be trained in GMT**. Appendix G-II of the NIH guidelines describes four levels of physical containment **BL1, BL2, BL3, and BL4**, representing facilities in which experiments ranging from **low to high potential hazard** may be conducted. For each BL, the guidelines specify the following:

- Standard microbiological practices,
- Special practices,
- Containment equipment, and
- Laboratory facilities.

Table 11.2 summarizes the facility requirements at the four BLs (for details, see Box 11.1).

Table 11.2 Summary of Biosafety Level Requirements	Biosafety Level			
	1	2	3	4
Isolation[a] of laboratory	No	No	Yes	Yes
Room sealable for decontamination	No	No	Yes	Yes
Ventilation				
Inward airflow	No	Desirable	Yes	Yes
Controlled ventilating system	No	Desirable	Yes	Yes
HEPA-filtered air exhaust	No	No	Yes/No[b]	Yes
Double-door entry	No	No	Yes	Yes
Airlock	No	No	No	Yes
Airlock with shower	No	No	No	Yes
Anteroom	No	No	Yes	—
Anteroom with shower	No	No	Yes/No[c]	No
Effluent treatment	No	No	Yes/No[c]	Yes
Autoclave				
On site	No	Desirable	Yes	Yes
In laboratory room	No	No	Desirable	Yes
Double-ended	No	No	Desirable	Yes
Biological safety cabinets	No	Desirable	Yes	Yes
Personnel safety monitoring capability[d]	No	No	Desirable	Yes

[a]Environmental and functional isolation from general traffic.
[b]Dependent on location of exhaust.
[c]Dependent on agent(s) used in the laboratory.
[d]For example, window, closed-circuit television, two-way communication.
Reprinted with permission from WHO Laboratory Biosafety Manual, 2004. Third edition, retrieved from http://www.who.int/csr/resources/publications/biosafety/en/Biosafety7.pdf.

BOX 11.1 PHYSICAL CONTAINMENT FOR STANDARD LABORATORY EXPERIMENTS

Appendix G of the **NIH guidelines** identifies strict adherence to good microbiological practices as being the first principle of containment; hence, all personnel directly or indirectly associated with experiments involving recombinant or synthetic nucleic acid molecules should be trained in good microbiological techniques. The four levels of physical containment Biosafety Levels 1 through 4 as described in Appendix G are summarized below:

Biosafety Level 1:

- *Standard microbiological practices:*
  - Access to the laboratory is limited or restricted at the discretion of the Principal Investigator (PI).
  - Work surfaces are decontaminated once a day, all liquid and solid wastes are decontaminated before disposal.
  - Mouth pipetting is prohibited.
  - Eating, drinking, smoking, or storing food in the refrigerators is prohibited.
  - Procedures are performed carefully to prevent formation of aerosols.
  - Good hygiene including washing hands and wearing protective clothes is encouraged.
- *Special practices:*
  - Contaminated materials to be decontaminated at a site away from the laboratory are transported in durable, leak-proof containers with closed lids.
  - An insect and rodent control program is required.
- *Containment equipment:*
  - Generally, not required for BL1
- *Laboratory facilities:*
  - The laboratory should be designed to be easily cleaned.
  - Benchtops should be resistant to water, acid/alkali/organic solvents and should have sinks for hand-washing.

Biosafety Level 2:

- *Standard microbiological practices:*
  - As described for BL1 and
  - Experiments of lesser biohazard can be conducted concurrently in demarcated areas of the laboratory.
- *Special practices:*
  - As described for BL1 and
  - PI limits access to the laboratory and establishes policies and procedures whereby persons entering the laboratory are aware of the hazard and meet any specific entry requirements (such as immunization).
  - The hazard-warning sign with the universal biosafety symbol (Fig. 11.1) with details of the agent used, contact information of the PI, and any special requirements for entry, are to be posted on the access door.
  - Protective clothing used exclusively in the laboratory is required; gloves are to be used to prevent skin contamination with experimental organisms.
  - Only needle-locking hypodermic syringes are used, placed in puncture-proof containers after use, and decontaminated before disposal.
  - A biosafety manual is prepared and adopted for safety of personnel.
  - Baseline serum samples of all laboratory and at-risk personnel should be collected and stored in accordance with institutional policy.
- *Containment equipment:*
  - Biological safety cabinets (class I or II) (Fig. 11.2) or other appropriate personal protective devices are used.
- *Laboratory facilities:*
  - As described for BL1 and
  - An autoclave required for decontamination.

(Continued)

BOX 11.1 (CONTINUED)

FIGURE 11.1

Biohazard warning sign for laboratory doors.

Reprinted with permission from WHO Laboratory Biosafety Manual, 2004. Third edition, retrieved from http://www.who.int/csr/
resources/publications/biosafety/en/Biosafety7.pdf.

Biosafety Level 3:

- *Standard microbiological practices:*
  - As described for BL2 and.
  - Persons below 16 years of age are not permitted entry.
- *Special practices:*
  - As described for BL2 and
  - Laboratory doors are kept close when experiments are in progress.
  - Laboratory clothing that protects street clothes is to be worn in the laboratory, removed when exiting the laboratory, and decontaminated prior to laundry or disposal.
  - Molded surgical masks or respirators are worn in rooms containing experimental animals.
  - If animals housed with conventional caging system, personnel must wear protective devices that includes wrap-around gowns, head covers, gloves, shoe covers, and respirators; personnel shall shower on exit from areas where these devices are required.
  - Alternatively, laboratory animals shall be housed in partial-containment caging systems; no animals other than the experimental animals are allowed.
  - Vacuum lines are protected with high efficiency particulate air (HEPA) filters and liquid disinfectant traps.

(Continued)

BOX 11.1 (CONTINUED)

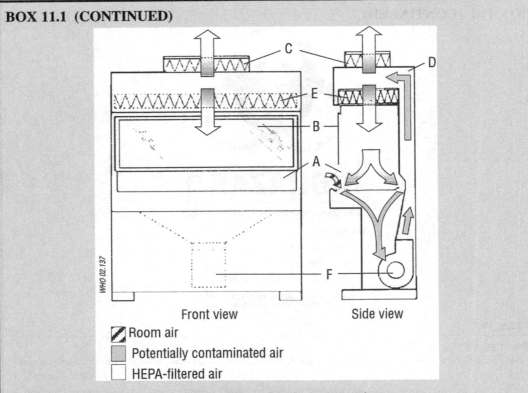

WHO 02.137

Front view   Side view

▨ Room air
▧ Potentially contaminated air
☐ HEPA-filtered air

FIGURE 11.2 Schematic representation of a Class II biological safety cabinet.

(A) Front opening; (B) sash; (C) exhaust HEPA filter; (D) rear plenum; (E) supply HEPA filter; (F) blower.

Reprinted with permission from WHO Laboratory Biosafety Manual, 2004. Third edition, retrieved from http://www.who.int/csr/resources/publications/biosafety/en/Biosafety7.pdf.

- Spills and accidents which result in potential exposure to modified organisms are immediately reported to the Biological Safety Officer, Institutional Biosafety Committee (IBSC) and to the NIH Office of Science Policy. Written records are to be maintained on appropriate medical evaluation, surveillance, and treatment provided.
- *Containment equipment:*
  - Biological safety cabinets (class I, II, or III) or other appropriate personal protective devices (such as special protective clothing, masks, gloves, respirators, centrifuge safety cups, sealed centrifuge rotors, containment cages for animals) are used.
- *Laboratory facilities:*
  - Laboratory to be separated from open areas within the building and accessed through two sets of doors; physical separation of high containment laboratory from other laboratories or activities, may be provided by a double-door clothes change room with showers, airlock, or other double-door access features.
  - Interior surfaces of walls, floors, and ceilings are water resistant for easy cleaning, should be capable of being sealed for decontaminating the area.

(Continued)

BOX 11.1 (CONTINUED)

- Access doors are self-closing.
- The HEPA-filtered exhaust air from Class I or II biological cabinets is discharged directly to the outside or through the building exhaust system.

Biosafety Level 4:

- *Standard microbiological practices:*
  - As described for BL3
- *Special practices:*
  - As described for BL3 and
  - Access to the facility is limited by means of secure locked doors; accessibility is restricted to authorized personnel and is supervised and managed by the PI, Biological Safety Officer, or person responsible for the physical security of the facility. A log of entry and exit of personnel is maintained. All personnel are advised of potential biohazards and are to comply with instructions on entry and exit procedure. Protocols for emergency situations are established.
  - Biological material to be removed in an intact state are to be sealed in a primary nonbreakable container, enclosed and sealed in a secondary nonbreakable container, and removed from the facility through a disinfectant dunk tank, fumigation chamber, or an airlock designed for the purpose.
  - Any other material to be removed from the facility are to be autoclaved or decontaminated before exiting the maximum containment laboratory.
  - Personnel enter and exit the facility only through clothing change and shower rooms; shower every time they exit the facility.
  - Street clothing is removed and kept in an outer changing room. Complete laboratory clothing (may be disposable) is provided and to be used by all personnel entering the facility. When exiting, the laboratory clothing is removed in an inner changing room before proceeding to the shower area. The clothing is decontaminated prior to laundering or disposal.
  - Supplies and material are brought into the facility through a double-door autoclave, fumigation chamber, or airlock.
- *Containment equipment:*
  - All procedures within the maximum containment facility are conducted in the Class III biological safety cabinet (Fig. 11.3), or in a Class I or II biological safety cabinet used in conjunction with a one-piece positive pressure personnel suits ventilated by a life-support system.
- *Laboratory facilities:*
  - The maximum containment facility is to be housed in a separate building or a clearly demarcated and isolated zone within a building. Access to the facility requires outer and inner change rooms separated by showers for entry and exit of personnel, and double-door autoclave, fumigation chamber or airlock for passage of materials, supplies, and equipment.
  - Internal surfaces of walls, floors, and ceilings of the facility should be water, acid, and alkali resistant; the facility should be sealable for fumigation, animal and insect proof. Drains in the floor contain traps filled with suitable chemical disinfectant and are connected directly to the liquid-waste decontamination system. Sewer and other ventilation lines contain HEPA filters.
  - Benchtops have seamless surfaces impervious to acids, alkalis, organic solvents, and moderate heat; construction of the facility should have adequate space for accessibility for cleaning.
  - Access doors are self-closing and locking.
  - An individual supply and exhaust air ventilation system that maintains pressure differentials and directional airflow ensures that airflow inwards from areas outside the facility toward areas of highest potential risk within the facility. The supply and exhaust airflow is monitored by manometers to assure inward (or zero) airflow at all times.

(Continued)

BOX 11.1 (CONTINUED)

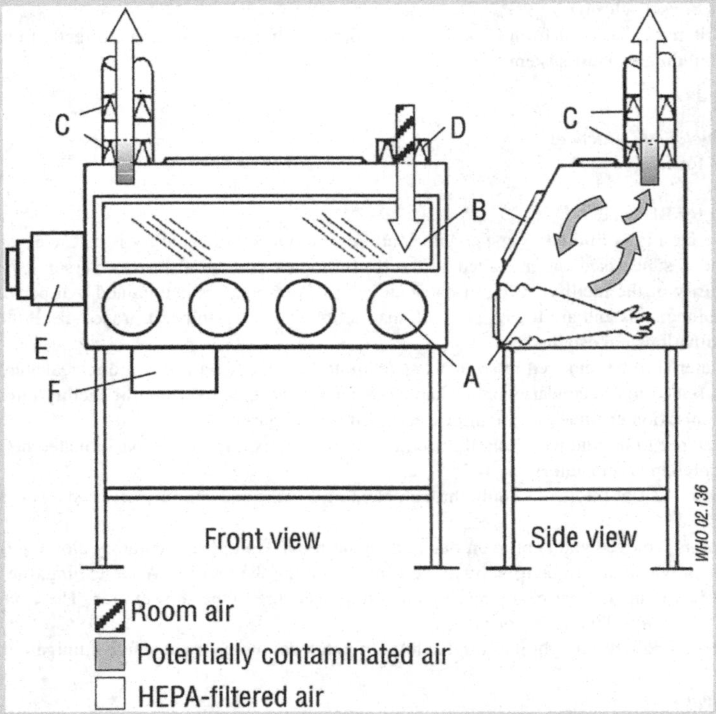

FIGURE 11.3 Schematic representation of a Class III biological safety cabinet (glove box).
(A) glove ports for arm-length gloves; (B) sash; (C) double-exhaust HEPA filters; (D) supply HEPA filter; (E) double-ended autoclave or pass-through box; (F) chemical dunk tank. Connection of the cabinet exhaust to an independent building exhaust air system is required.

Reprinted with permission from WHO Laboratory Biosafety Manual, 2004. Third edition, retrieved from http://www.who.int/csr/resources/publications/biosafety/en/Biosafety7.pdf.

- Exhaust air from the facility is filtered through HEPA filters before discharge to the outside.
- A specially designed suit area may be provided in the facility entry into which is through an airlock fitted with airtight doors. The air pressure within the suit area is maintained greater than that of adjacent areas. Personnel who enter this area shall wear a one-piece positive pressure suit ventilated by life-support system. A chemical shower is provided to decontaminate the surface of the suit before the worker exits the area.

NIH Guidelines for Research Involving Recombinant or Synthetic Nucleic Acid Molecules (NIH Guidelines) retrieved from http://osp.od.nih.gov/sites/default/files/NIH_Guidelines_0.pdf.

11.3.2 **BIOLOGICAL CONTAINMENT**

The growth and dissemination of organisms are naturally limited. For ensuring safety, biological containment takes advantage of these natural barriers such as:

1. The infectivity or **host specificity** of a vector or virus
2. Its **spread and survival** in the environment.

Appendix I of the NIH guidelines describes *Biological containment* strategies for recombinant or synthetic nucleic acid molecules. The *vector* (plasmid, organelle, or virus) for the recombinant or synthetic nucleic acid molecule and *the host* (bacterial, animal, or plant cell), in which the vector is propagated, are taken together as a *Host–Vector system* for consideration of biological containment. Selection of a Host–Vector system aims to minimize:

1. Survival of the vector in its host outside the laboratory and
2. Transmission of the vector from the propagation host to other nonlaboratory hosts.

Host–Vector 1 Systems provide moderate level of containment. The **EK1 system** has *E. coli* **K-12 (or derivatives) as the host**, and the vectors include **nonconjugative plasmids** (e.g., pSC101, Co1E1) and **variants of bacteriophage lambda**.

Host–Vector 2 Systems (EK2) provide a **high level of biological containment** with escape of the recombinant or synthetic nucleic acid molecule to other organisms under specified conditions being $< 1/10^8$.

11.4 **PHYSICAL AND BIOLOGICAL CONTAINMENT FOR RESEARCH INVOLVING PLANTS**

The BLs 1 through 4 are applicable to microorganisms, but Appendix P of the NIH guidelines specifies physical and biological containment conditions for experiments involving **recombinant or synthetic nucleic acids in plants, plant-associated microorganisms, and small animals**. The plants include, but are not limited to, mosses, liverworts, macroscopic algae, and vascular plants including terrestrial crops, forest, and ornamental species. Plant-associated microorganisms include those that have a benign or beneficial effect as also those that cause diseases, and include viroids, virusoids, viruses, bacteria, fungi, protozoans, small algae, as well as microbes being modified for association to plants. Plant-associated small animals include arthropods and nematodes, tests on which require the use of plants. **The purpose of the containment is to prevent unintentional transmission of recombinant or synthetic nucleic acid molecule containing plant genomes (nuclear or organellar DNA), or release of modified organisms associated with plants**. Appendix P-II establishes four levels referred to as BL1-Plants (P), BL2-P, BL3-P, and BL4-P, which specify the use of plant tissue culture rooms, growth chambers within laboratory facilities, or experiments performed on open benches. Appendix P-III specifies Biological Containment Practices if botanical reproductive structures are produced that can potentially be released. For further details, see Box 11.2.

BOX 11.2 PHYSICAL CONTAINMENT FOR EXPERIMENTS INVOLVING PLANTS

Appendix P of the **NIH guidelines** supersedes Appendix G (Physical Containment) when the research plants are of a size, number, or have growth requirements that preclude use of the containment conditions outlined in Appendix G. The containment principles in Appendix P are based on the premise that the organisms pose no health threat to humans or higher animals and that the purpose of the containment is to minimize the possibility of unanticipated deleterious effects on organisms and the ecosystem. The physical containment levels describe greenhouse practices and special greenhouse facilities for physical containment.

Biosafety Level 1—Plants (BL1-P):

- *Standard practices:*
  - *Greenhouse access:*
    - Limited or restricted, at the discretion of the Greenhouse director when experiments are in progress.
    - Personnel shall be required to follow standard greenhouse procedures.
  - *Records:*
    - Record shall be maintained of experiments in progress in the facility.
  - *Decontamination and inactivation:*
    - Experimental organisms shall be rendered biologically inactive by appropriate methods prior to disposal.
  - *Control of undesired species and motile macroorganisms:*
    - Appropriate methods shall be adopted to control undesired species of weeds, rodents, arthropod pests, and pathogens.
    - If macroorganisms are released in the greenhouse, precautions are to be taken to minimize escape from the greenhouse facility.
  - *Concurrent experiments conducted in the greenhouse:*
    - Provided the work is conducted in accordance with BL1-P practices, experiments involving other organisms that require containment level lower than BL1-P may be conducted.
- *Greenhouse design:*
  - The floor of the greenhouse may be of gravel or other porous materials, impervious (concrete) walkways are recommended.
  - The greenhouse may be vented with windows or other openings in walls or roof, screens are recommended as barriers to contain or exclude pollen, microorganisms, or small flying animals.

Biosafety Level 2—Plants (BL2-P):

- *Standard practices:*
  - *Greenhouse access:*
    - As described for BL1-P
    - Personnel should be aware of and follow BL2-P practices and procedures.
  - *Records:*
    - As described for BL1-P and
    - The PI shall report any greenhouse accident involving inadvertent release or spill of microorganisms to the Greenhouse Director, Institutional Biosafety Committee, the NIH Office of Science Policy, and other appropriate authorities. Written records are to be maintained on any such accident.
  - *Decontamination and inactivation:*
    - As described for BL1-P and
    - Decontamination of run-off water is not generally required, although periodic cleaning to remove any organisms potentially entrapped by the gravel is to be done.
  - *Control of undesired species and motile microorganisms:*
    - As described for BL1-P
  - *Concurrent experiments conducted in the greenhouse:*
    - As described for BL1-P

(Continued)

BOX 11.2 (CONTINUED)

- *Signs:*
    - A sign shall be posted to indicate that a restricted experiment is in progress, shall indicate the name of the responsible person, the plants in use, and any special requirements for using the area.
- *Transfer of materials:*
    - Materials containing experimental organisms brought into or removed from the greenhouse facility in a viable state shall be transferred in a closed nonbreakable container.
- *Greenhouse design:*
    - As described for BL1-P.
    - An autoclave shall be available for treatment of contaminated greenhouse materials.

Biosafety Level 3—Plants (BL3-P):

- *Standard practices:*
    - *Greenhouse access:*
        - As described for BL1-P.
        - Personnel should be aware of and follow BL3-P practices and procedures.
    - *Records:*
        - As described for BL2-P.
    - *Decontamination and inactivation:*
        - All experimental materials including water shall be sterilized in an autoclave or rendered biologically inactive by appropriate methods before disposal (except those that are to remain in a viable or intact state for experimental purposes).
    - *Control of undesired species and motile microorganisms:*
        - As described for BL1-P.
        - Arthropods and other motile macroorganisms shall be housed in appropriate cages; when appropriate to the organism, experiments shall be conducted in the cages.
    - *Concurrent experiments conducted in the greenhouse:*
        - Involving organisms that require containment lower than BL3-P may be conducted concurrently provided BL3-P practices are followed.
    - *Signs:*
        - As described for BL2-P and
        - If organisms used have a recognized potential for causing detrimental impacts on managed or natural ecosystems, their presence should be indicated on a sign posted on the greenhouse access door.
        - If there is a risk to human health, a sign with the universal biosafety symbol shall be posted.
    - *Transfer of materials:*
        - A sealed nonbreakable secondary container shall be used for experimental material brought into or removed from the greenhouse facility in a viable state.
        - At the time of transfer, the surface of the secondary container shall be decontaminated by passage through a chemical disinfectant or fumigation chamber or any method found effective.
    - *Protective clothing:*
        - Disposable clothing (such as solid front or wrap around gowns, scrub suits, or other appropriate clothing) shall be worn if deemed necessary by the Greenhouse Director.
        - Such clothing shall be removed before exiting the facility and decontaminated before laundering or disposal.
- *Greenhouse design:*
    - The greenhouse floor shall be of concrete or other impervious material with provision to collect and decontaminate liquid run-off.
    - Windows shall be sealable, glazing shall be resistant to breakage; internal walls, ceilings, and floors shall be resistant to penetration by liquids to facilitate cleaning and decontamination; benchtops and other surfaces should be seamless, resistant to acids, alkali, organic solvents, and moderate heat; a foot, elbow, or automatically operated sink should be located near the exit for hand washing.

(Continued)

BOX 11.2 (CONTINUED)

- The greenhouse shall be a closed self-contained structure, separated from areas open to unrestricted flow of traffic; it shall be surrounded by a security fence or protected by security measures.
- An autoclave (double door recommended) shall be available for decontaminating materials within the facility.
- An individual supply and exhaust air ventilation shall be provided that maintains pressure differentials and directional airflow (assures inward, or zero, airflow from areas outside the greenhouse).
- Exhaust air shall be filtered through HEPA filters prior to discharge.

Biosafety Level 4—Plants (BL4-P):

- Standard practices:
  - Greenhouse access:
    - As described for BL3-P and
    - Personnel shall enter and exit the greenhouse facility only through the clothing change and shower rooms and shall shower each time they exit the facility; airlocks are used only for emergency exits; all reasonable efforts taken to ensure that viable propagules are not transported from the facility in an emergency.
    - Prior to entry, personnel should read and follow instructions on BL4-P procedures.
  - Records:
    - As described for BL2-P and
    - A record and time-log is kept of all people entering or exiting the facility.
  - Decontamination and inactivation:
    - As described for BL3-P and
    - Water that comes in contact with the experimental material (such as run-off water) shall be collected and decontaminated before disposal; all equipment and materials used will be decontaminated as in standard microbiological practices.
  - Control of undesired species and motile microorganisms:
    - As described for BL3-P
  - Concurrent experiments conducted in the greenhouse:
    - Experiments involving organisms that require containment less than BL4-P may be conducted concurrently.
  - Signs:
    - As described for BL3-P
  - Transfer of material:
    - As described for BL3-P and
    - Supplies and materials shall be brought into the facility through a double-door autoclave, fumigation chamber, or airlock that is fumigated between uses.
  - Protective clothing:
    - Street clothing is removed and kept in an outer changing room. Complete laboratory clothing (may be disposable) is provided and to be used by all personnel entering the facility. When exiting, the laboratory clothing is removed in an inner changing room before proceeding to the shower area. The clothing is decontaminated prior to laundering or disposal.
- Greenhouse design:
  - The maximum containment greenhouse shall consist of a separate building or a clearly demarcated area; should be able to maintain negative pressure; surrounded by a security fence or similar security measures.
  - Outer and inner change rooms separated by a shower shall be provided for entry and exit of personnel; doors should be self-closing; windows closed and sealed; glazing shall be resistant to breakage; ceilings and floors shall be resistant to penetration by liquids to facilitate cleaning and decontamination; benchtops and other surfaces should be seamless, resistant to acids, alkali, organic solvents, and moderate heat.
  - A double-door autoclave, fumigation chamber, or ventilated airlock shall be provided for passage of materials, supplies, and equipment.
  - An individual supply and exhaust air ventilation shall be provided that maintains pressure differentials and directional airflow (assures inward, or zero, airflow from areas outside the greenhouse).
  - Exhaust air shall be filtered through HEPA filters prior to discharge.

11.5 PHYSICAL AND BIOLOGICAL CONTAINMENT FOR RESEARCH INVOLVING ANIMALS

Appendix Q of the NIH guidelines deals with the requirements for containment and confinement for research involving whole animals. The guideline covers both animals whose genome has been altered by stable integration of recombinant or synthetic nucleic acid molecules into the germ line (**transgenic animals**), as well as **modified microorganisms tested on whole animals**. The animals covered in the guidelines include, but are not limited to, cattle, swine, goats, horses, sheep, and poultry. As in the case of plants, four levels of containment are established, referred to as BL1- Animals (N), BL2-N, Bl3-N, and BL4-N. For further details, see Box 11.3.

BOX 11.3 PHYSICAL CONTAINMENT FOR EXPERIMENTS INVOLVING ANIMALS

Appendix Q of the **NIH guidelines** supersedes Appendix G (Physical Containment) when the animals are of a size or have growth requirements that preclude the use of the physical containment described in Appendix G. For experiments that require prior approval of the IBSC that utilize facilities described in Appendix Q, the IBSC shall include at least one scientist with expertise in animal containment principles. The institute shall establish a health surveillance program for personnel working with viable microorganisms carrying recombinant or synthetic DNA that require BL 3 or greater.

Biosafety Level 1—Animals (BL1-N):

- *Standard practices:*
  - *Animal facility access:*
    - The containment area shall be locked; access shall be limited or restricted when experiments are in progress; the area shall be patrolled/monitored at frequent intervals.
    - The containment area shall be in accordance with state and Federal laws and animal care requirements.
    - All genetically engineered neonates shall be permanently marked within 72 hours of birth (or if size does not permit, the containers shall be marked); transgenic animals should contain distinct and biochemically assayable DNA sequences that allow distinction between modified and nonmodified animals; a double barrier shall separate male and female animals unless reproductive studies are part of the study.
- *Animal facilities:*
  - Animals shall be confined to securely fenced areas or enclosed animal rooms to minimize the possibility of theft or unintended release.

Biosafety Level 2—Animals (BL2-N):

- *Standard practices:*
  - *Animal facility access:*
    - As described for BL1-N and
    - The Animal Facility Director shall establish procedures to ensure personnel who enter are advised of potential hazards and meet specific requirements (such as vaccination); animals of the same or different species, not involved in the experiment, shall not be permitted.
  - *Decontamination and inactivation:*
    - Materials to be decontaminated elsewhere are to be placed in closed durable leak-proof containers, needles and syringes in puncture-proof containers to be autoclaved before disposal.
  - *Signs:*
    - Warning signs incorporating the universal biosafety symbol to be posted on access doors containing details of special provisions (such as vaccinations) for entry, agents and animal species involved in the experiments, and details of the Animal Facility Director.

(Continued)

BOX 11.3 (CONTINUED)

- *Protective clothing:*
  - Protective coating to be worn in the animal area, to be removed in nonlaboratory areas, gloves to be worn and care to be taken to avoid skin contamination.
- *Records:*
  - Any incident involving spills or inadvertent exposure or release of modified microorganisms shall be reported to the Animal Facility Director, Institutional Biosafety Committee, the NIH Office of Science Policy, and other appropriate authorities. Written records are to be maintained on any such accident, and if necessary, the area shall be decontaminated.
  - When appropriate, base line serum samples of animal care workers and at-risk personnel may be collected and stored.
- *Transfer of materials:*
  - Advance approval for transfer of material shall be obtained from the Animal Facility Director; biological material shall be transferred in a sealed nonbreakable primary container, sealed in a second nonbreakable container, both of which are to be disinfected before removal; unless inactivated, packages are to be opened in a facility having equivalent or higher physical containment.
- *Other:*
  - As described for BL1-N and
  - Appropriate steps to be taken to prevent horizontal transmission or exposure of personnel; eating, drinking, smoking is not permitted in the work area.
- *Animal facilities:*
  - As described for BL1-N and
  - Surfaces shall be impervious to water and resistant to acids, alkalis, organic solvents, moderate heat, easy to clean; windows that open shall be fitted with fly screens; special attention to be taken to prevent entry and exit of arthropods.

Biosafety Level 3—Animals (BL3-N):

- *Standard practices:*
  - *Animal facility access:*
    - As described for BL2-N
  - *Decontamination and inactivation:*
    - As described for BL2-N and
    - Special safety testing, decontamination procedures, and IBSC approval require for transfer of agents or tissue/organ specimens from a BL3-N to a facility of lower containment classification.
    - Liquid effluent from the facility shall be decontaminated by heat treatment prior to release to the sanitary system.
  - *Signs:*
    - As described for BL2-N
  - *Protective clothing:*
    - Full protective clothing shall be worn in the animal area; personnel are required to shower before exiting the BL3-N facility; protective clothing shall not be worn outside the containment area and will be decontaminated before laundering or disposal.
    - Appropriate respiratory protection shall be worn in the containment rooms.
  - *Records:*
    - As described for BL2-N and
    - A permanent record book shall maintain details of experimental animal use and disposal.
  - *Transfer of materials:*
    - As described for BL2-N and
    - Special safety testing, decontamination procedures, and IBSC approval require for transfer of agents or tissue/organ specimens from a BL3-N to a facility of lower containment classification.
  - *Other:*
    - As described for BL2-N

(Continued)

BOX 11.3 (CONTINUED)

- *Animal facilities:*
  - As described for BL2-N and
  - The animal containment area shall be separated from other areas; access doors shall be self-closing; passage through two sets of doors and clothes change room equipped with integral showers and airlock.
  - An exhaust air ventilation system shall be provided that creates a directional airflow; that draws air into the animal rooms vacuum lines shall be protected with HEPA filters; liquid effluent from containment rooms shall be decontaminated before discharge into the sanitary system.

  Biosafety Level 4—Animals (BL4-N):

- *Standard practices:*
  - *Animal facility access:*
    - As described for BL3-N and
    - Individuals below 16 years of age shall not be permitted to enter the animal area.
    - Personnel shall enter and exit through the clothing change and shower rooms, and use the airlocks in case of an emergency.
  - *Decontamination and inactivation:*
    - As described for BL3-N and
    - All contaminated liquid and solid wastes and wastes from the animal rooms shall be decontaminated before disposal.
  - *Signs:*
    - As described for BL3-N
  - *Protective clothing:*
    - Street clothes shall be removed and kept in the outer changing room; complete laboratory clothing (may be disposable) shall be provided for all personnel entering the animal facility, which is to be removed and placed in bins in the inner changing room while exiting the facility; clothing is decontaminated before laundering or disposal; personnel shall shower each time they exit the containment facility.
    - A ventilated head-hood or a one-piece positive pressure suit shall be worn by personnel entering rooms that contain experimental animals when appropriate.
  - *Records:*
    - As described for BL3-N and
    - A permanent record and time-log of entry and exit of personnel is to be maintained.
  - *Transfer of materials:*
    - As described for BL3-N and
    - Supplies and materials needed in the animal facility shall be brought in by way of the double-door autoclave, fumigation chamber, or airlock appropriately decontaminated between use.
  - *Other:*
    - As described for BL3-N and
    - Animal-holding areas shall be cleaned at least once a day and decontaminated immediately if spilling of viable materials occurs.
    - An essential adjunct to the reporting, surveillance system is the availability of a facility for quarantine, isolation, and medical care of personnel with potential or known laboratory associated diseases.
- *Animal facilities:*
  - As described for BL3-N and
  - The BL4-N shall have a double barrier to prevent release of recombinant or synthetic nucleic acid molecule containing microorganisms to the environment such that even if the barrier of the inner facility is breached, the outer barrier will prevent release into the environment; physical separation of the animal containment area is by double-door clothes change room equipped with showers and airlock.
  - All equipment and floor drains shall be equipped with minimally 5-in.-deep traps; ducted exhaust air ventilation shall be provided that is filtered through double HEPA filters and creates a directional airflow that draws air into the laboratory.

11.6 GOOD LABORATORY PRACTICE

By definition *"Good Laboratory Practice embodies a set of principles that provide a framework within which laboratory studies are planned, performed, monitored, recorded, reported, and archived"* (Dolan, 2007). The primary purpose of GLP is to **ensure uniformity, consistency, and reliability of safety tests (nonclinical) for pharmaceuticals, agrochemicals, aroma and color food/feed additives, cosmetics, detergents, novel foods, nutritional supplements for livestock, and other chemicals.** These safety tests are used to generate data on various parameters from physicochemical properties to toxicity (nonclinical) for use of regulatory authorities in order to make risk/safety assessments. Originally, GLP regulations were intended for toxicity testing only and were reserved for laboratories undertaking animal studies for preclinical work. **GLP is now followed in all laboratories where research or marketing studies are to be submitted to regulatory authorities** such as the FDA. **Establishment of GLP is mandatory to evaluate safety or toxicity of products intended to undergo clinical trials.**

Historically, GLP was introduced in several countries (including the United States in 1978) in response to a scandal involving an American industrial product safety testing laboratory in Illinois, the Industrial Bio-Test (IBT) Laboratory. This laboratory performed more than one-third of all toxicology testing in the United States in the 1950s to 1970s, but was found guilty of extensive scientific misconduct, resulting in indictment and convictions of several of its staff in the early 1980s. As data generated by IBT had been used by regulatory authorities for marketing licenses, the United States Environmental Protection Agency was forced to pull several pesticides from the market pending reevaluation of its safety data.

The **Organization for Economic Cooperation and Development** *Principles of Good Laboratory Practice (GLP)* was first developed in 1978 by an Expert Group led by the United States with experts from Australia, Austria, Belgium, Canada, Denmark, France, the Federal Republic of Germany, Greece, Italy, Japan, the Netherlands, New Zealand, Norway, Sweden, Switzerland, the United Kingdom, the Commission of the European Communities, the WHO, and the International Organization for Standardization. The GLP was formally recommended for use in Member countries in 1981. A more comprehensive document specifying the Principles of GLP was brought out by the OECD in 1992 (revised in 1997) (OECD, 1998) and has since been adopted by several countries and incorporated in national regulatory policies and documents.

Compliance with GLP requires that:

1. The tests should be conducted by **qualified personnel**.
2. Each study should have a **Study Director responsible** for the overall conduct of the tests.
3. The laboratory study and the accompanying data should be **audited by a Quality Assurance Unit.**
4. All laboratory activities must be performed in accordance with **written and filed management-approved Standard Operating Procedures (SOPs).** SOPs should cover policies, administration, equipment operation, technical operation, and analytical methods.
5. All control and test articles and reagents must be **identified, characterized, and labeled** with information regarding **source, purity, stability, concentration, storage conditions, and expiration date**.
6. The equipment must be **maintained, calibrated, and must be designed to meet analytical requirements**.

Compliance with GLP has served to harmonize test methods across nations, facilitating generation of **mutually acceptable data**, thus avoiding duplication of tests, and saving time and
resources.

KEY TAKEAWAYS

The primary purpose of **GLP** is to ensure uniformity, consistency, and reliability of safety
tests (nonclinical) for pharmaceuticals, agrochemicals, aroma and color food/feed additives,
cosmetics, detergents, novel foods, nutritional supplements for livestock, and other chemicals. **Establishment of GLP is mandatory to evaluate safety or toxicity of products
intended to undergo clinical trials.**

11.7 SUMMARY

Crucial to the research and development of new applications of genetically modified organisms
derived by the transfer of synthetic or recombinant nucleic acid molecules are measures to prevent
hazards (to laboratory personnel as well as to other persons, animals, and the ecosystems) from
being realized. Guidelines prepared by the NIH and the WHO have helped establish processes and
systems that build on GMT in order to ensure biosafety. The guidelines form an integral part of
normative policies and regulation of genetically modified organisms in countries using recombinant
DNA technology. Both the NIH and the WHO guidelines recommend classification of biological
agents based on their potential to cause harm to humans, animals, and the environment. Four BLs
are recommended to handle organisms of increasing risk potential. Recommended for each level
are standard microbiological practices as well as facilities for physical and biological containment
of genetically modified organisms (microbes, plants, or animals). In order to harmonize toxicity
testing and generation of mutually acceptable preclinical data that may be used for decisions
regarding regulation including commercialization, several countries have adopted the principles of
GLPs. These principles establish a framework and a minimum standard for the conduct of tests,
and documentation and analysis of data.

REFERENCES

Dolan, K. (2007). *Laboratory animal law: Legal control of the use of animals in research.* Oxford: Blackwell
    Publishing Limited.
NIH Guidelines for Research Involving Recombinant or Synthetic Nucleic Acid Molecules (NIH Guidelines)
    retrieved from http://osp.od.nih.gov/sites/default/files/NIH_Guidelines_0.pdf.
OECD (1998). *OECD principles on good laboratory practice (as revised in 1997). OECD series on principles
    of good laboratory practice and compliance monitoring number 1.* Paris: Environment Directorate,
    Organisation for Economic Co-operation and Development. Retrieved from http://www.oecd.org/official
    documents/publicdisplaydocumentpdf/?cote=env/mc/chem(98)17&doclanguage=en.
WHO Laboratory Biosafety Manual 2004 Third edition, retrieved from http://www.who.int/csr/resources/
    publications/biosafety/en/Biosafety7.pdf.

RECOMBINANT DNA SAFETY CONSIDERATIONS IN LARGE-SCALE APPLICATIONS AND GOOD MANUFACTURING PRACTICE

12

Standards should not be forced down from above but rather set by the production workers themselves.
-Taiichi Ohno, Japanese industrial engineer, author of "Toyota Production System: Beyond Large-scale Production"

CHAPTER OUTLINE

12.1 Introduction 274
12.2 Safety Considerations for Industrial Applications of Organisms Derived by Recombinant DNA Techniques 274
12.3 Good Industrial Large-Scale Practice 275
12.4 Good Developmental Principles 278
12.5 Safety Considerations for Field/Market Release of GMOs and/or Their Products 279
    12.5.1 Safety Considerations for Field Release of Genetically Modified (GM) Crops 279
    12.5.2 Safety Considerations for Field Release of Genetically Modified Animals 282
    12.5.3 Safety Considerations for Marketing of Foods from Genetically Modified Organisms.... 283
    12.5.4 Safety Considerations for Market Approval of Biopharmaceuticals........................... 284
    12.5.5 Safety Considerations for Market Approval of Biosimilars 286
12.6 Good Manufacturing Practices 286
12.7 Summary 289
References 289
Further Reading 290

An Introduction to Ethical, Safety and Intellectual Property Rights Issues in Biotechnology.
DOI: http://dx.doi.org/10.1016/B978-0-12-809231-6.00012-0

AquAdvantage Salmon.

(Photo courtesy: AquaBounty Technologies)

12.1 INTRODUCTION

Within 3 years of the creation of the first recombinant DNA molecule, the first **biotechnology industry**, Genentech, was founded by Herbert Boyer and Robert Swanson to commercially exploit the technology (see Chapter 4: Recombinant DNA Technology and Genetically Modified Organisms). The initial products of this industry were proteins of pharmaceutical importance (for instance, insulin and human growth factor) obtained by cloning human genes for the proteins in bacteria. With further development of the technology, transgenic plants [genetically modified (GM) crops such as GM soybean, maize, or canola] and transgenic animals (such as the AquAdvantage[(R)] salmon and RIDL mosquitoes of Oxitec) were developed for commercial cultivation and have been released (or are awaiting regulatory approval for release) into the environment (see Chapter 4: Recombinant DNA Technology and Genetically Modified Organisms). Although risk analysis as part of the regulatory requirement for commercial release of GM organisms (GMOs) ensures that these organisms pose minimal risk to humans, animals, and the environment (see Chapter 10: Risk Analysis), care needs to be exercised while handling these organisms. This chapter discusses Good Industrial Large-Scale Practice (GILSP) and Good Manufacturing Practice (GMP) as applied to GM microbes, plants, animals, and products derived from GMOs.

12.2 SAFETY CONSIDERATIONS FOR INDUSTRIAL APPLICATIONS OF ORGANISMS DERIVED BY RECOMBINANT DNA TECHNIQUES

The first product from a GMO was marketed in 1982 (Humulin, human insulin expressed in bacteria, developed by Genentech and licensed to Eli Lilly and Co.) and was soon followed by a number of other pharmaceutically important proteins (see Chapter 4: Recombinant DNA Technology and Genetically Modified Organisms). Production volumes of over thousands of liters of bacterial culture could be achieved through **extensive experience of scale-up from laboratory level, to pilot scale, and finally, manufacturing levels of foods by fermentation processes using bacteria, yeasts, and fungi**. Traditional fermentation processes in industries (such as the beer, wine, cheese, and other fermented food industries) use microorganisms that are well characterized and considered to be of low-risk. Therefore, such industries require only minimal controls and containment procedures. When rDNA was first introduced, the major concern was regarding potential hazards, such as allergenicity, toxicity, or other effects, on humans and animals. In order to avoid persistence of escapees in the environment, **initial research and production systems using rDNA technology were confined to a strain of gut bacteria, *Escherichia coli* K12**. This strain had been cultured in laboratories for several decades and had lost several genes present in the wild-type strains of *E. coli* necessary for colonizing the human gut. These included the cell surface K antigen, part of the lipopolysaccharide side chain, resistance to lysis by complement in human serum and to phagocytosis by white blood cells, and an adherence factor that enabled the bacteria to stick to epithelial cells of the gut. The *E. coli* K12 bacteria were also incapable of synthesizing certain ingredients necessary for growth (had to be supplied in the culture medium) or repairing DNA by recombination (hence, readily killed by exposure to ultra violet rays present in sunlight). In short, **the *E. coli* K12 strain was incapable of causing allergies, disease, or survival outside the laboratory**.

Pioneering efforts in understanding safety issues in the nascent biotech industry and in implementing processes that would ensure safety in the application of rDNA technology were made by the Ad hoc Group of government experts created by the Committee for Scientific and Technological Policy of the **Organization for Economic Cooperation and Development** (OECD; http://www.oecd.org/). In 1986, OECD Council decided to make public a report prepared by the Ad hoc Group and to adopt recommendations made in the report to ensure safety in applications of rDNA organisms in industry, agriculture, and the environment (OECD, 1986). The report established a concept of **GILSP** applicable to **low-risk organisms used in industrial production**. Key concepts to GILSP are as follows:

1. Risk assessment of the recombinant organism to ensure that it is as safe as the low-risk host organism
2. Identification and adoption of safe practices.

12.3 GOOD INDUSTRIAL LARGE-SCALE PRACTICE

Potential risks to the environment of the applications of rDNA organisms are minimized due to a **"step-by-step assessment during the research and development process,"** that is from laboratory scale, to pilot scale level, to finally, industrial level. For rDNA microorganisms and cell cultures, the **criteria for GILSP** suggested by the OECD include the following:

- **Host**:
  - *Nonpathogenic*: Hosts containing the recombinant nucleic acid should be identified and established to be nonpathogenic. They should also not produce any toxins or allergens.
  - *No adventitious agents*: The hosts should not harbor any viruses or mycoplasma.

- *Extended history of safe use*: Sufficient documented experience of safe use of the host organism should be available. Safe use could also be established by laboratory/pilot-scale fermentations under conditions of minimal containment.
- *Built-in environmental limitations*: Should permit optimal growth in industry but limited survival in the environment, such as strains sensitive to ultraviolet light or requiring supplements in growth media of substances not found in nature. Any surviving microbes should have minimal adverse environmental consequences.
- **Vector/Insert**:
  - *Well characterized and free from known harmful sequences*:
    - — *Vector*: Knowledge of the derivation and construction of the vector and subsequent experimental confirmation of the construct is necessary to ensure that the vector is free from sequences that result in phenotypes harmful to humans or the environment such as production of toxins or factors that promote pathogenicity.
    - — *Insert*: Source and function of the DNA being inserted and the point of insertion should be known.
  - *Limited in size*: The vector/insert should be as limited in size as possible in order to decrease the probability of carrying unwanted genes and other sequences.
  - *Should be poorly mobilizable*: The rate at which it may be transferred from the original recipient to other organisms should be low, for example, by eliminating transfer functions of plasmids, or by integration into host chromosome.
  - *Should not transfer any resistance markers to microorganisms not known to acquire them naturally*: Genetic markers conferring resistance to substances, such as antibiotics, herbicides, or heavy metals, are often used in rDNA technology to select the transformed organisms from untransformed hosts. Use of these markers should take into consideration the possibility of spread and the impact of the marker on the environment.
- **rDNA organism**:
  - *Nonpathogenic*: The rDNA organism is expected to be nonpathogenic as the gene product has no known role in pathogenicity and the host is nonpathogenic.
  - *As safe in an industrial setting as host organism*: The rDNA organism should have limited survival or have no adverse consequence to humans and the environment.

The report recognized that there may be some circumstances under which physical containment may be warranted, as when pathogenic organisms are used, or genes coding for harmful products are inserted. Under such circumstances, industry safety programs rely on the two approaches of:

- **biological containment**—takes advantage of natural barriers that limit the survival and multiplication of the organism in the environment, and/or transmission of the genetic information to other organisms;
- **physical containment**—which uses (1) equipment, (2) operating practices, and (3) design of facility to protect personnel handling the organisms and the environment outside the facility.

Physical containment for large-scale uses of organisms containing recombinant or synthetic nucleic acids have been addressed in Appendix K of the guidelines proposed by the National Institutes of Health (see Box 12.1).

BOX 12.1 NIH GUIDELINES: PHYSICAL CONTAINMENT FOR LARGE-SCALE APPLICATIONS OF ORGANISMS CONTAINING RECOMBINANT OR SYNTHETIC NUCLEIC ACIDS

Appendix K of the NIH guidelines (NIH Guidelines, 2016) specifies physical containment guidelines for addressing the biological hazard associated with research or production involving greater than 10 L of culture of viable organisms containing recombinant or synthetic nucleic acid molecules. Appendix K supersedes Appendix G, *Physical Containment*, in cases when culture volumes are in excess of 10 L. The guideline establishes four levels of physical containment commensurate with the assessed degree of hazard to health or to the environment posed by the organism based on experience with similar organisms that have not been modified, and on GILSP. The four levels of containment are referred to as Good Large-Scale Practice, BL1-Large Scale, BL2-Large Scale, and BL3-Large Scale.

Good Large-Scale Practice: This level is recommended for large-scale research or production involving organisms that are generally regarded as safe, nonpathogenic, nontoxigenic, and derived from host organisms that have an extended history of safe use. In order to ensure safety, measures taken include the following:

- Institutional codes of practice shall be formulated and implemented
- Personnel are trained to handle modified organisms
- Basic hygiene and safety measures such as hand washing, prohibition of eating, drinking, and smoking in work area
- Discharges containing viable organisms are treated as per environmental regulations
- An emergency response plan shall include provisions for handling spills.

Biosafety Level 1 (BL1)-Large Scale: This level is recommended for organisms that qualify for BL1 containment at the laboratory scale and do not qualify for Good Large-Scale Practice.

- Spills and accidents that result in exposure to the modified organisms are immediately reported to the Laboratory Director. Medical evaluation, surveillance, and treatment are provided as appropriate and documented.
- Cultures of viable modified organisms shall be handled in a closed system (e.g., closed culture vessels)
- Culture fluids shall not be removed from the closed vessel unless viable organisms have been inactivated
- Exhaust gases shall be treated by filters
- Emergency plans including handling large losses of culture on an emergency basis required by the Institutional Biosafety Committee and Biological Safety Officer.

Biosafety Level 2 (BL2)-Large Scale: This level is recommended for organisms that qualify for BL2 containment at the laboratory scale.

- Spills and accidents that result in exposure to the modified organisms are immediately reported to the Biological Safety Officer, Institutional Biosafety Committee, NIH Office of Science Policy, and other appropriate authorities. Medical evaluation, surveillance, and treatment are provided as appropriate and documented.
- As in BL1-Large Scale, closed systems shall be used for the propagation and growth of viable modified organisms, which shall not be opened unless sterilized by a validated procedure.
- A sign with the universal biosafety symbol shall be posted on each closed system and primary containment equipment.
- Emergency plans including handling large losses of culture on an emergency basis required by the Institutional Biosafety Committee and Biological Safety Officer.

Biosafety Level 3 (BL3)-Large Scale: This level is recommended for organisms that qualify for BL3 containment at the laboratory scale.

- As in BL1-Large Scale and
- The controlled area shall have a separate entry, double door with airlocks, or change room separating the controlled area from the rest of the facility. A shower facility shall be provided. Entry to the controlled area shall be restricted to authorized personnel only and will be only through the double doors.

(Continued)

BOX 12.1 (CONTINUED)

- An effective insect and rodent program shall be maintained; the controlled area shall be decontaminated in accordance with standard procedures in the event of a spill or accident.

    Currently, organisms that require BL4 containment in the laboratory scale are not permitted for large-scale applications.

Reference

NIH Guidelines for Research Involving Recombinant or Synthetic Nucleic Acid Molecules (NIH Guidelines), 2016, Retrieved from http://osp.od. nih.gov/sites/default/files/NIH_Guidelines_0.pdf.

An internal survey carried out in the OECD countries in 1988 revealed that the underlying principles of the GILSP concept had been adopted in national guidelines in several countries and was being considered for implementation by others. In 1992, the OECD brought out an updated follow-up to the 1986 publication (OECD, 1992). This document introduced the concept of "**Good Developmental Principles**" (GDP) for the design of small-scale field research with plants and microorganisms with newly introduced traits.

KEY TAKEAWAYS

Criteria suggested by OECD for rDNA GILSP microorganisms and cell cultures are as follows:

- **Host organism**—should be nonpathogenic, should not harbor adventitious agents, should have extended history of safe use or built-in environmental limitations
- **Vector/Insert**—should be well characterized, free from harmful sequence, as limited in size as possible, poorly mobilizable, should not transfer resistance markers to microorganism not known to acquire them naturally
- **rDNA organism**—should be nonpathogenic, safe in industrial settings, limited survival with minimal adverse effects in environment.

12.4 GOOD DEVELOPMENTAL PRINCIPLES

The development of a GMO for use in the environment generally goes through **three stages**. The **first stage consists of experiments done in the laboratory and the glasshouse/greenhouse**. Both nationally and internationally, codes have been developed for ensuring safety under laboratory conditions. But the OECD's Group of National Experts on Safety in Biotechnology felt that it was necessary to "*develop general principles that would identify a generic approach to the safety assessment of low—or negligible risk small-scale field research*" (OECD, 1992) **which represents the second stage of development. The third stage of development is the release of the variety/production**. The GDP were developed to address this need.

KEY TAKEAWAYS

Three stages of product development are as follows:

- **Stage 1**—Laboratory/greenhouse
- **Stage 2**—Basic field research and small-scale field research (*principles of GDP applied here*)
- **Stage 3**—Applied large-scale field trials and production/release.

Key safety factors identified for ensuring safety in experiments include:

1. *The characteristics of the organism*: That is, the introduced gene/genetic material. In many cases, the organism may be safe under a wide range of environmental conditions, but it could be possible to grow organisms known to cause adverse effects under confinement or by exercising mitigation methods.
2. *The characteristics of the research site*: Research sites selected for field trials should meet the objectives of the experiment and should take into account important ecological and/or environmental considerations related to safety; the climatic conditions; appropriateness in terms of geographical location and proximity to specific biota that may be affected; and the size of the site.
3. *The experimental conditions*: In order to obtain scientifically acceptable and environmentally sound data, the experiment should be designed carefully. This necessitates attention to the formulation of a hypothesis and statement of objectives, precise experimental protocols including planting density and treatment patterns, collection, and analysis of experimental data to draw conclusions based on statistical significance.

The principles of GDP facilitate the design and conduct of field experiments so that

- the experimental GM plants remain reproductively isolated from unmodified plants grown outside the experimental area
- GMOs or their genes will not be released into the environment beyond the experimental site, or
- even without reproductive isolation, the plants will not cause unintended, uncontrolled adverse effects.

12.5 SAFETY CONSIDERATIONS FOR FIELD/MARKET RELEASE OF GMOs AND/OR THEIR PRODUCTS

Recombinant DNA technology has been used to produce food, industrial chemicals, and medicinal products from GM microorganisms, cell lines, plants, and animals. Safety considerations for each application are distinct and dependent on the host, the gene transferred, and the application itself. Discussed in the following sections are the issues associated with each category of applications:

12.5.1 SAFETY CONSIDERATIONS FOR FIELD RELEASE OF GENETICALLY MODIFIED (GM) CROPS

Commercial cultivation of GM crops began in 1986 with regulatory approval being given for the cultivation of herbicide resistant tobacco. Since then, GM crops have been grown in 28 countries in around 180 million hectares. GM varieties exist for major crops such as soybean, maize, canola, cotton and are being developed for rice and several others (see Chapter 4: Recombinant DNA Technology and Genetically Modified Organisms). Risk analysis of the GM crop and of foods derived from such crops conducted as part of the regulatory process ensures that commercially cultivated GM crops pose no safety issues to humans, animals, or to the environment (see Chapter 10: Risk Analysis). However, one category of GM plants that **warrant special consideration regarding biosafety are plants modified for the purpose of producing recombinant proteins for pharmaceutical or industrial use (plant molecular farming (PMF)**, also known as pharming, see Chapter 4: Recombinant DNA Technology and Genetically Modified Organisms). This type of application raises concerns regarding aspects of **transgene spread in the environment**, or **accidental contamination of the food/feed chains as these plants are not meant for food/feed use**. Host systems used in PMF include **food plants** (such as maize, soybean, potato, oilseed rape, tomato, banana, and rice), **nonfood plants** (such as tobacco), **noncultivated plants** (such as duckweed, *Arabidopsis*), and **cultured plant cells** (such as carrot, tobacco, and tomato). **The choice of production platform has a significant impact on biosafety in PMF**.

The use of food crops as production systems has been particularly controversial due to the **risk of GM crops inadvertently entering the food chain**. An instance illustrative of the problem of accidental contamination occurred in 2002 when farmers in Nebraska planted conventional soybeans for human consumption in a field previously used to test GM maize producing a pig vaccine by a biotech company, ProdiGene. US Department of Agriculture (USDA) inspectors found 500,000 bushels of soybean contaminated with GM maize stalks and leaves as the firm had neglected to remove volunteer corn plants that sprouted alongside the soybean. ProdiGene was ordered to pay a US$ 250,000 fine, and buy and destroy all the contaminated soybean. Earlier that year, the company had been asked to destroy 155 acres of corn in Iowa contaminated with GM corn producing Trypsin (pancreatic serine protease), not approved for consumption by humans or animals (Fox, 2003). **The USDA has since enforced a zero-tolerance standard, whereby plants grown for pharmaceutical or industrial products (and not approved for food/feed) have to remain distinct from the food system** (USDA, 2006). Many countries recommend the use of nonfood plants or cell cultures for pharming.

Physical and biological containment may be considered on a case-by-case basis as a viable option to limit adverse impacts on the environment or contamination of food/feed systems (Breyer et al., 2009) as discussed below:

Physical containment:

Several plants, such as tobacco, potato, and tomatoes, can be grown in glasshouses, greenhouses, plastic tunnels, and other forms of physical containment. Although this option is effective in preventing contamination of food systems, practical difficulties include **issues of scale-up** and **additional financial resources**.

Spatial containment:

This option aims to minimize pollen transmission of traits from the GM to non-GM crops while being more flexible in terms of scale-up. Strategies used include the following:

- *Dedicated land*: Pharming is conducted in regions where similar crops are not grown or locations considerably distant from nonmodified crops so as to eliminate the risk of gene flow. This option is not always feasible due to unfavorable agroclimatic conditions.

- *Restricted use*: This option is to restrict pharming to a designated area for a specified number of growing seasons, during which nonmodified crops for food/feed would not be grown.
- *Buffer and border zones*: Pharming could also employ strategies used to grow other GM crops. A minimum isolation distance (buffer zone) which depends on the biology of the crop plant (self-pollinated/wind/insect pollinated) could be set up around the GM crop. Alternatively, borders of non-GM plants could be planted around the GM crop stand to "trap" pollen from the GM plants. These strategies may not, however, ensure zero contamination.

Biological containment:

Several strategies based on many different biological principles have been suggested:

- *Plastid transformation*: Here, the transgene is inserted in the chloroplast genome rather than the nuclear genome. Several advantages of this technique include the ability to control gene insertion more precisely, higher rates of transgene expression and protein accumulation, but most significantly, in higher plants, it prevents pollen transmission of the transgene as pollen grains lack chloroplasts.
- *Male sterility*: Natural and induced male sterility has been used by plant breeders to control crossing and may be used to prevent pollen transmission of modified genes.
- *Genetic Use Restriction Technologies (GURT)*: Although this genetic system has been much criticized by the social media as being a technique developed by multinational seed companies to control the seed market and to enforce intellectual property rights, GURT could be used to ensure that the trait is not carried forward to the next generation either because the seeds are sterile (as in V-GURT or "terminator technologies") or the trait is not expressed in the progeny (as in T-GURT or "traitor technologies") unless treated with an inducer (see Chapter 5: Relevance of Ethics in Biotechnology).

Other Biological Containment Strategies:

Several other containment mechanisms may be developed in future which exploit natural mechanisms, such as apomixes (production of seed without pollination); cleistogamy (self-pollination and fertilization within flowers that are closed); or targeted/spatial and temporal gene expression so the expressed products are present only in specific organs such as roots, seeds, or edible plant parts.

Temporal confinement:

Temporal confinement can be achieved by either physical or biological methods, such as timing the crop for PMF at different times to prevent overlap with the food/feed crop; or to have only transient expression of the introduced genes (as the gene is not stably integrated in the host genome, it will not be heritable).

Additional considerations:

In order to avoid issues of contamination of nonmodified plant products with products from pharming, care is to be exercised in the handling and transport of products, the cleaning of equipment (preferably, dedicated equipment to be used), and the personnel employed. Also warranting attention is waste management: residual material left on the field/storage areas and the by-products of the processing. Adequate postmarket management measures such as inspection of the cultivation site and monitoring of the product is necessary to ensure that no adverse effects occur.

KEY TAKEAWAYS

Safety considerations for field release of GM crops for nonfood products to prevent contamination of food/feed include the following:

- *Physical containment*: growing plants in glasshouse, greenhouses, plastic tunnels
- *Spatial containment*: minimizing pollen transmission of traits from modified to non-GM crops
  - *Dedicated land*
  - *Restricted use*
  - *Buffer/border zones*
- *Biological containment*:
  - *Plastid transformation*: prevents pollen transmission of traits
  - *Male sterility*
  - *GURT*
  - *Other Biological techniques*: apomixes, cleistogamy, targeted spatial gene expression
- *Temporal confinement*: timing the modified crop to prevent overlap with the food/feed crop, transient gene expression.

12.5.2 SAFETY CONSIDERATIONS FOR FIELD RELEASE OF GENETICALLY MODIFIED ANIMALS

Genetically engineered animals have biomedical applications, such as production of human proteins, drugs, vaccines, and replacement tissues, as well as applications in agriculture, such as production of food and animal welfare, in addition to reducing the impact on the environment due to better utilization of resources (Gottlieb & Wheeler, 2011). Management practices employed to mitigate assessed environmental risks of transgenic animals include maintaining the animals in specialized facilities that minimize contact with people, other animals, insects, and infectious agents. There have been till date, **no GE animals placed in the market in the European Union**, but in the United States, the **Food and Drug Administration (FDA)** has approved several **transgenic animals for the commercial production of pharmaceutical compounds and for food uses**. These include:

- goats that produce an anticoagulant, ATryn (antithrombin), in milk (2009)
- rabbits that produce a drug for treating angioedema (2014)
- AquAdvantage salmon for food (2015)
- Chicken that produce a drug kanuma (sebelipase alfa) in eggs (used to treat people with a rare genetic condition that prevents the body from breaking down fatty molecules in cells) (2015)

In the pipeline are GM mosquitos, *Aedes aegypti* (OX513A), produced by Oxitec for vector control strategy for preventing mosquito-borne viral diseases including Zika, dengue, chikungunya, and yellow fever. Although the FDA released the final Environment Assessment submitted by Oxitec based on a field trial conducted in Florida Keys and the final Finding of No Significant Impact on August 5, 2016, the GE mosquitos are yet to be approved for commercial use (FDA, 2016).

The FDA places the onus of ensuring safety in commercial use of transgenic on the sponsor. In the case of the GE mosquitoes, the company Oxitec is responsible for ensuring all local, state, and federal requirements met before conducting field trials. The company AquaBounty Technologies which produces AquAdvantage salmon has put in place physical and biological containment strategies to ensure that the modified salmon does not impact natural salmon population (see Box 4.1).

Containment strategies for GM animals would be concomitant with the phenotype of the animal, the nature of the activity, and the assessed risk due to the modification. Physical containment should prevent animals from escaping into the wider environment and will typically consist of pens, cages, and other enclosures. Double fencing may sometimes be appropriate given the level of risk. Aquatic animals should be kept in tanks fitted with filters sufficient to prevent escape of eggs or the smallest fingerlings. Access to the containment facility should be restricted and monitored. Disposal of waste and carcasses from the facility should be handled with care. Additional biological containment could be effected by reproductive sterility (e.g., by polyploidy as in the case of AquAdvantage salmon or genetic sterility as in the case of the RIDL mosquitos).

KEY TAKEAWAYS

Safety considerations for field release of transgenic animals:

- *Physical containment*: pens, cages, water tanks with water filters; double fencing; waste disposal done with care
- *Biological containment*: sterility due to polyploidy or genetic sterility

---

12.5.3 SAFETY CONSIDERATIONS FOR MARKETING OF FOODS FROM GENETICALLY MODIFIED ORGANISMS

Highly polarized views exist with regard to foods produced from GM plants and animals, with opponents referring to them as "Frankenfoods" and proponents insisting that they are **substantially equivalent to** and inherently **do not pose more risk than foods from unmodified organisms**. Many scientific organizations believe that genetic engineering is merely an extension of breeding techniques and is **no more unsafe as the genetic manipulations of conventional breeding methods.** For instance, the American Association for the Advancement of Science in a statement issued in 2012 pointed out that the EU had invested more than €300 million in research on the biosafety of GMOs and concluded in its report based on more than 130 research projects over 25 years involving more than 500 independent research groups, that biotechnology, in particular **GMOs, are not per se more risky than conventional breeding technologies** (AAAS, 2012). Subsequently, the US FDA issued guidance on voluntary labeling of foods from genetically engineered sources stating that under the federal FD&C Act, "*the FDA can only require additional labeling of foods derived from GE sources if there is a material difference—such as different nutritional profile—between the GE product and its non-GE counterpart*" (FDA, 2015). US polls on GE food labeling show that the majority (89%) favor mandatory labeling with only 6% opposed to it. These views were widespread across demographic lines: Democrats (92% favor, 2% oppose), Independents

(89% favor, 7% oppose), and Republicans (84% favor, 7% oppose) (The Mellman Group, Inc., 2015). On July 29, **2016, President Obama signed into law a bill that will require labeling of GM ingredients being marketed in the United States**—food packages would need to carry a text label, a symbol or an electronic code readable by a smartphone whether the food contains GMOs (Jalonick, 2016). Earlier Vermont state laws had made it mandatory for GMO foods to be labeled as "produced with genetic engineering."

The European Union has enforced consumer "right to know" laws for GM foods through **Regulation (EC) No. 1830/2003** that mandates the **traceability and labeling of GMOs** and of **food and feed products produced from GMOs** (European Union, 2003). Under this regulation, products such as flour, oils, and syrups have to be labeled as GM if derived from GM crops. However, products produced with GM technology (for instance, cheese produced using GM enzymes, as well as meat, milk, or eggs from animals fed on GM feed) do not have to be labeled. Traceability requirements enable tracking GMOs and GM food/feed products at all stages of the supply chain. This means that all operators involved (such as the farmers/producers of food or feed) must provide customers with information regarding the product, or ingredients in the product, having been derived from GMOs. Also, a record of transactions within the supply chain is to be maintained by all operators and made available for a period of 5 years.

Several nations have also introduced **"GM-free"** labels that indicate that specific measures have been taken to strictly exclude the presence or use of GMOs in the food/feed products.

KEY TAKEAWAYS

Safety considerations for GM foods:

- The FDA considers food from GM animals to be not per se more risky than that from conventionally bred animals and only requires voluntary labeling of foods derived from GE sources
- The EU mandates traceability and labeling of GMOs and food/feed products produced from GMOs
- Several nations have introduced "GM-free" labels to indicate that the product contains no GM ingredients.

12.5.4 SAFETY CONSIDERATIONS FOR MARKET APPROVAL OF BIOPHARMACEUTICALS

Biopharmaceuticals [also known as "biologics" or "biologic(al) medicinal products"] are medicinal products manufactured by biotechnology methods from living organisms or their products and include all recombinant proteins, monoclonal antibodies, vaccines, blood/plasma-derived products, nonrecombinant culture-derived proteins, and cultured cells and tissues. Although technical differences exist in the manner in which biologics are regulated in different regions such as the United States and Europe, efforts have been made to harmonize requirements for market approval of this class of medicines to ensure patient safety (Kingham, Klasa & Carver, 2014). This is because regulatory authorities world over recognize that biologics (unlike chemical drugs) are largely complex in structure and susceptible to variation during manufacture.

In the United States, for market approval, the sponsor of a chemical (nonbiologic) drug must submit a New Drug Application (NDA) that shows that the drug is safe and effective. But in the case of a biological product, the **Biologics License Application (BLA)** must prove that the product is *"safe, pure, and potent."* This means that for approval biologics must undergo **laboratory and animal testing to define their pharmacologic and toxicologic effects and prove clinical benefit in human clinical trials**. For **nonclinical studies for biologics**, the FDA has adopted the International Conference on Harmonization (ICH) of Technical Requirements for Registration of Pharmaceuticals for Human Use **S6 guidelines** (ICH, 2011). In order to conduct clinical tests, the sponsor must first have an **Investigational New Drug (IND) Application** in effect (an IND generally goes into effect 30 days after the application, unless on review the FDA places the trial on hold, for instance because it deems it to place trial patients in unreasonable risk). **The IND should therefore provide sufficient proof of safety in preclinical trials for the conduct of a clinical trial**. The FDA has adopted the guideline **for Good Clinical Practice (E6 guideline)** developed by the ICH (1996) for the conduct of clinical trials in order to protect clinical trial subjects and to ensure the integrity of data collected during the trial (see Section 5.6 in Chapter 5: Relevance of Ethics in Biotechnology). **In addition to the nonclinical and clinical data, the BLA should also contain a full description of manufacturing methods for the product; stability data substantiating the expiration date; product samples along with summary of test results for the batch from which derived; as well as details of address of manufacturing unit, labeling, packaging, and enclosures**.

In Europe, the regulatory authority Committee for Medicinal Products for Human use (CHMP) of the European Medicines Agency (EMA) defines biologics "largely by the method of manufacture." The CHMP has also adopted the **ICH S6 as guideline for preclinical testing** of biologics. Subsequent to the preclinical trials, clinical trials have to be conducted before a **Market Authorization Application (MAA)** can be made. **Clinical trials of biologics must comply with Directive 2005/28/EC on Good Clinical Practice and the ICH E6 guideline** adopted by CHMP. The principles for clinical trials detailed in the directive and guideline ensure that the rights, safety, and well-being of trial subjects take precedence over the interests of science and society and the ethical principles of the World Medical Association's Declaration of Helsinki (see Box 5.1 in Chapter 5: Relevance of Ethics in Biotechnology) are upheld. Under the **Clinical Trials Directive**, information regarding trials must be recorded in the European database of clinical trials accessible only to other competent authorities, the EMA and the European Commission. The MAAs for biologics in addition to providing standard information described in the Medicines Directive, also has **special information requirements**, such as: details of manufacturing process; origin and history of starting materials; should demonstrate that the active substance complies with special measures for preventing transmission of animal spongiform encephalopathies; should demonstrate that cell banks if used are stable; provide information on adventitious agents that may be present; describe origin, criteria, procedures for collection, transportation, and storage of starting material if medicines derived from blood or plasma; base vaccine production on a seed lot system and established cell banks if possible; and describe the manufacturing facilities and equipment. The manufacture of biologics is expected to comply with **GMPs** during all clinical trial phases and after market approval.

KEY TAKEAWAYS

United States:

- BLA must prove product is safe, pure, and potent
- Nonclinical trials to comply with ICH S6 Guideline
- Sponsor should have an Investigational NDA in effect for conduct of clinical trials
- Clinical trials should comply with Good Clinical practice, the ICH E6 Guideline

Europe:

- The CHMP of EMA also adopts ICH S6 for preclinical testing
- Clinical trials must comply with Directive 2005/28/EC and ICH E6 Guideline
- MAA should also have details of manufacturing process for the biologic

12.5.5 SAFETY CONSIDERATIONS FOR MARKET APPROVAL OF BIOSIMILARS

Biosimilars (also known as "follow on biologics" or "subsequent entry biologics") are medicinal products similar to an original "innovator" biologic that can be manufactured when the original product's patent expires. Biosimilars can therefore be produced by different companies and very often use different starting materials as these companies may not have access to the original cell line, or the exact fermentation or purification method used by the originator. In order to be approved for commercial production, companies have to demonstrate to the regulatory authorities that the product is **"similar" in terms of safety and efficacy to the reference product** (hence, the term "biosimilar"). The approval process for biosimilars is not the same as for generic versions of small molecule drugs that are products of easily defined synthetic or semisynthetic processes. As biologics are complex and as mentioned earlier, prone to variations, product quality, and integrity will differ for each manufacturer. Regulatory authorities such as the EMA, the FDA, and Health Canada each have issued specific guidance on the requirements for the approval of biosimilars (Blank et al., 2013). **The approval procedure is based on the demonstration of "comparability" of the structure and function(s), pharmacokinetic profiles and pharmacodynamics effects/efficacy of the biosimilar, to the approved biologic**. Biosimilars, as do biologics, also present some risk of adverse reactions or unwanted immune reactions to the medicine. In order to ensure patient safety, the introduction of biosimilars requires a specifically designed **pharmacovigilance plan**. The EMA needs a risk management plan to be submitted along with the market approval application and requires that the company provides regular safety update reports after the product is in the market. In the United States, a drug approved for marketing has to be revaluated for its safety and efficacy once every 6 months for the first 2 years, and subsequently every year, reports of which are to be filed with the FDA.

12.6 GOOD MANUFACTURING PRACTICES

GMPs are systems that provide proper design, monitoring and control of manufacturing processes and facilities, and thereby assure the identity, strength, quality, and purity of drug products.

They are often referred to as Current GMPs, indicating that manufacturers are to use state-of-the-art, or the most modern technologies and systems. The primary aim of GMP is to diminish risks inherent in production of pharmaceuticals, such as cross-contamination and false labeling.

The first draft text of GMP was prepared in 1967 by the World Health Organization (WHO) and published as an Annex to its 22nd report in 1968. Subsequent developments have resulted in revisions and incorporation of the concept of validation (available online at http://www.who.int/medicines/areas/quality_safety/quality_assurance/production/en/). The main principles of GMP for pharmaceutical products were updated and published in 2014 as **Annex 2** of the **WHO Technical Report 986** (WHO, 2014). Biologics warrant special considerations and **GMP for biologics was first published in 1992**. The current updated version has been published as **Annex 3** of the **WHO Technical Report 996** in 2016 (WHO, 2016). These documents serve as guidance for national regulatory authorities and for manufacturers of pharmaceutical products and could be incorporated into national legal requirements. The GMP for biologics address manufacturing procedures that involve growth of microorganisms and eukaryotic cells; extraction of substances from biological tissues; recombinant DNA and hybridoma techniques; and propagation of microorganisms in embryos or animals. For more details, see Box 12.2.

BOX 12.2 WHO GOOD MANUFACTURING PRACTICES FOR BIOLOGICAL PRODUCTS

The WHO GMP for biological products applies to the manufacture of medicinal products including *"allergens, antigens, vaccines, certain hormones, cytokines, monoclonal antibodies (mAbs), enzymes, animal immune sera, products of fermentation (including products derived from rDNA), biological diagnostic reagents for in vivo use and advanced therapy medicinal products (ATMPs) used for example in gene therapy and cell therapy"* (WHO, 2016). Special considerations and precautions are warranted in the manufacture of these products because unlike other medicinal products manufactured by defined and mostly consistent chemical or physical methods, biologics are derivatives of biological processes which may be inherently variable. Quality Risk Management (QRM) principles are, therefore, especially important to this class of medicines and extend across all stages of the manufacturing process including: material sourcing and storage; manufacture and packaging; quality control; quality assurance, storage, and distribution activities.

Personnel: Only trained personnel with adequate scientific experience should handle the different steps in the manufacture of biologics.

Starting materials: The source, origin, and suitability of active substances starting materials, buffers and media, and other components should be documented and the information retained for at least 1 year after the expiry date of the finished product as it may be useful in investigating adverse events if it occurs. All suppliers should be initially qualified and identity tests performed on each batch of supplies without adversely affecting the quality of the product in order to prevent contamination or cross-contamination. Sterilization of starting material if required should be done with heat whenever possible. Risk of contamination of the starting material during passage through the supply chain should be assessed.

Seed lots and cell banks: Appropriate controls over sourcing, testing, transport, and storage should be exercised when human or animal cells are used as feeder cells in the manufacture process. In order to prevent genetic drift due to passaging, a system of master seed lots or cell banks (MCB) and working cell banks (WCB) should be set up. The number of passages between the seed lot or cell bank and the finished product should be consistent with the marketing authorization application. Establishment and handling of the MCB and WCB should be performed under appropriate conditions, and during establishment, no other infectious agent should be simultaneously handled in the same area or by the same personnel. Appropriate quarantine and release procedures should be followed for the MCB and WCB.

(Continued)

BOX 12.2 (CONTINUED)

Also, the seed banks should be handled in a manner as to minimize risk of contamination, alteration, or cross-contamination. Storage and handling conditions should be defined, access to the material restricted to authorized personnel, and records maintained as to location and identity.

Premises and equipment: Quality risk management (QRM) principles should be adhered to in the handling of preparations containing live microorganisms or viruses, which includes avoiding the handling of organisms in areas used for processing of other pharmaceutical products. The use of closed systems to improve sepsis, and containment, should be considered wherever possible. Adequate attention is to be given to cleaning and sanitation measures.

Containment: Airborne dissemination of live microorganisms and viruses including those from personnel is to be avoided. Drainage systems should be designed to allow decontamination or effective neutralization of effluents and minimize risk of cross-contamination. For handling of pathogenic organisms of Biosafety risk group 3 or 4, and/or spore forming organisms, dedicated production areas should be used. Air-handling systems should be designed, constructed, and maintained to prevent crosscontamination between different manufacturing areas. Areas where Biosafety risk group 3 or 4 organisms are handled should always be under negative air pressure.

Clean rooms: The WHO GMP for sterile pharmaceutical products defines and establishes requirements for clean areas for the manufacture and aseptic fill of sterile products. Specific guideline is also available for the production of vaccines. The degree of environmental control of particulate and microbial contamination of the production area would depend on the potential level of contamination in the starting material and risks to the finished product.

Production: Typically biologics would require conditions, media, and reagents that promote growth of cells or microbes in axenic conditions; hence, effective technical and organizational measures are to be taken to prevent contamination and cross-contamination. This includes the design of the facility as well as the processes involved that need to be in keeping with QRM principles.

Labeling: Information to be provided on the container (inner) label as well as the packaging (outer label) should be readable and legible and the content approved by the national regulatory authority. Care is to be taken for the label to remain attached under different storage conditions, including ultralow temperatures of the product.

Validation: The handling of live material, and cleaning, is the major aspects of biological product manufacturing that require validation. It plays an important part in production consistency, control of critical process parameters, and product attributes. A QRM approach is to be adopted to determine the scope and extent of validation.

Quality control: Special consideration is to be given to the nature of the materials being sampled for quality control and testing. Samples for postrelease use belong to two categories: the reference samples and the retention samples, which for finished products may be presented as fully packaged units. Reference samples of biological starting materials should be retained under recommended storage conditions for at least a year. Retention samples of finished product should be stored in their final packaging for at least a year after the expiry date under the recommended storage conditions. The traceability, proper use, and storage of reference standards should be ensured, defined, and recorded. All analytical methods used in quality control of biological products should be well characterized, validated, and documented; the fundamental parameters of validation include linearity, accuracy, precision, selectivity/specificity, sensitivity, and reproducibility.

Documentation (batch-processing records): Processing records of regular production batches should provide a complete account of the manufacturing activities. Manufacturing batch records are to be retained for at least a year after the expiry date of the batch of the biological product.

Use of animals: Animals may be used for the manufacture or quality control of biological products. Live animals are to be avoided in the production area unless otherwise justified. Embryonated eggs if applicable are allowed in the production area. If extraction of tissues or organs is required, special care is to be taken to prevent contamination of the production area. Areas used for performing tests should be well separated from areas used for manufacturing and should have a separate ventilation system. The animals are to be properly housed and care to be taken to prevent and monitor infections.

Reference

WHO (2016). *WHO good manufacturing practices for biological products*. Annexe 3, WHO Technical Report Series 996. Retrieved from http://www.who.int/medicines/publications/pharmprep/WHO_TRS_996_annex03.pdf?ua=1.

12.7 SUMMARY

One of the most significant aspects of recombinant DNA technology and one that has spurred innovation in the field is the possibility to commercially exploit the technology. Safety considerations that prevent harm to human and animal health and to the environment could often be different at the laboratory scale and at a level required for commercialization. Most large-scale applications rely on the modified organism being no more dangerous than the nonmodified host. In instances where the risk assessment of the host organisms indicates a possibility for causing disease or unforeseen adverse effects, physical and/or biological containment may offer solutions. This chapter examined the mechanisms that ensure safety in large-scale applications of GMOs such as GILSP for GM microorganisms, and physical and biological containment appropriate for field release of GM crops and transgenic animals. The chapter also discussed GMPs for the production of medicinal products from biological sources.

REFERENCES

AAAS (October 20, 2012) *Statement by the AAAS Board of Directors on labelling of genetically modified foods.* American Association for the Advancement of Science. Retrieved from https://www.aaas.org/sites/default/files/AAAS_GM_statement.pdf.

Blank, T., Netzer, T., Hildebrandt, W., Vogt-Eisele, A., & Kaszkin-Bettaf, M. (2013). Safety and toxicity of biosimilars — EU versus US regulation. *Generics and Biosimilars Initiative Journal, 2*(3), 144–150. Retrieved from http://gabi-journal.net/safety-and-toxicity-of-biosimilars-eu-versus-us-regulation.html.

Breyer, D., Goossens, M., Herman, P., & Sneyers, M. (2009). Biosafety considerations associated with molecular farming in genetically modified plants. *Journal of Medicinal Plants Research, 3*(11), 825–838. Retrieved from http://www.academicjournals.org/article/article1380379852_Breyer%20et%20al.pdf.

European Union (2003) Regulation (EC) No 1830/2003 of the European Parliament and of the Council of 22 September 2003 concerning the traceability and labelling of genetically modified organisms and the traceability of food and feed products produced from genetically modified organisms and amending directive 2001/18/EC. Retrieved from http://eur-lex.europa.eu/LexUriServ/LexUriServ.do?uri = OJ:L:2003:268:0024:0028:EN:PDF.

FDA (November 19, 2015) *FDA takes several actions involving genetically engineered plants and animals for food.* [FDA News Release] Retrieved from http://www.fda.gov/NewsEvents/Newsroom/PressAnnouncements/ucm473249.htm.

FDA (August 5, 2016) *Oxitec Mosquito.* [Webpage] Retrieved from http://www.fda.gov/AnimalVeterinary/DevelopmentApprovalProcess/GeneticEngineering/GeneticallyEngineeredAnimals/ucm446529.htm.

Fox, J. L. (2003). Puzzling industry response to ProdiGene fiasco. *Nature Biotechnology, 21*, 3–4.

Gottlieb, S., & Wheeler, M. B. (2011). *Genetically engineered animals and public health: Compelling benefits for health care, nutrition, the environment, and animal welfare.* Washington, D.C: Biotechnology Industry Organization. Retrieved from https://www3.bio.org/foodag/2011_ge%20animal_benefits_report.pdf.

ICH (1996) *Guideline for good clinical practice E6(R1).* International Conference on Harmonisation of Technical Requirements for Registration of Pharmaceuticals for Human Use. Retrieved from http://www.ich.org/fileadmin/Public_Web_Site/ICH_Products/Guidelines/Efficacy/E6/E6_R1_Guideline.pdf.

ICH (2011) *Preclinical safety evaluation of biotechnology-derived pharmaceuticals S6(R1).* International Conference on Harmonisation of Technical Requirements for Registration of Pharmaceuticals for Human Use. Retrieved from http://www.ich.org/fileadmin/Public_Web_Site/ICH_Products/Guidelines/Safety/S6_R1/Step4/S6_R1_Guideline.pdf.

Jalonick, M.C. (July 29, 2016) (Reporter) *Obama signs bill requiring labelling of GMO foods*. [Associated Press] Retrieved from http://www.usnews.com/news/business/articles/2016-07-29/obama-signs-bill-requiring-labeling-of-gmo-foods.

Kingham, R., Klasa, G., & Carver, K. H. (2014). Key regulatory guidelines for the development of biologics in the United States and Europe. In W. Wang, & M. Singh (Eds.), *Biological drug products: Development and strategies* (First Edition, pp. 75−109). John Wiley and Sons. Retrieved from https://www.cov.com/~/media/files/corporate/publications/2013/10/chapter4_key_regulatory_guidlines_for_the_development_of_biologics_in_the_united_states_and_europe.pdf.

OECD (1986) *Recombinant DNA safety considerations*: Recommendations of the council concerning safety considerations for applications of recombinant DNA organisms in industry, agriculture and the environment. Retrieved from https://www.oecd.org/sti/biotech/40986855.pdf.

OECD (1992) *Safety considerations for Biotechnology*. Organization for Economic Co-operation and Development. Retrieved from https://www.oecd.org/sti/biotech/2375496.pdf.

The Mellman Group, Inc., (November 23, 2015) *U.S. polls on GE Food Labeling*. Retrieved from http://www.centerforfoodsafety.org/issues/976/ge-food-labeling/us-polls-on-ge-food-labeling#.

USDA (2006) *Permitting genetically engineered plants that produce pharmaceutical compounds*. BRS factsheet. Retrieved from https://aglearn.usda.gov/customcontent/APHIS-BRS-RegProcess/images/brs_fs_pharmaceutical_02-06.pdf.

WHO (2014) *WHO good manufacturing practices for pharmaceutical products: main principles*. Annexe 2, WHO technical report series 986. Retrieved from http://www.who.int/medicines/areas/quality_safety/quality_assurance/TRS986annex2.pdf?ua=1.

WHO (2016) *WHO good manufacturing practices for biological products*. Annexe 3, WHO technical report series 996. Retrieved from http://www.who.int/medicines/publications/pharmprep/WHO_TRS_996_annex03.pdf?ua=1.

FURTHER READING

Richman D.A. (n.d.) FDA basics for biotech drugs, biologics and devices. [Web Article] King and Spalding. Retrieved from http://www.kslaw.com/library/pdf/richman_fda_basics_biotech.pdf.

RELEVANCE OF INTELLECTUAL PROPERTY RIGHTS IN BIOTECHNOLOGY

13

"Intellectual property is a key aspect for economic development."
-Craig Venter, American biotechnologist and entrepreneur.

CHAPTER OUTLINE

13.1 Introduction 292
13.2 Intellectual Property Rights 293
13.3 Role of Intellectual Property Rights in Trade... 293
13.4 International Conventions and Treaties for Protection of IPRs 295
    13.4.1 Paris Convention 296
    13.4.2 World Intellectual Property Organization 296
13.5 Intellectual Property Rights Issues in Biotechnology 297
    13.5.1 Issues of Patentability 298
    13.5.2 Issues of Ownership... 302
    13.5.3 Issues of Enforcement 302
    13.5.4 Issues of Sharing of Costs and Benefits... 306
    13.5.5 Issues of Ethics... 307
13.6 Summary 308
References 308

An Introduction to Ethical, Safety and Intellectual Property Rights Issues in Biotechnology.
DOI: http://dx.doi.org/10.1016/B978-0-12-809231-6.00013-2

Louis Pasteur (1822—1895)

13.1 INTRODUCTION

Within a few years of the creation of the first genetically modified organism using recombinant DNA technology, Herbert Boyer and Robert Swanson founded Genentech (in 1976) with the aim of commercially exploiting the new technology. Since then, major developments in the technology have been funded and new commercially viable applications explored by the biotechnology industry. Increasing global competition, high investments in research and development (R&D), and the need for highly skilled personnel has created an environment in which biotechnology companies feel the need for exclusive marketing rights and protection of their innovation. Applying existing instruments of intellectual property rights such as patents (for microbes/plants/animals) or copyrights (for protein and nucleic acid sequences) directly to the products of biotechnology may not be possible in most instances. Being products of nature these are not considered as patentable matter by most countries. The challenge is to evolve mechanisms that allow disclosure of the technology so as to encourage innovation, while simultaneously allowing developers exclusive rights to commercially exploit their innovation in order to recover R&D and associated costs of development, and make a profit in order to sustain further R&D efforts. This chapter discusses some of the issues in the protection of intellectual property rights (IPRs) in biotechnology.

13.2 INTELLECTUAL PROPERTY RIGHTS

Generally described as being "intangible," IPRs are nevertheless **similar to other legal property rights** in that they can be **sold, leased, gifted, licensed, or waived** and allow the holder **exclusive rights to use or commercially exploit the product of his/her intellect**. Different forms of IPRs are recognized such as:

- Patents which protect inventions
- Copyrights which protect works of literature and art
- Trademarks
- Registered (industrial) design
- Layout designs of Integrated Circuits
- Trade secrets
- Geographical Indications

Except for copyright which is global in nature (applicable in all nations that are members of the Berne convention), **IPRs are territorial, confined to the country in which awarded**. These are **monopoly rights** which mean that the consent of the rights owner is required for use. IPRs are **awarded by the state for a limited period**, except for trademarks and geographical indications which have an indefinite life provided they are renewed by paying official fees, and trade secrets which have an infinite life and do not have to be renewed (see Box 13.1 for the salient features of different forms of IPR).

13.3 ROLE OF INTELLECTUAL PROPERTY RIGHTS IN TRADE

Historically, IPRs **evolved as a means of encouraging the introduction of new technologies, while sustaining trade**, by the award of monopolies of specified duration. In the 14th century, in United Kingdom (UK), monopolies on trade were banned, but an exception was made for inventions. **Patents emerged as exemptions to the ban on monopolies and were granted by the Crown** with the aim of introducing new technology to UK. **The term of the patent during which the holder could enjoy exclusive marketing rights was specified as 14 years, which was the estimated time required to train two consecutive apprentices in the art of manufacturing the invention**. Patents were thus a *quid pro quo* arrangement—exclusive marketing rights, in exchange for disclosure of the invention. **IPRs gained momentum subsequent to the industrial revolution** in the 18th and 19th centuries. The printing press for instance made mass production and duplication of literary works easy, giving birth to copyright laws. Trade at an international level increased after the industrial revolution due to improvements in manufacturing skills and shipping. It also led to a better exchange of ideas and ushered in an era in which innovation dictated marketability. IPRs in its current form have their origin in the hosting of international exhibitions of inventions in 1873 by the Government of the Empire of Austria-Hungary. As a result of inventors from the United States demanding some form of protection and marketing exclusivity in return for exhibiting their inventions in Europe, Austria passed a law securing temporary protection to all foreign participants for their inventions, trademarks and industrial designs. In 1883, the *International Convention for the Protection of Industrial Property* was signed in Paris by 11 states.

BOX 13.1 INTELLECTUAL PROPERTY RIGHTS (IPRs)

Intellectual Property is defined broadly as *"legal rights which result from intellectual activity in the industrial, scientific, literary and artistic fields"* (WIPO, 2004). Intellectual Property Rights are traditionally divided into:

- Industrial property rights
- Copyright and related rights

Summarized in the table below are the key aspects of different intellectual property instruments:

Type of IPR	What Does it Cover	Term of the IPR	International Conventions
Patent	Inventions	20 years from date of filing	Paris Convention
Copyright and related Rights	Forms of creativity primarily concerned with mass communication, e.g., books, paintings or drawings, music, poems	Life of the author and not less than 50 years after the death of the author	Berne Convention
Traditional Cultural Expressions (TCEs)	Music, musical instruments, stories, art, handicrafts, words, names and insignia, performances, textiles, carpet, jewelry designs, forms of architecture		Undertaken in cooperation with UNESCO
Trademark (™), Service Marks, Collective Marks, and Certification Marks	Signs that individualizes the goods of a given enterprise and distinguishes them from the goods of its competitors	Usually 10 years, renewed indefinitely on payment of additional fees	Madrid Agreement
Industrial Designs	Formal or ornamental appearance for mass produced items (original ornamental or nonfunctional features that result from design activity)	Usually 15 years, renewal is usually subject to payment of renewal fees	Hague Agreement
Integrated Circuits	Layout designs (topographies) of integrated designs	Usually 15 years	
Geographical Indications (GI)	Names and symbols which indicate a certain geographical origin of a given product—applied to products whose quality and characteristics are attributable to their geographical origin	Indefinitely	TRIPS Agreement
Appellations of Origin (AO)	Special kind of indication of source	Indefinitely	Lisbon Agreement
Protection against unfair competition, Trade secrets	Confidential business information which provides an enterprise a competitive edge (manufacturing or industrial secrets and commercial secrets)	Indefinitely, until public disclosure of the secret occurs	

Reference
WIPO (2004). WIPO intellectual property handbook, Reprinted 2008. Retrieved from http://www.wipo.int/edocs/pubdocs/en/intproperty/489/wipo_pub_489.pdf.

After the Second World War, at the United Nations Conference on Trade and Employment, the creation of an International Trade Organization was attempted. Although that did not materialize, 23 nations entered into a multilateral agreement, ***General Agreement on Tariffs and Trade (GATT)***. The agreement was signed on October 30, 1947 and took effect on January 1, 1948. The purpose of GATT was the reduction of tariffs and elimination of trade barriers on a "reciprocal and mutually advantageous basis." Between its inception in 1947 and its final round in 2001 (Doha round), the

GATT held nine rounds of discussions on tariffs and nontariff measures to foster international trade and increased its membership to 159 countries. In the eighth Uruguay Round Agreements which lasted from 1986 to 1994, the founding members established the *World Trade Organization (WTO)* effective January 1, 1995. The WTO which has its headquarters in Geneva is vested with the responsibility of implementing, administering and monitoring GATT. Its main function is to *"ensure that trade flows as smoothly, predictably and freely as possible"* (WTO, 2014). The WTO also oversees the General Agreement on Trade in Services, Intellectual Property Agreement and has a major role in resolving trade quarrels under the Dispute Settlement Understanding. Since over three quarters of the WTO members are developing or least-developed countries, the WTO assists in trade policy issues through technical assistance and training programs.

The WTO sets minimum standards for the different forms of IPRs through administering the *Agreement on Trade-Related Aspects of Intellectual Property Rights (TRIPS)*. This agreement requires WTO members to provide *copyright* and *related rights* (covering the rights of performers, producers of sound recordings, broadcasting organizations); *trademarks* (including service marks); *geographical indications* (including appellations of origin); *industrial designs*; *patents* (including new plant varieties); *layout designs of integrated circuits*; and *undisclosed information* (including trade secrets and test data) (WTO, 2016).

The three main features of the TRIPS agreement are as follows:

- *Standards*: In each area of intellectual property covered by the TRIPS Agreement, the Agreement establishes the minimum standards of protection to be provided by each member.
- *Enforcement*: The Agreement lays down general principles applicable to all IPR enforcement procedures, and the procedures and remedies that must be available so that right holders can effectively enforce their rights.
- *Dispute settlement*: Disputes between WTO members regarding TRIPS obligations are subject to the WTO dispute settlement procedures.

KEY TAKEAWAYS

- Patents developed as exemptions to the ban on monopolies in the 14th-century UK
- IPRs gained momentum during the industrial revolution in the 18th and 19th centuries
- 1883—the *International Convention for the Protection of Industrial Property* was signed in Paris by 11 states.
- 1947—*General Agreement on Tariffs and Trade (GATT)*, a multilateral agreement between 23 nations
- 1995—*World Trade Organization (WTO)* established in the Uruguay Round Agreements of GATT.

13.4 INTERNATIONAL CONVENTIONS AND TREATIES FOR PROTECTION OF IPRs

Historically, IPRs have been protected by two major international conventions: The *International Convention for the Protection of Industrial Property* (which came to be known as the *Paris Convention)* and the *Berne Convention for the Protection of Literary and Artistic Works* (known as the *Berne Convention*). The Berne convention is an international agreement governing copyright, which was first accepted in Berne in 1886.

13.4.1 PARIS CONVENTION

The *Paris Convention* signed in 1883 has since been revised 6 times: at Brussels in 1900, Washington in 1911, The Hague in 1925, London in 1934, Lisbon in 1958, and Stockholm in 1967. It was amended in 1979 (http://www.wipo.int/treaties/en/ip/paris/). The number of members in the convention has swelled from the initial 11 to 176 (in 2014). The Paris Convention established three main provisions as follows:

- *Right to national treatment*—signatories are obligated to extend the same rights under the law to nonresidents and nonnationals as citizens of a country.
- *Right to priority*—a 12-month period to file a Paris Convention application is allocated to a patent applicant so that the initial priority date in the home country may be recognized as priority date in applications filed in countries that are party to the Paris Convention.
- *Common rules* and a unified framework for implementation of the convention.

There are several other treaties that are open only to members of the Paris Convention, including:

1. *Patent Cooperation Treaty (PCT)*: a multilateral treaty that entered into force in 1978. The PCT facilitates obtaining priority for an invention in any of the member countries by designating them in the PCT application without having to file separate patent applications in each country of interest.
2. *Budapest Treaty*: signed in 1973 and amended in 1980, this treaty governs registration of deposits of microorganisms and cell-lines in approved culture collections for the purpose of applying for a patent in any country that is party to the treaty.
3. *UPOV (Union pour la Protection des Obtentions Vegetales, International Union for the Protection of New Varieties of Plants)*: an international convention that provides a common basis for the examination of plant varieties for determining whether it merits protection under the UPOV.
4. *Madrid Agreement and Madrid Protocol*: makes it possible to protect a mark in multiple countries by obtaining an international registration that has effect in each of the designated contracting parties.
5. *Hague Agreement*: first adopted in 1925, the agreement establishes a system that allows industrial designs to be protected in multiple countries or regions.
6. *Lisbon Agreement*: facilitates the means of obtaining protection for an appellation of origin (AO) in the contracting parties through a single registration.

The Paris Convention is administered by the World Intellectual Property Organization (WIPO).

13.4.2 WORLD INTELLECTUAL PROPERTY ORGANIZATION

The origins of WIPO go back to the International Bureau of Intellectual Property that was created in 1893 by the unification of the secretariats of the Paris Convention and the Berne Convention. The "Convention Establishing the World Intellectual Property Organization" met in Stockholm in

1967, and the WIPO Convention entered into force in 1970. In 1974, the WIPO became a specialized agency in the United Nations system of organizations.

The objectives of WIPO are as follows:

- To promote the protection of intellectual property throughout the world through cooperation among the states, and where appropriate through collaboration with other international organizations
- To ensure administrative cooperation among the intellectual property Unions created by the Paris and Berne Conventions and several subtreaties signed by members of the Paris Union

Since January 1, 1996, WIPO has an Agreement with the WTO (*1995 WIPO–WTO Cooperation Agreement*) that provides for cooperation between the International Bureau of the WIPO and the Secretariat of the WTO in respect of assistance to the developing countries; notification and collection of intellectual property laws and regulations of WTO members; and notification of emblems of States and international organizations.

KEY TAKEAWAYS

International efforts for protection of IPRs are as follows:

- 1883—Paris Convention (*International Convention for the Protection of Industrial Property*) established
  - *Right to national treatment*
  - *Right to priority*
  - *Common rules*
    - *Patent Cooperation Treaty (PCT)*
    - *Budapest Treaty*
    - *UPOV*
    - *Madrid Agreement and Madrid Protocol*
    - *Hague Agreement*
    - *Lisbon Agreement*
- 1886—Berne Convention (*Berne Convention for the Protection of Literary and Artistic Works*)
- 1970—WIPO (created by unification of the secretariats of the Paris Convention and the Berne Convention)
- 1974—WIPO becomes a specialized agency in the United Nations system of organizations
- 1995—WIPO has an Agreement with the WTO (*1995 WIPO–WTO Cooperation Agreement*) to help developing nations with IPR issues among other duties.

13.5 INTELLECTUAL PROPERTY RIGHTS ISSUES IN BIOTECHNOLOGY

Innovations in biology have been perceived as being different from other industrial innovations for several reasons: New strains of microbes isolated from soil or water, or new varieties of plants and animals bred by conventional methods, are not patent eligible in most countries, being mere "discoveries," or "products of nature." Gene or protein sequences cannot be protected by copyrights, as they are not "original" or "creative" works. Also, establishing the "inventiveness" of new organisms has been challenging. However, there have been several innovations in biology deserving of IPR protection (for a review, see Maxham, 2015).

It is necessary that a patent application disclose sufficient details of the invention to allow a person skilled in the art to make the invention without undue experimentation. In the case of rare microorganisms, cells, or cell-lines that form part of the claim for a process or method, a written description was deemed insufficient since the public would not have access to the biological material crucial for enabling them to use the invention. Patent Offices in most countries therefore required inventors to deposit the claimed biological material (cells, cell-lines, or organisms) in a depository that can be accessed by the public. This meant that inventors would have to make a deposit in each country/region that they wanted protection for their invention. In 1977, the WIPO adopted the ***Budapest Treaty on the International Recognition of the Deposit of Microorganisms for the Purposes of Patent Procedure*** (WIPO, 1977). The Budapest Treaty allows inventors from contracting States to make a *"deposit of a microorganism with any 'international depositary authority,' irrespective of whether such authority is in or outside the territory of the said State."* **This eliminates the need to deposit in each country in which protection is sought**. On May 1, 2016, there were 45 such authorities (see WIPO, 1977). The number of submissions to the international depositary authority is over 96,000 till August 2016 (http://www.wipo.int/ipstats/en/).

13.5.1 ISSUES OF PATENTABILITY

Patents are IPRs that protect inventions (see Box 13.2) and provide one of the most powerful instruments in IPR. However, protecting innovations in biology using patents has been controversial primarily because living systems are considered products of nature, not conforming to conventional patent subject matter.

13.5.1.1 Patenting organisms

Perhaps the earliest record of patents given to living systems were those relating to the manufacture of beer awarded to **Louis Pasteur** (Federico, 1937). Pasteur held patents in at least four different countries, France, England, Italy, and the United States. Two patents granted in the United States in 1873 were for improvements in the process of manufacture and preservation of beer; one of the claims in the second patent is for *"Yeast, free from organic germs of disease, as an article of manufacture"* (see Federico, 1937).

New strains of microbes, isolated from soil or water, or new varieties of plants and animals bred by conventional methods, are not patent eligible in most countries The United States Patent and Trademark Office (USPTO) does however provide for **plant patents by Title 35 United States Code (U.S.C.),** Section 161 which states: *"Whoever invents or discovers and asexually reproduces any distinct and new variety of plant, including cultivated sports, mutants, hybrids, and newly found seedlings, other than a tuber propagated plant or a plant found in an uncultivated state, may obtain a patent therefor, subject to the conditions and requirements of title"* (USPTO, 2015a). The decision of the US Supreme Court to allow **novel organisms created by the use of modern biotechnology to be eligible for patents** prompted the formation of a new class of companies, the biotech companies, and the development of bioengineered products.

BOX 13.2 PATENTS

Patents are a form of intellectual property rights granted for inventions which, in exchange for a full disclosure of the invention, allow an inventor to exclude others from commercially exploiting the invention for a specified period of time. More specifically, a patent grant confers on the patent holder *"the right to exclude others from making, using, offering for sale, or selling"* the invention. (Note that, the patent does not confer the right to make, use, offer for sale or sell the invention, only to exclude others from doing so.) In order to qualify for award of a patent, an invention should typically be:

- novel (should not be available in documented literature referred to as "prior art"),
- should involve an inventive step (i.e., it should not be obvious to a person skilled in the art; not anticipated by prior art; ideally the invention should represent a technical advancement), and
- should have a utility or be capable of industrial application.

Patent laws in most countries recognize different types of patents such as:

1. *Utility patents* (product, or process patents): granted for the invention of a new and useful process, machine, article of manufacture, or composition of matter, or any new and useful improvement thereof
2. *Design patents*: granted for the invention of new, original, and ornamental design for an article of manufacture; and,
3. *Plant patents*: granted to anyone who invents or discovers and asexually reproduces any distinct and new variety of plant.

Most countries also specify the types of inventions that are not eligible for patents. Typically, these include:

- an invention that is based on the mere discovery of scientific principles, or the formulation of an abstract theory, or discovery of any living or nonliving substances occurring in nature
- an invention that is frivolous or which claims anything that is obviously contrary to well-established natural laws
- an invention the primary or intended use or commercial exploitation of which could be contrary to public order or morality or which cause prejudice to human, animal or plant life or health, or harm to the environment
- a mere discovery of a new form of a known substance which does not result in enhancement of the known efficiency of that substance, or the mere discovery of any new property or new use for a known substance, or of the mere use of a known process, machine or apparatus, unless such known process results in a new product or employs at least one new reactant
- a substance obtained by mere admixture resulting only in the aggregation of the properties of the components thereof or a process for producing such substances.
- a mere arrangement or rearrangement or duplication of known devices each functioning independently of one another in a known way
- an invention which, in effect, is traditional knowledge or which is an aggregation or duplication of known properties of traditional knowledge component(s)
- a mathematical or business method, or a computer program per se or algorithm, a mere scheme or rule or method of performing mental act or method of playing a game.

For socioethical reasons and not allowing patents to inhibit the transfer of benefits of new developments in science and technology from reaching the public, patents are not given for:

- any process for medicinal, surgical, curative, prophylactic, diagnostic, therapeutic, or other treatment of humans or animals to render them free of disease or to increase their economic value or that of their products
- a method of agriculture or horticulture
- inventions relating to atomic energy.

In Europe and several countries (like India, which has an agrarian economy, and agriculture provides livelihood for over 70% of the population) patents are not given for:

- plants and animals in whole or part thereof, other than microorganisms, but including seeds, varieties, and species
- biological processes for production or propagation of plants and animals.

(Continued)

BOX 13.2 (CONTINUED)

Patents are awarded by patent offices in each country/region in response to a patent application which includes sufficient details of the invention to enable a person with ordinary skill in the art to make and use it without difficulty. Patent rights are enforceable by law. Patents can be revoked, or compulsorily licensed depending on various circumstances. Under the European Patent Convention, post grant, patents may be opposed by an opposition procedure before the European Patent Office (EPO) that establishes that the subject matter of the patent is not patentable.

Utility and plant patents have a term of 20 years from date of filing. On expiry, the invention becomes public knowledge and free to make, use, and sell by anyone.

Four theories on the principle purpose of patents are:

- *"Invention-Inducement Theory: The anticipation of receiving patents provides motivation for useful invention.*
- *Disclosure Theory: Patents facilitate wide knowledge about and use of inventions by inducing inventors to disclose their inventions when otherwise they would rely on secrecy.*
- *Development and Commercialization Theory: Patents induce the investment needed to develop and commercialize inventions.*
- *Prospect Development Theory: Patents enable the orderly exploration of broad prospects for derivative inventions"* (Nelson & Mazzoleni, 1997).

Reference

Nelson, R.D., & Mazzoleni, R. (1997). Economic theories about the costs and benefits of patents. In: Intellectual Property Rights and the Dissemination of Research Tools in Molecular Biology: Summary of a workshop held in the National Academy of Science, February 15–16, 1996. Retrieved from http://www.ncbi.nlm.nih.gov/books/NBK233535/.

In 1972, microbiologist **Ananda Chakrabarty** filed a patent application for an invention to treat oil spills. Chakrabarty had isolated bacterial strains of *Pseudomonas* which were able to degrade hydrocarbons in petroleum to simpler substances which could be used as food by several organisms. Each of these strains could metabolize only one or few of the different hydrocarbons found in petroleum, so a mixture of bacterial strains (consortium) was required to treat oil spills. The problem was the consortium was seldom stable due to differences in growth and competition between the strains. Chakrabarty found that the genes for the enzymes which degrade hydrocarbons were present on small self-replicating circular DNA molecules known as plasmids present in bacterial cells. Using simple genetic techniques, Chakrabarty was able to mobilize different plasmids into a single bacterial strain. This modified bacterial strain was far more efficient at tackling oil spills than the consortium. The patent application included 36 claims related to the invention that were

1. for the modified bacterium;
2. for the method of producing the bacterial strain (process claims); and
3. for the inoculum composed of the bacterium and carrier material like straw that allowed the inoculum to float on the surface of water (process claims).

The patent examiner accepted the process claims but rejected the claim for the modified bacterium on the grounds that living things were not patentable subject matter under **Section 101 of Title 35 U.S.C**. While the Board of Patent Appeals agreed with the patent examiner, on appeal, the Court of Customs and Patent Appeals ruled in Chakrabarty's favor, stating, *"the fact that the microorganisms are alive is without significance for purposes of the patent law."* The Court concluded that the new bacteria were **not "products of nature"** as the **bacteria containing the**

plasmids are created and are not naturally occurring. In the now famous *Diamond v. Chakrabarty* case, Sidney Diamond, Commissioner of Patents and Trademarks appealed to the US Supreme Court for a writ of certiorari. The case was heard in March 1980, and the verdict given in June that year. The court ruled 5–4 in favor of Chakrabarty, concluding that a *"live human-made micro-organism is patentable subject matter under 35 U.S.C.* Section 101... *Respondent's organism constitutes a 'manufacture' or 'composition of matter' within that statute"* (United States Supreme Court, 1980). Chief Justice Burger in delivering the verdict quoted the congressional report accompanying the 1952 Patent Act that Congress intended statutory subject matter to include *"anything under the sun that is made by man."* The patent was granted by the USPTO on March 31, 1981. Although the modified bacteria in this case were not created using recombinant DNA technology, subsequently, a number of different transgenic organisms have been found to be eligible for protection by patents in the United States and elsewhere (see Chapter14: Patenting of Life Forms).

13.5.1.2 Patenting genes

Initial patents granted for DNA sequences were for those that specified the amino acid sequences of proteins of therapeutic applications such as insulin and growth hormone. These **patents were important to the nascent biotech industry** as a means of recovering cost of R&D, and generating funds for further growth. Subsequent developments in human genomics and the use of **gene sequences for testing and diagnostics** (see Chapters 1 and 7) provided new business opportunities in healthcare. Ideological differences in the IPR protection to be accorded to DNA sequences were apparent even while the human genome was being deciphered in the 1990s. Robert Cook-Deegan and Christopher Heaney in a paper published in 2010 recount that the *"battle over patents was part of a larger war over how to conduct the Human Genome Project"* (Cook-Deegan & Heaney, 2010). The differences in perceptions on the role of public and private sectors on the use of patents became apparent when J. Craig Venter, then with the National Institutes of Health (NIH), publicly announced in June 1991 that DNA sequences were the subject of two patent applications filed one day before the publication of the information. The sequences in question were expressed sequence tags (ESTs) of around 600 protein coding sequences from human brain, which could potentially be used for developing diagnostic kits. In 1991, NIH under Director Bernadine Healy who hoped to encourage rapid developments of products for disease treatment, filed patent applications for around 2750 ESTs from human genome. The patent applications were filed in the United States and through the Patent Cooperation treaty in Europe, Australia, Canada and Japan (Gold & Gallochet, 2001). The ensuing NIH EST patent controversy had the next NIH director, Harold Varmus, consult patent scholars for guidance. Their advice was that patents should be pursued only if it would advance commercialization without hindering science, but in their considered opinion, prospects for commercialization were dim but opportunities for impeding science were real. The controversial NIH EST patents were withdrawn in 1994, but only after they had been rejected in Europe. The policy of the NIH and others involved in the Human Genome Project codified in the 1997 **Bermuda Principles** was that the DNA sequence information should be released into publicly available databases within 24 hours of being generated with no restrictions on use (NHGRI, 2013). One fall out of the EST controversy was that it drew the attention of venture capitalists to business possibilities in genomics and private enterprise in sequencing and patenting genes (for a review see Cook-Deegan & Heaney, 2010). But **whether gene sequences are patentable subject matter has**

itself been a subject of much debate and discussion. Patent laws do not consider laws of nature and natural phenomenon, fundamental principles, and abstract ideas to be eligible for patents. In a much publicized case *Association for Molecular Pathology v. Myriad Genetics, Inc*. (see Box 14.2 in Chapter14: Patenting of Life Forms), some of the questions raised by the Justices of the Supreme Court during the hearing on April 15, 2013, were as follows:

- "How much modification must occur during the extraction process to make an otherwise natural product patent eligible?
- Is deciding where to cut DNA to isolate a gene considered an invention?
- Does isolated DNA have a new function/use that is different from DNA inside the body?
- Is cDNA materially different from genomic DNA?
- Will there be incentive for companies to invest in new discovery if they are not assured patent protection?" (Genetics Generation, 2015).

In a landmark **decision in 2013** putting to rest speculation on whether genes were patent eligible, the **US Supreme Court unanimously ruled that genes were not eligible for patents**, and that separating a gene from its surrounding genetic material is not an act of invention. The court did however find **cDNA patent eligible** as it is created in the laboratory and does not exist in nature.

13.5.2 ISSUES OF OWNERSHIP

When patents are obtained for genes, cell-lines, or diagnostic kits developed from cells, tissues, or organs from patients, often without their knowledge or consent, apart from ethical issues, it also raises issues of ownership. A case illustrative of this is *Moore v. Regents of California* (Box 13.3) tried in the US Supreme Court in 1990 that established the legal status of ownership of (discarded) human cells, tissues, and organs. In its verdict, the court **rejected Moore's claim to ownership** and felt that extending property rights to include medical samples would unnecessarily complicate medical research and practice. This rationale has **prevailed in the instance of the cell line developed from cancerous tissues** from Henrietta Lacks (**the HeLa cell line**). Although several patents have been obtained using the HeLa cell line, it has not been legally possible to make royalty payments to her descendants (see Box 13.4).

13.5.3 ISSUES OF ENFORCEMENT

Patent rights are enforced by law. Infringement of a patent consists of the unauthorized making, using, offering for sale or selling any patented invention (USPTO, 2015b). In cases of infringement, patentees can sue for relief in the appropriate federal court. Patent infringement is of **special concern to biotech companies since biotech products are difficult to create but easy to replicate** as they are often living organisms/cells capable of self-replication. This is **more so in the area of plant biotechnology**: "*Agrichemical companies devote significant resources toward investigating and prosecuting farmers for alleged seed patent violations. ... As of December 2012, Monsanto had filed 142 alleged seed patent infringement lawsuits involving 410 farmers and 56 small farm businesses in 27 states... DuPont, the world's second largest seed company hired at least 45 farm investigators in 2012 to examine planting and purchasing records of Canadian farmers to take*

BOX 13.3 MOORE V. REGENTS OF CALIFORNIA

This was a landmark case tried in the Supreme Court of California in 1990 that established the legal status of ownership of (discarded) human cells, tissues, and organs. The case was filed by John Moore on the basis that he was denied a share of the profits arising from the commercialization of cell lines derived from his organs and tissues.

Moore in 1976 was diagnosed with hairy cell leukemia and was treated for the same in the University of California, Los Angeles (UCLA) Medical Centre by Dr. David W. Golde and his team. In October 1976, Golde recommended removal of the spleen (splenectomy) in order to slow down the progress of Moore's disease. Prior to the operation, Golde obtained from Moore a signed written form consenting to the splenectomy and the disposal of any severed tissue or member by cremation. It appears however that Golde was aware of the financial benefit that could be derived from the tissues and body fluids from cancer patients, but failed to reveal this to Moore. Between 1976 and 1983, Moore visited the UCLA Medical Centre several times for treatment recommended by Golde. On each visit, samples of blood, serum, bone marrow, skin, and sperm were collected. In 1983 however, Moore was asked to sign a form that granted the UCLA rights that either he or his heirs might have to any cell line or potential products derived from blood or bone marrow derived from him. Moore did not sign this form and instead consulted an attorney. From further investigation, it emerged that Golde and his coworkers had been doing research on Moore's cancer and had established a cell line from Moore's T-lymphocytes sometime in 1979 which could be exploited commercially. Golde had withheld this information from Moore and requisitioned the consent as part of the process of applying for a patent in 1983. On March 20, 1984, US Patent 4438032 was issued for the cell line. The patent named Golde and coworker Shirley G. Quan as inventors and the Regents of the University of California as the assignee. Meanwhile, Golde had, with the help of the Regents, entered into an agreement conferring exclusive rights to the cell line and products derived from it with a company called Genetics Institute. In turn Golde became a paid consultant with 75,000 shares of common stock and he and the Regents were paid at least $330,000 over the next 3 years. The compensation was increased by $110,000 when Sandoz joined the agreement in 1982.

Named in the suit given by Moore were Dr. David W. Golde, the Regents of University of California, Dr. Shirley G. Quan, Genetics Institute Inc., and Sandoz Pharmaceutical Corporation. In 1988, the California Court of appeals ruled in favor of John Moore, implying that patients should have the power to control the use of their own tissue. The decision was appealed by the University of California and Golde. In 1990, the Supreme Court of California rejected Moore's claim of property rights to his discarded cells or any profit derived from them. For one, it did not accept that the cells were unique to Moore, or that his spleen should be protected as property in order to protect his privacy or dignity as that had already been addressed by the consent form he had signed. Any property in question was not the cells from Moore's body, but the cell line derived from the cells. The court felt that extending conversion of property (appropriation for own use without clear legal authority to do so) laws to include cells, tissues, and organs would make medical research complicated given the millions of samples handled by medical laboratories. Moore could possibly sue his doctor for nondisclosure of the financial implications, but he could hardly consent to the procedure and yet reserve the right to sell his body parts.

Dorney, M.S. (1990). Moore v. the Regents of the University of California: Balancing the need for biotechnology innovation against the right to informed consent. *Berkeley Technology Law Journal: 5(2) Article 4.* Retrieved from http://scholarship.law.berkeley.edu/cgi/viewcontent.cgi?article=1087&context=btlj.

---

samples from their fields to send to DuPont for genetic analysis" (Center for Food Safety & Save our Seeds, 2013). Highly publicized, many of the infringement suits have been portrayed as a David versus Goliath scenario with **farmers having to take on the might of multinational seed companies** as in the ***Monsanto Canada Inc. v. Schmeiser*** and the ***Bowman v. Monsanto Co.*** cases (see Box 13.5). The development of Genetic Use Restriction Technologies ("terminator" and "traitor" technologies, see Chapter 5: Relevance of Ethics in Biotechnology) would have facilitated the prevention of patent infringements as farmers would have had to buy the seed from companies every season; the farm saved seed would not have the same characteristics.

BOX 13.4 HENRIETTA LACKS AND THE HELA CELL LINE

The HeLa cell line was developed in the 1950s from a particularly aggressive strain of cervical cancer cells taken during a routine biopsy from a 30-year-old mother of five, Henrietta Lacks. She was treated for the disease by Dr. George Gey in the colored ward of The Johns Hopkins Hospital. As the head of tissue research, Dr. George Gey was at that time attempting to establish immortal cell lines that could be used in medical research. Taking tissue samples from poor and/or American Negro cancer patients being treated in the colored ward without informed written consent was not unusual and neither doctors nor their patients were aware of any of the ethical issues involved. Henrietta Lacks was diagnosed in 1950, and died within a year at the age of 31. But her cells lived on and became the first human cell line to be established. The reason why these cells were immortal while several others failed to grow in culture is not entirely understood. In addition to cervical cancer caused the Human Papilloma Virus (HPV) (multiple copies of HPV genome were later found in the HeLa cell line), Henrietta Lacks also had syphilis which probably suppressed her immune system. In any case, her cells pioneered research that led to a better understanding of the causes and treatment of human cancers. In 1952, the Tuskegee Institute set up a laboratory to supply the cell line to other researchers and laboratories which soon spawned a company named Microbiological Associates that supplied HeLa cells for profit. The cells were used in laboratories around the world and have been crucial for the development of vaccines, for instance, the polio vaccine. Although initially the cells were known by the pseudonym Helen Lane, as a tribute to George Gey who died of pancreatic cancer in 1970, the cells were correctly identified as having been sourced from Henrietta Lacks. The Lacks family learnt for the first time that her cells were still alive when scientists at the Johns Hopkins Hospital approached them for tissue samples for a genetic analysis. They soon learnt that her cells had also been commercialized. Henrietta's children found this upsetting—for one, they wondered how their mother's soul could rest in peace if her cells were still around; for another, the family was poor and could ill afford health insurance, and yet their mother's cells were part of a multibillion-dollar industry. Henrietta Lacks story caught public attention in 2010 with the publication of an award-winning book, which stayed on *the New York Times* best-seller list for 2 years, *"The Immortal Life of Henrietta Lacks"* by Rebecca Skloot. The book highlighted the fact that the development of the cell-line, the genetic and molecular analysis, and the commercial applications developed from the cell line had all been done without the knowledge or consent of Henrietta Lacks or her family, nor had they been the beneficiaries of the developments. The family felt their privacy had been violated, so much so that her grandchildren contacted the European Molecular Biology Laboratory and asked them to withdraw a paper that they had published on the genetic make-up of HeLa cells. The scientists withdrew the paper with apologies to the family. Meanwhile, Rebecca Skloot set up The Henrietta Lacks Foundation in 2010 by donating a portion of the book's proceeds and donations from the reading public and scientists who would like to do something in return for the family. The foundation provides scholarship funds and health coverage to Henrietta's descendants (Skloot, 2011). Also, in 2013, Skloot and the Lacks family held meetings with the NIH Director Francis Collins, and NIH Deputy Director for Science, Outreach, and Policy Kathy Hudson, along with scientists and ethicists from Johns Hopkins to discuss modalities for publishing genetic information and future applications using the cells. What was agreed upon was that Lacks' genome data will be accessible only to those who apply for and are granted permission; two representatives of the Lacks family would serve on the NIH group responsible for reviewing applications for controlled access to the HeLa cells. Also, researchers who use the data would include an acknowledgment to the Lacks family in their publication. There was however no legal provision for monetary compensation to the family. Nevertheless, the agreement is viewed as *"a moral and ethical victory for a family long excluded from any acknowledgment and involvement in genetic research their matriarch made possible"* (Caplan, 2013).

Today, prior informed consent is routinely sought for taking tissue samples.

References
Caplan, A. (2013, 8 August) (NBC News contributor) *NIH finally makes good with Henrietta Lacks' family*. Retrieved from http://www.cnbc.com/id/100946766.
Skloot, R. (2011). The immortal life of Henrietta lacks. A readers guide. Retrieved from http://rebeccaskloot.com/wp-content/uploads/2011/03/HenriettaLacks_RGG.pdf.

BOX 13.5 PATENT IINFRINGEMENT IN GENETICALLY MODIFIED CROPS

1. Monsanto Canada Inc. v. Schmeiser:

This case tried in the Supreme Court of Canada in 2004 attracted public attention to issues of patent infringement. The legal argument in the case pertains to whether growing a genetically modified plant would constitute "use" of a patent obtained for a modified plant cell. Percy Schmeiser, a farmer from Saskatchewan, Canada apparently noticed canola plants growing in a ditch in his field which were resistant to the herbicide RoundUp (glyphosate) which he had used to clear weeds. He then sprayed an additional 3—4 acres of the crop in the same field with RoundUp and found that around 60% survived. He collected the seeds separately and in 1998, used the seed in around 1000 acres. He also sold the seed from the 1998 crop. Monsanto discovered that 95%—98% of the 1998 crop had the herbicide resistant gene patented by them. The company insisted that Schmeiser sign a license agreement and pay them the $35 per acre license fee as well as a portion of the profit. Schmeiser refused and claimed that he had not planted the modified seeds in 1997 and that it was probably a contamination. Since it was on his property, as a farmer he had the right to save and plant seed from his property. Monsanto sued Schmeiser for patent infringement, filing its case in the Canadian Federal Court in August 1998. When out of court settlements failed, Schmeiser filed a counterclaim against Monsanto in 1999 for $10 million for libel, trespass and contaminating his fields. The case attracted tremendous publicity and was quoted by anti-GM activists as an instance of gene pollution and harassment of farmers by multinational seed companies.

The Federal Court of Canada tried the case in June 2000 and ruled in favor of Monsanto as the Court felt that Schmeiser either knew or "ought to have known" that the herbicide resistant crop he grew in 1998 was RoundUp Ready canola. Schmeiser did not put up a defense of accidental contamination for the 1998 crop. Before the trial, all claims prior to 1998, including the manner in which the 1997 plants appeared in Schmeiser's fields were dropped. The ruling was upheld in the Federal Court of Appeals in 2002. It also held the argument put forward by Schmeiser that he had only grown the plants but not used herbicide on the crop irrelevant from the point of patent infringement as a patent prohibits unauthorized use in any manner, not merely its use for its intended purpose. Request for the case to be heard in the Supreme Court was granted in May 2003 and the trial began in January 2004. On May 21, 2004, the Supreme Court ruled 5—4 in favor of Monsanto, it dismissed the argument the "use" of patented genes or cells applied only in the context of the isolated form. It held Schmeiser guilty of infringement as a few contaminating seeds could not have resulted in a commercial level of purity of 95%—98% RoundUp Ready canola in the 1998 crop. He had moreover sold the seeds from that crop. Schmeiser however was saved from paying damages, as he did not spray the crop with herbicide and had not therefore profited from the invention.

2. Bowman v. Monsanto Co. et al.

This case, tried in the US Supreme Court in 2013, examined the issue of patent exhaustion (that gives the purchaser of a patented article, or any subsequent owner, the right to use or resell that article) with respect to genetically modified crops.

Monsanto sells its patented RoundUp Ready soybeans resistant to its herbicide RoundUp subject to a licensing agreement that permits farmers to grow the seed in only one growing season. The harvest may be sold for various purposes but cannot be saved for use as seed for raising the next crop. Vernon Bowman, a farmer from Indiana, purchased RoundUp Ready soybean from Monsanto to raise his first crop and sold the entire crop to a grain elevator. He then purchased commodity soybeans from the grain elevator and used it to seed his second crop. He figured that since most of the farmers grew the herbicide resistant soybeans, the crop would also show the trait, and he could save on the cost of company seeds. He saved the seeds from his crop each year, adding more from the grain elevator if necessary, and harvested eight crops in this manner. Each time he sprayed the crop with RoundUp to rid it of weeds and any soybean that was not resistant to the herbicide. When Monsanto got wind of it, the company sued Bowman for patent infringement. Bowman argued patent exhaustion in defense as the grains were the subject of a prior authorized sale from local farmers to the grain elevator. The argument was rejected by the District Court, and Bowman was asked to pay $84,456 in damages to Monsanto. On appeal, the verdict was upheld by the Federal Circuit on the grounds that Bowman had *"created a newly infringing article."* The court explained that the right to use a patented article following

(Continued)

BOX 13.5 (CONTINUED)

an authorized sale *"does not include the right to construct an essentially new article on the template of the original, for the right to make the article remains with the patentee."* In response to a writ of certiorari, the Supreme Court affirmed in its verdict given in May 2013, that the exhaustion doctrine did not enable Bowman to make additional patented soybeans without Monsanto's permission.

Berkeley Technology Law Journal (2005) *Monsanto Canada Inc. v. Schmeiser.* Berkley Technology Law Journal: 20(1) Article 18. Retrieved from http://scholarship.law.berkeley.edu/cgi/viewcontent.cgi?article=1507&context=btlj; Supreme Court of the United States (2013) *Vernon Hugh Bowman, Petitioner v. Monsanto Company et al.* On writ of certiorari to the United States Court of Appeals for the Federal Circuit. Retrieved from https://www.supremecourt.gov/opinions/12pdf/11-796_c07d.pdf.

13.5.4 ISSUES OF SHARING OF COSTS AND BENEFITS

Related to notions of ownership are issues of sharing of costs and benefits from the working of a patent. An illustrative example is that of the enzyme **Taq polymerase and the Yellowstone National Park**. Taq polymerase is crucial to the polymerase chain reaction (PCR) invented by Kary Mullis of Cetus Corporation in 1984. The enzyme was obtained from a temperature tolerant bacterial strain, *Thermus aquaticus* (Taq), which had been isolated from the Yellowstone National Park by Thomas Brock and Hudson Freeze of Indiana University in 1966 and submitted to the American Type Culture Collection (ATCC). PCR spawned an extraordinary number of patents (over 600 in the United States alone) covering various aspects of the technology (Carroll & Casimir, 2003). Patents covering the basic method originally owned by Cetus Corporation were sold to Hoffmann-LaRoche for US$300 million. Other patents include those for polymerases used in the application (such as Taq polymerase, Tth polymerase, Tma polymerase); devices used in PCR (thermocyclers, solid supports); reagents (primers, buffers, internal standards); and several applications using the technology (Reverse-Transcriptase PCR, nested PCT, Real Time PCR). Annual sales for the licensees of PCR equipment and supplies based on Taq polymerase were estimated to be around US$200 million and growing in 1997 (cited in Kate, Touche, & Collis, 1998). However, the Yellowstone National Park did not receive a share of the financial benefits arising from the use of Taq and PCR (Kate et al. 1998). This experience led the National Park Service to examine options for benefit-sharing in 1996, so as to maximize benefits to resource conservation that could result from research on biological samples acquired from the park. One such Cooperative Research and Development Agreement (CRADA) was signed in 1997 between Yellowstone National Park and a Delaware based company, Diversa Corporation, which sources materials from extreme environments for research on molecular biology and functional genomics. Under this agreement, the company provided the Park with an up-front payment of US$100,000 million to be offset against future royalty payments received under the agreement. In return, Diversa received nonexclusive access to the genetic resources in the unique habitats of the Yellowstone National Park. In accordance with Title 36 of the Code of Federal Regulations, ownership of all specimens collected, including those removed to Diversa's laboratories, is retained by the Yellowstone National Park (Federal Government), but, as CRADA did not deal with IPRs, the company is free to patent any innovations and to sell the resulting products (Kate et al., 1998).

Ideological differences exist between Western and Eastern approach to IPR - Eastern philosophies consider nature as a shared resource, and property rights do not carry the status and importance attached to it as in Western societies. In the developing countries especially, it is felt that a strong patent regime would lead to the increase in price of essential commodities, such as drugs and inputs for agriculture (seeds), thereby increasing the cost of living and proving to be a burden for the poor. Other examples of issues of cost and benefit-sharing involve the use of traditional knowledge in the development of patentable products, and are discussed in Chapter 16: Protection of Traditional Knowledge Associated with Genetic Resources.

13.5.5 ISSUES OF ETHICS

With respect to intellectual property, the two main ethical issues revolve around (1) the **morality of patenting life**; and (2) whether it is **morally acceptable to restrict access to scientific advances that could potentially benefit a large number of people** if freely available.

13.5.5.1 Morality of patenting life forms

In a letter addressed to the US President Jimmy Carter a few weeks after the verdict on the *Diamond vs. Chakrabarty* case, the general secretaries of the three largest US religious denominations, the National Council of Churches, the Synagogue Council of America, and the US Catholic Council, raised the following questions and concerns: *"Who shall determine how human good is best served when new life forms are being engineered? Who shall control genetic experimentation and its results which could have untold implications for human survival? Who will benefit and who will bear any adverse consequences, directly or indirectly?"* (cited in Cook-Deegan & Heaney, 2010). To many, it was **morally abhorrent that new life forms could be created and owned by individuals,** they considered it to be **contrary to religious sentiments** that subscribed to Divine creation and held that all living organisms were gifts of nature. Also, several people held the opinion that **patenting human genes was tantamount to patenting human life and was therefore immoral**.

13.5.5.2 Restricting access to innovations

The use of patents to protect innovations has been promoted by industrialized nations as being necessary to spur further development as without this mechanism inventors would keep the details secret. However, patents are also monopolistic rights, excluding others from making or using the invention. Many view this as being **contrary to the common good**. For instance, in the developing countries, **patented seeds of crop plants increase inputs costs for farmers**, while **patented drugs put healthcare out of the reach of the common man**. For **diagnostics, patents on gene sequences appear to be inhibiting the development of new products** (see Chapter 15: Patents in Biopharma). Ethical concerns that gene patents could **stifle research** and **clinical use of genomics** was what prompted the leaders of the Human Genome Project to place sequence data in **public databases thus rendering them impossible to patent**. In several jurisdictions, such as the European Patent Office (EPO), legal and ethical concerns are to be addressed in the patent application for genes and DNA sequences.

KEY TAKEAWAYS

IPR issues in Biotechnology:

- Issues of patentability:
  - Patenting organisms—whether it constitutes a "manufacture" or "composition of matter"
  - Patenting genes: whether it constitutes "invention" and not "discovery"
- Issues of ethics:
  - Morality of patenting life
  - Whether it is morally acceptable to restrict access to scientific advances that could potentially benefit a large number of people if freely available
- Issues of ownership: who owns cells/tissues/organs from which patentable products are developed
- Issues of sharing of costs and benefits
- Issues of enforcement: unlike machines, inventions in biology are able to self-replicate

13.6 SUMMARY

IPRs have been closely tied to trade as it affords protection to innovators from unfair competition and trade practices. Patents especially have been deemed necessary to foster a spirit of innovation and to encourage the development of new technologies. The argument is that in the absence of such protection, inventors would keep their inventions secret in order to maintain a market edge, thus inhibiting further development of the innovation. A patent ensures that the innovator enjoys exclusive marketing rights while others gain access to the information. Living organisms and their parts such as the biomolecules (proteins or nucleic acids) isolated from them were not considered eligible for patents as they were products of nature. With the development of genetic engineering and the possibility to create organisms and products (such as cDNA) not found in nature, the patent system has now been extended to include biology and biotechnology. Patenting of life forms is however different from that of mechanical inventions and pose unique problems such as: establishing the "inventive" nature, and difficulties in preventing infringements due to the ability of the organism to self-replicate.

REFERENCES

Carroll, P., & Casimir, D. (2003). PCR patent issues. In J. M. S. Bartlett, & D. Stirling (Eds.), *Methods in molecular biology* (Vol. 226, pp. 7−14). Totowa, NJ: Humana Press Inc, PCR Protocols.

Center for Food Safety & Save our Seeds (2013) *Seed Giants vs. U.S. Farmers*. Retrieved from http://www.centerforfoodsafety.org/files/seed-giants_final_04424.pdf.

Cook-Deegan, R., & Heaney, C. (2010). Patents in genomics and human genetics. *Annual Review of Genomics and Human Genetics., 11*, 383−425. Retrieved from http://www.ncbi.nlm.nih.gov/pmc/articles/PMC2935940/.

Federico, P. J. (1937). Louis Pasteur's patents. *Science, 86*(2232), 327. Retrieved from http://science.sciencemag.org/content/86/2232/327.

Genetics Generation (2015) *Patenting genes* [Webpage] Retrieved from http://knowgenetics.org/patenting-genes/.

Gold, E. R., & Gallochet, A. (2001). The European Biotech Directive: past as prologue. *Eur.Law J.*, *7*, 331–366. Retrieved from https://www.researchgate.net/publication/227903993_The_European_Biotech_ Directive_Past_as_Prologue.

Kate, K.T., Touche, L., & Collis, A. (1998) *Benefit-sharing case study: Yellowstone National Park and Diversa Corporation.* Submission to the Executive Secretary of the Convention on Biological Diversity by the Royal Botanic Gardens, Kew. Retrieved from https://www.cbd.int/doc/case-studies/abs/cs-abs-yellow-stone.pdf.

Maxham, A. (2015). *Gene revolution. Center for protection of intellectual property, school of law.* George Mason University. Retrieved from http://cpip.gmu.edu/wp-content/uploads/2015/11/Maxham-The-Gene-Revolution.pdf.

NHGRI (2013) 1997: *Bermuda Meeting affirms principle of data release.* [Webpage] Retrieved from https://www.genome.gov/25520385/online-education-kit-1997-bermuda-meeting-affirms-principle-of-data-release/.

United States Supreme Court (1980) *Diamond v. Chakrabarty No.79–136.* Retrieved from http://caselaw.find law.com/us-supreme-court/447/303.html.

USPTO (2015a) *General information about 35 U.S.C. 161 Plant Patents* [Website] Retrieved from https://www.uspto.gov/patents-getting-started/patent-basics/types-patent-applications/general-information-about-35-usc-161.

USPTO (2015b) General information concerning patents [Website] Retrieved from https://www.uspto.gov/patents-getting-started/general-information-concerning-patents#heading-28.

WIPO (1977) Summary of the Budapest Treaty on the International Recognition of the Deposit of Microorganisms for the Purposes of Patent Procedure (1977) [Website]. Retrieved from http://www.wipo.int/treaties/en/registration/budapest/summary_budapest.html.

WTO (2014) *World Trade Organization.* Retrieved from https://www.wto.org/english/res_e/doload_e/inbr_e.pdf.

WTO (2016) *Overview: the TRIPS Agreement* [Webpage] Retrieved from https://www.wto.org/english/tratop_e/trips_e/intel2_e.htm.

PATENTING OF LIFE FORMS

14

A handful of global corporations, research institutions, and governments could hold patents on virtually 100,000 genes that make up the blueprints of the human race, as well as the cells, organs, tissues that comprise the human body. They may also own similar patents on tens of thousands of microorganisms, plants, and animals, allowing them unprecedented power to dictate the terms by which we and future generations will live our lives.

-Jeremy Rifkin, author, in The Biotech Century, 1998

Shall an invention be patented or donated to the public freely? I have known some well-meaning scientific men to look askance at the patenting of inventions, as if it were a rather selfish and ungracious act, essentially unworthy. The answer is very simple. Publish an invention freely, and it will almost surely die from lack of interest in its development. It will not be developed and the world will not be benefited. Patent it, and if valuable, it will be taken up and developed into a business.

-Elihu Thomson, engineer and inventor, founder of General Electric, American Institute of Electrical Engineers, Thomson-Houston Electric Company

CHAPTER OUTLINE

14.1 Introduction 312
14.2 Criteria for Award of Patent 312
    14.2.1 Prerequisites for Patentability 313
    14.2.2 Essential Requirements for Patentability 314
14.3 Patenting Cells and Cell Lines 316
14.4 Patenting Genes and DNA Sequences 318
    14.4.1 Sequences Used for Diagnostic Testing 319
    14.4.2 Sequences Used as Research Tools 319
    14.4.3 Sequences Used in Gene Therapy 321
    14.4.4 Sequences Used in Production of Therapeutic Proteins 321
14.5 Patenting of Animals 321
14.6 Protection of Plant Varieties 323
    14.6.1 Patents 323
    14.6.2 Sui Generis Forms of Protection of Plant Varieties: International Union for the Protection of New Varieties of Plants (*Union internationale pour la protection des obtentions végétales*, UPOV) 324
    14.6.3 Geographical Indications 325
14.7 Summary 326
References 326
Further Reading 327

An Introduction to Ethical, Safety and Intellectual Property Rights Issues in Biotechnology.
DOI: http://dx.doi.org/10.1016/B978-0-12-809231-6.00014-4

Charles Darwin's 1837 sketch, his first diagram of an evolutionary tree
From his *First Notebook on Transmutation of Species* (1837) on view at the Museum of Natural History in Manhattan. Charles Darwin (1809–82) *[Public Domain, Wikimedia Commons]*

14.1 INTRODUCTION

The theory of evolution as proposed by Charles Darwin in his 1859 book *On the Origin of Species* is that all species on earth have arisen through a series of mutations from preexisting living organisms. Species that had mutations which made them better adapted to changing environmental conditions were able to reproduce more efficiently than others and survived, while species unable to compete for resources went extinct (summarized as the "survival of the fittest"). Living organisms are products of nature and consequently cannot be patented. This premise changed in the 1980s when it became apparent that genetic engineering techniques could create novel organisms not seen in nature. Biologists today are beginning to design new life forms with synthetic biology and are foreseeing the addition of a new kingdom **"Synthetica"** to the tree of life which currently has the kingdoms Archaea, Bacteria, and Eucarya (Ginsberg, 2010). As these are creations of the human intellect, these new organisms are technically patent eligible. This chapter examines the issues associated with the patenting of life forms and components derived from living cells.

14.2 CRITERIA FOR AWARD OF PATENT

Intellectual Property Rights are **territorial**, limited to the country/region where awarded. Intellectual property laws are similar in different parts of the world in that **patents are awarded** for inventions which satisfy conditions of novelty, inventiveness, and utility or industrial application (see Chapter 13: Relevance of Intellectual Property Rights in Biotechnology). However, despite agreements such as the TRIPS agreement that attempt to harmonize rules across jurisdictions, there are sufficient differences in interpretation so as to result in disparate outcomes.

14.2.1 PREREQUISITES FOR PATENTABILITY

For an innovation involving a biological system to be considered eligible for a patent, it has to be first established as an invention and not a discovery, as patents are only given for inventions. Also, as listed in Box 13.2, there are several inventions that are not eligible for patents as they are considered frivolous, or contrary to natural laws, or morality.

14.2.1.1 "Discovery" v. "Invention"

In the United States, since the award of a patent for a modified microorganism in the *Diamond v. Chakrabarty* case (1980), it is accepted that "anything under the sun that is made by man" is eligible for patents, and that it is "living" is not significant (see Chapter 13: Relevance of Intellectual Property Rights in Biotechnology). The only issue that needs to be resolved for patent eligibility is the distinction between discovery and invention. **Pure products of nature would be a discovery** (cannot be patented under 35 U.S.C. Section 102), but **if human endeavor is able to impart a new form, new quality, or at least one new property to the original product existing in nature, it would be patentable**.

In Europe, in 1973, the European Patent Office (EPO) was established by a multilateral treaty, the **European Patent Convention (EPC)**. It provides an autonomous legal system that grants European patents which are essentially a group of nationally enforceable and nationally revocable patents. In order to accommodate legal protection of biotechnological inventions, a new **EU Directive 98/44/EC was adopted in July 1998** (EUR-Lex, 1998). While the directive recognized, *"no prohibition or exclusion exists in national or European patent law (Munich Convention) which precludes a priori the patentability of biological matter"*, Article 4 of the Directive 98/44/EC (same as Article 53(b) of EPC) finds *"plant and animal varieties"* and the *"essential biological processes for the production of plants or animals"* not patentable. Nevertheless, Article 2 of the directive does establish, *"biological material which is isolated from its natural environment or produced by means of a technical process may be the subject of an invention even if it previously occurred in nature."* Gene patents have however been a controversial issue in Europe; the Directive contains some confusing and even contradictory statements about human gene patents (Gold & Gallochet, 2001).

In Australia, IP Australia (https://www.ipaustralia.gov.au/) which administers Australia's intellectual property **does not consider building blocks of living matter, such as DNA and genes as a discovery when they** *"have been identified for the first time and copied from their natural source and then manufactured synthetically as unique materials with a definite industrial use."* In the context of gene patents, IP Australia's Manual (IP Australia, 2002) provides specific guidance on the difference between discovery and invention, *"The discovery of a micro-organism, protein, enantiomer or antibiotic in nature* **can be claimed in its isolated form** *or as substantially free of (perhaps, specified) impurities. Also,* **a gene can be claimed as the gene per se** *(as long as the claim does not include within its scope the native chromosome of which the gene forms part) or as the recombinant or isolated or purified gene."* In 2015 however, the Australian High Court **clarified** that mere isolation of nucleic acids is not sufficient to support its characterization as a "manner of manufacture" required under Section 18(1)(a) of the Australian Patent Act (see *D'Arcy v Myriad Genetics Inc. & Anor [2015]* in *Box 14.2*).

14.2.1.2 Should not be contrary to "ordre public"

A significant component of European patent law articulated in Directive 98/44/EC Article 6 is that *"inventions shall be considered unpatentable where their commercial exploitation would be contrary to ordre public or morality"*. This includes *"processes for **cloning human beings**; processes for **modifying the germ line genetic identity of human beings**; uses of **human embryos for industrial or commercial purposes**; processes for **modifying the genetic identity of animals** which are likely to cause them suffering without any substantial medical benefit to man or animals, and also animals resulting from such processes."* The patent laws must be applied to respect the dignity and integrity of a person. This means, *"the principle that the human body, at any stage in its formation or development including germ cells, and the simple discovery of one of its elements or one of its products, including the sequence or partial sequence of a human gene, cannot be patented"* (Article 5).

14.2.2 ESSENTIAL REQUIREMENTS FOR PATENTABILITY

Novelty, inventiveness, and utility or industrial application are the three essential requirements for any invention to be considered patent eligible. While this is usually easy to establish for mechanical inventions, in the case of innovations in biology, they are subject to interpretation and current understanding of living systems.

14.2.2.1 Novelty

The novelty criteria in patent law is based on an examination of "prior art." For patenting genetically modified organisms and products, it is necessary to establish that they do not exist in nature in that specific form. In the *Diamond v. Chakrabarty* case, it had been possible to establish novelty as the modified bacterium was **not seen in nature**. However, in the case of *Funk Brothers Seed Co. v. Kalo Inoculant Co.*, argued in the US Supreme Court in 1948, the Court ruled that a simple mixture of different bacterial strains could not be considered "novel" (Justia website). The case was one of patent infringement in which the patentee had found that certain strains of nitrogen fixing bacteria, each of which was specific for a particular legume crop, could be mixed, and thus sold for use in multiple crops. The patent was being exploited by Kalo, but when similar packets were sold by Funk, Kalo sued for patent infringement. In the hearing, the Court held that the properties of inhibition or of non-inhibition of bacterial strains were the *"work of nature.... the aggregation of species fell short of invention within the meaning of the patent statutes"* and therefore reversed the patent.

With respect to **patents on genes**, there have been several different viewpoints. The public opinion has often veered toward considering **genes as naturally occurring entities** present in living organisms; **hence gene sequences are not novel**, and identifying them may be considered as discoveries not inventions. However, **scientists maintain that the genes and DNA sequences used in various applications, such as diagnosis (gene testing) or gene therapy, are modified, different from nature, novel, and therefore patentable**. With many gene sequences currently being available in public databases, **in order to be patentable, a sequence should not have been used or described**

anywhere in the world. Ensuring that human genome sequences would not be patent eligible and should benefit all of humanity was one of the motivating factors for setting up of public databases that made this information available to researchers all around the world within 24 hours of being sequenced.

14.2.2.2 Inventive step/nonobviousness step

The criterion of inventiveness in European patent law (or nonobviousness in US patent law) is **based on what a person in the inventor's field is expected to know already**. In the context of patent prosecution, if the definition of obviousness is too lax, a patent claiming obvious subject matter may be allowed (patent owner can exclude previously accessible information from the market). But if too stringent, many biotechnologists may not be able to get patents necessary to start a business (Varma & Abraham, 1996). In biological systems, the **central dogma** dictates that the information encoded by the DNA is transcribed and translated into protein (see Chapter 1: Genes, Genomes, and Genomics). **The question arises whether the colinearity of information lends itself to obviousness**. In other words, if the protein sequence is known, is the gene sequence obvious (or vice versa). Two cases that exemplify the issue are *In re Bell* (Federal Circuit, 1993) and *In re Deuel* (Federal Circuit, 1995).

In the case of *In re Bell*, the applicants appealed the United States Patent and Trademark Office (PTO) Board of Patent Appeals and Interferences affirmation of the **patent examiner's rejection of the claims made by them** on Preproinsulin-Like Growth Factors (IGF) I and II on the **grounds of obviousness**. The claims were for the nucleic acid molecules containing human sequences that code for IGF I and II, which are single chain serum proteins that play a role in somatic cell growth. The prior art in the case referred to a publication disclosing the amino acid sequence, and to another that described a general method of cloning genes for which at least a short amino acid sequence of encoded protein is known. The method involved synthesizing a small stretch of nucleotides (corresponding to the known amino acid sequence) to be used as a probe to fish out the gene of interest. The prior art described the use of the method for cloning a histocompatibility gene and used amino acids specified by unique codons to synthesize the probe. The patent examiner rejected the claims in the patent application on the grounds that the prior art had sufficiently demonstrated that a skilled person could have isolated the IGF I and II sequences. However, Bell argued that the PTO had not established a prima facie case of obviousness as it had failed to show how the prior art references could teach or suggest the claimed invention. Because of the degeneracy of the genetic code, several different combinations of nucleic acids could code for a protein (Bell pointed out that 10^{36} nucleic acids might potentially code for IGF; he did not claim for all nucleic acid sequences that could code for IGF, only the one that codes for human IGF). Moreover, none of the amino acids he had used to design the probe had unique codons, hence prior art did not apply. The Federal Circuit concluded that the combination of prior art references did not render the claimed invention obvious, and *reversed* the Board's decision.

In re Deuel's patent application concerned the isolation and purification of cDNA molecules encoding heparin binding growth factors which had also been **found to be unpatentable on the ground of obviousness**. Prior art quoted in this case was a reference that disclosed a group of growth factors and a reference to a technique of isolating DNAs or cDNAs by screening a DNA/cDNA library with a gene probe. The case was also *reversed on examination* by the Federal Circuit.

14.2.2.3 Industrial applicability/utility requirement

In order to satisfy 35USC Section 101, an invention must have a specific and credible utility. For example, in **In re Fisher**, tried in the Court of Appeals for the Federal Circuit in September 2005, the court **affirmed the rejection of a patent claim for five purified nucleic acid sequences that encode proteins and protein fragments in maize.** The claimed sequences are known as **"expressed sequence tags (ESTs),"** with putative use in research, but the claim was rejected because it *"did not disclose a specific and substantial utility for them"* (Federal Circuit, 2005). The verdict rendered hundreds of pending patent applications for nucleic acid sequences worthless.

KEY TAKEAWAYS

Prerequisites for patents in biology

- Should not be a "discovery"—must be isolated from natural environment or produced by means of a technical process
- Should not be contrary to "ordre public"—cloning of human beings, modifying germ line, or causing harm to animals or humans are not patentable.

Criteria for patenting:

- Should have Novelty: should not exist in prior art
- Should have an Inventive/nonobvious step: should not be obvious to a person skilled in the art
- Should have a Utility or industrial applicability

14.3 PATENTING CELLS AND CELL LINES

In Europe, for patent prosecution, **cells are deemed to be novel** if they have been **isolated or purified from their natural environment or reproduced outside the body.** However, the patentability of **human embryonic stem (hES) cells** has been questioned and interpreted variously on *grounds of morality*. The EU Directive 98/44/EC prohibits patenting of *"uses of human embryos for industrial or commercial purposes"*. The position of the EPO is that patent protection can only be given if the cells had been obtained by means other than the destruction of a human embryo (Faure-Andre, 2015).

In the United States, until 2012, the USPTO granted patents protecting human pluripotent cells and hES cell lines. However, subsequent to decisions by the Supreme Court in cases such as **Association for Molecular Pathology v. Myriad Genetics, 2013** (see Box 14.1), and *Mayo Collaborative Services v. Prometheus Laboratories Inc.*, 2012 (see Box 15.6 in Chapter 15: Patents in Biopharma), questions have been raised on patentability of embryonic stem cells (whether they are only **"products of nature"**). Consequently, according to the Guidelines issued by the USPTO on December 16, 2014, **cells, "isolated" from its natural "human induced" environment, are not considered different from those occurring naturally** (USPTO, 2015b). A cell can only be considered to be patentable if it is **"significantly different"** in structure, function, or other aspects from natural cells.

BOX 14.1 ASSOCIATION FOR MOLECULAR PATHOLOGY V. MYRIAD GENETICS, INC., 2013

Association for Molecular Pathology v. Myriad Genetics, Inc., 569 U.S., 133S.Ct. 2107, 106 USPQ2d 1972 (2013)

Myriad Genetics, a start-up company founded in 1991 out of University of Utah, along with the National Institute of Environment Health Sciences and University of Utah, filed a patent for BRCA1 gene which they had isolated in 1994 and for BRCA2 the following year. In 1996, Myriad launched a diagnostic product "BRACAnalysis" which detects mutations in the two genes which puts women at a higher risk for breast cancer and ovarian cancer (see Chapter 7: Genetic Testing, Genetic Discrimination, and Human Rights). The company has since launched several commercial molecular diagnostic products including three personalized medicine products and two prognostic products and has been successful, growing from a start-up to a publicly traded company with over 1000 employees. Myriad's business model is based on the premium price that the company is able to collect during the 20-year life of its patents. But this means that the company has to be vigilant about patent infringement by competitors. In 1998, Myriad issued cease and desist letters on the basis of patent infringement to the University of Pennsylvania's Genetic Diagnostic Laboratory to stop testing patient samples for BRCA. With many diagnostic laboratories opposed to exclusive licensing of diagnostic tests, the Association of Molecular Pathology (AMP) along with the researchers at University of Pennsylvania, Columbia, Yale, Emory, and New York University built a case challenging the validity of gene patents, specifically the use of gene sequences to diagnose propensity to cancer. Initially tried at the District Court for the Southern District of New York, the verdict that the challenged claims were not patent eligible was appealed by Myriad at the United States Court of Appeals of the Federal Circuit. The Federal Circuit ruled that the isolated DNA does not exist in nature and are therefore patentable. AMP appealed to Supreme Court which remanded the case to the Federal Circuit. The Federal Court however did not change its opinion. A petition was again filed in 2012 with the Supreme Court for a certiorari with respect to the second Federal Circuit Decision. The unanimous verdict of the Supreme Court pronounced on June 13, 2013, was that merely isolating genes that are found in nature does not make them patentable—*Myriad did not create or alter either the genetic information encoded in the BRCA1 and BRCA2 genes or the genetic structure of the DNA*. The court did however find that the cDNA that was used in the diagnosis was patentable.

In a related case in Australia, *D'Arcy v Myriad Genetics Inc & Anor [2015] HCA35*, Cancer Voices Australia and Yvonne D'Arcy challenged the patents owned by Myriad on the ground that the claim involved a mere isolation of the nucleic acid without the necessary *"manner of manufacture"* as prescribed in 18 (1)(a) of the Patent Act. The Federal Court of Australia dismissed the challenge. On appeal, the Full Federal Court also dismissed the appeal holding that an isolated nucleic acid was chemically, structurally and functionally different from a nucleic acid inside a human cell. As a result, the invention was a manner of manufacture because an isolated nucleic acid with the characteristics specified in Myriad Genetics claims resulted in an artificially created state of affairs for economic benefit. Yvonne D'Arcy was granted an appeal in the High Court of Australia in September 2014. On October 7, 2015, the High Court of Australia made a unanimous decision that an isolated nucleic acid coding for mutant BRCA1 protein indicated in susceptibility to breast and ovarian cancer is not a "patentable invention" within the meaning of 18(1)(a) of the Australian Patents Act 1990 (Slezak, 2015). The decision reversed earlier court rulings on the disputed claims filed by Myriad Genetics.

References

Association for Molecular Pathology et al. v. Myriad Genetics, Inc., et al. certiorari to the United States Court of Appeals for the Federal Circuit No. 12–398. Argued April 15, 2013—Decided June 13, 2013. Retrieved from https://supreme.justia.com/cases/federal/us/569/12-398/.

Slezak, M. (2015, October 7) [Reporter] Genes can't be patented, rules Australia's High Court. *New Scientist*. Retrieved from https://www.newscientist.com/article/gene-patents-struck-down-by-australias-high-court/.

Early patents in cells and cell lines claim methods for isolating a human cell population enriched in Hematopoietic Stem Cells (HSC) to use in bone transplantation (for review, see Martin-Rendon & Blake, 2007). Patents have also been filed for the use of HSC for therapeutic applications other than bone marrow transplantation. The first patent granted for hES cells was in 1998 and is held by the Wisconsin Alumni Research Fund (WARF), with James Thomson as inventor (Martin-Rendon & Blake, 2007).

KEY TAKEAWAYS

Patenting cells and cell-lines:

- In Europe, cells and cell-lines can be patented provided no embryos have been sacrificed for developing the line
- In the US isolated cells are patentable only if they are significantly different from naturally occurring ones

14.4 PATENTING GENES AND DNA SEQUENCES

Concerns over patenting of human genes in 2001 as a result of the Human Genome Project and other sequencing efforts prompted the then NIH Director, Harold Varmus, and the then NHGRI Director, Francis Collins, to write to the USPTO to **implement stricter criteria for biotechnology patents** (NHGRI, 2014). The USPTO issued utility examination guidelines at that time were as follows:

*"A patent claim directed to an isolated and purified DNA molecule could cover, e.g., **a gene excised from a natural chromosome or a synthesized DNA molecule. An isolated and purified DNA molecule that has the same sequence as a naturally occurring gene is eligible for a patent because (1) an excised gene is eligible for a patent as a composition of matter or as an article of manufacture** because that DNA does not occur in that isolated form in nature, or (2) **synthetic DNA preparations are eligible for patents because their purified state is different from the naturally occurring compound"*.

USPTO (2001)

This view has since evolved to more stringent requirements for patent eligibility since the U.S. Supreme Court ruling in 2013 in the *Association for Molecular Pathology v. Myriad Genetics, Inc.* case (see Box 14.1).

The Nuffield Council on Bioethics distinguishes four applications of DNA sequences in relation to patent claims (Nuffield Council on Bioethics, 2002):

1. *Sequences used for diagnostic testing*: knowledge of the gene sequence is used for detecting the presence of a faulty gene that is the cause for a disease
2. *Sequences used as research tools*: knowledge of the gene sequence can help in identification of potential targets for new drugs and vaccines
3. *Sequences used in gene therapy*: replacing a faulty gene with a normal gene in the body
4. *Sequences used in production of therapeutic proteins*: to be used as medicines

For each category of DNA sequences, whether the sequences meet the eligibility criteria for patenting has been much discussed and subjected to interpretation.

14.4.1 SEQUENCES USED FOR DIAGNOSTIC TESTING

DNA sequences can be used for genetic testing useful in diagnosis and personalized medicine (see Chapter 7: Genetic Testing, Genetic Discrimination, and Human Rights). Since these sequences are derived from human genomic sequences, doubts have been raised as to whether it is patent eligible subject matter. Illustrative of this issue is *Association for Molecular Pathology v. Myriad Genetics, Inc.* tried in the US Supreme Court in 2013 (see Box 14.1). Prior to this case, the US Patent Office accepted patents on DNA sequences as a "**composition of matter**," hence Myriad held the patents for BRCA1 and BRCA2 genes. The decision in this case was, "*DNA's existence in an 'isolated' form alters neither this fundamental quality of DNA as it exists in the body nor the information it encodes. Therefore, **the patents at issue directed to 'isolated DNA' containing sequences found in nature are unsustainable as a matter of law and are deemed unpatentable under 35 U.S.C. §101**.*" Subsequently, in two other controversial cases, patents were found to be impermissible as the claims were deemed directed to a "**natural law**." In *Ariosa Diagnostics, Inc. v. Sequenom, Inc., 2015* (see Box 14.2), Federal Circuit said the claims "*are directed to a patent-ineligible concept*" because the "*method begins and ends with a natural phenomenon*". Second, it said, the claimed method did not "*transform the claimed naturally occurring phenomenon into a patent-eligible application*" of the phenomenon.

Similarly, in *Genetic Technologies Limited v. Merial LLC, 2016* (see Box 14.3), the Federal Circuit rejected the patentability of the method claim on account of its novelty residing in a newly discovered law of linkage disequilibrium between coding and noncoding regions of DNA.

14.4.2 SEQUENCES USED AS RESEARCH TOOLS

With the development of genomic projects, patent applications were made for partial DNA sequences such as the **Expressed Sequence Tags (ESTs)** and **Single Nucleotide Polymorphisms (SNPs)**. These sequences have **no immediate therapeutic or diagnostic use** but are of **value in developing a commercial product** such as medicine or vaccine, hence are considered as **research tools**. Very often at the time of patent application, the function of these sequences may not have been understood. The first patent for an EST was granted in 1998 to Incyte Pharmaceuticals Inc., but patents for ESTs have been highly controversial since Craig Venter working with NIH had first announced the filing of patent applications for ESTs in 1991 (reviewed in Cook-Deegan & Heaney, 2010). NIH as part of its policy had abandoned the applications in 1994, but the controversy had the effect of attracting private companies to this new business opportunity. Craig Venter himself left NIH to join a not-for profit organization, The Institute for Genomic Research, which focused on sequencing the genome, but would vest patent rights in a for-profit company, Human Genome Sciences. Also in the field was Incyte Pharmaceuticals Inc., a start-up that mainly did contract research for Genentech but soon began to work on human genome sequencing. Scientists and the public were concerned that private entities would sequence and patent genes thereby making them unavailable for others to use in development of applications without expensive licensing fees. Since many of the **patent applications were broad** with "**reach through**" **rights to future discoveries**, there were fears of patent "logjams" and "thickets." Cook-Deegan and Heaney (2010) write: "*Funding agencies and scientists realized*

BOX 14.2 ARIOSA DIAGNOSTICS, INC. V. SEQUENOM, INC., 2015

Sequenom Laboratories (https://www.sequenom.com/) is a San Diego, California based company that offers highly sensitive genetic tests for Noninvasive Prenatal Testing (NIPT) for common fetal chromosomal anomalies (for instance Down's syndrome: trisomy 21A) and carrier screening. Sequenom holds the patent filed in 1998 for NIPT which "*invention relates to a detection method performed on a maternal serum or plasma sample from a pregnant female, which method comprises detecting the presence of a nucleic acid of fetal origin in the sample. The invention enables non-invasive prenatal diagnosis including for example sex determination, blood typing and other genotyping, and detection of preeclampsia in the mother.*" The test uses PCR to amplify cell free DNA of fetal origin (cffDNA) present in the pregnant mother's blood using probes specific to paternally inherited DNA, and is safer that other techniques such as amniocentesis which could cause miscarriages. After Sequenom launched its product in the market, a number of other laboratories began offering the same test at reduced rates. Ariosa Diagnostics, Inc., Natera, Inc., and Diagnostics Center, Inc. received letters from Sequenom threatening them with patent infringement suits. The response of the companies was to ask the opinion of the court on legal rights by filing a declaratory judgment action against Sequenom. The district court examined the meaning of the patent claims and restrained Sequenom from threatening the legal rights of the other laboratories by a preliminary injunction. Sequenom asked the appellate court to review the case before the trial concluded by making an interlocutory appeal. The district court held the claims patent ineligible on the grounds, "*the claims at issue pose a substantial risk of preempting the natural phenomenon of paternally inherited cffDNA.*" Sequenom then appealed to the Federal Circuit. The Federal Court affirmed the judgment of the district court and held that the claims as invalid "directed to a patent ineligible concept" as the claimed method did not transform the claimed naturally occurring phenomenon into patent-eligible phenomenon. Sequenom's appeal for rehearing was denied in December 2015. In March 2016, Sequenom filed a certiorari petition, but this was also denied by the United States Supreme Court in June, 2016.

Brinckerhoff, C.C. (2015, June 15) Federal Circuit holds Sequenom diagnostic patent invalid under 101. *PharmaPatents* [Blog]. Retrieved from https://www.pharmapatentsblog.com/2015/06/15/federal-circuit-hold-sequenom-patent-invalid-under-101/.

BOX 14.3 GENETIC TECHNOLOGIES LIMITED V. MERIAL LLC, BRISTOL-MYERS SQUIBB COMPANY, 2016

Genetic Technologies Limited (http://gtglabs.com/) (GTG) is an Australian-based molecular diagnostic company founded in 1989, "offers predictive testing and assessment tools to help physicians proactively manage women's health." In the 1980s, Malcolm Simons discovered that in genomic DNA certain exons (coding regions of a gene) of certain genes are correlated with the introns (noncoding regions) of the same gene as well as introns in different genes, and other noncoding regions of the genome known as the "intergenic spacing sequences" (dubbed "junk DNA" since they have no apparent function). Simons found that these correlated regions show "linkage disequilibrium," that is they "tend to be inherited together more frequently than probability would dictate." This information could be used to isolate genes by amplifying and analyzing noncoding regions known to be linked to the coding region of the gene of interest. Between 1989 and 1992, Simons and GTG filed several patent applications for this discovery, one of which was granted as U.S. Patent No. 5,612,179 (referred to as the '179 patent). The claims in the patent were broadly defined, covering the method of detection of at least one coding region by amplifying noncoding regions. In 2011, GTG sued several pharmaceutical and biotechnology companies for patent infringement. Merial and Bristol-Myers Squibb Company (BMC) moved to dismiss the case arguing that the claims in the '179 patent were ineligible subject matter under 35 U.S.C.§ 101. The district court held the patent invalid for claiming a law of nature: "*A claim is unpatentable if it merely informs a relevant audience about certain laws of nature, even newly-discovered ones, and any additional steps collectively consist only of well-understood, routine, conventional activity already engaged in by the scientific community. The claim involved here, claim 1 of the '179 patent, does just that and no more.*" The verdict was affirmed by the United States Court of Appeals for the Federal Circuit in April 2016.

United States Court of Appeals for the Federal Circuit (2016, April 8) *Genetic Technologies Limited v. Merial L.L.C., Bristol-Myers Squibb Company*. Retrieved from http://www.cafc.uscourts.gov/sites/default/files/opinions-orders/15-1202.Opinion.4-6-2016.1.PDF.

that they needed systematic policies to cultivate a scientific commons, lest they lose control of their science to those wielding patents. Four examples of such collective action illustrate how norms of open science were put in place: (a) the public domain EST sequencing projects funded by Merck and the DOE, (b) the Bermuda Rules of sharing sequence data rapidly, (c) the SNP Consortium, and (d) the HapMap project."

Patent Offices in the United States, Europe and Japan have since **applied the criteria of utility more rigorously so as to allow only DNA sequences that demonstrate a "substantial, credible, and specific use" to be patented.** The Nuffield Council on Bioethics (2002) recommends a *"research exemption"* so that these patents will not hinder research.

14.4.3 SEQUENCES USED IN GENE THERAPY

The Nuffield Council on Bioethics (2002) opines, *"once a gene associated with a disease is identified, the use of the relevant DNA sequences in gene replacement therapy, to alleviate the effects of mutations in that gene is obvious...."* Therefore, they recommend that protection by product patents should seldom be permissible for DNA sequences used in gene therapy.

14.4.4 SEQUENCES USED IN PRODUCTION OF THERAPEUTIC PROTEINS

First generation gene patents were issued for full length DNA sequences that made proteins of use in medicine and were valuable as they helped recoup the cost of the research and development. The Nuffield Council on Bioethics (2002) does however recommend that these patents should be narrowly defined so that the rights to the DNA sequence extend only to the protein described.

KEY TAKEAWAYS

Patenting genes and DNA sequences:

- *Sequences used for diagnostic testing*: isolated DNA sequences are not patentable if similar to that in the living organism; cDNA is patentable
- *Sequences used as research tools*: ESTs and SNPs are not patentable if "substantial, credible and specific use" is not demonstrated
- *Sequences used in gene therapy*: Nuffield Council on Bioethics recommends that protection by product patents should seldom be permitted for DNA sequences used in gene therapy
- *Sequences used in production of therapeutic proteins*: most patented class of DNA sequences

---

14.5 PATENTING OF ANIMALS

Higher order living organisms, plants and animals, have not received the same treatment as have microorganisms in patent prosecution. In Europe for instance, since Article 4 of the Directive 98/44/EC (same as Article 53(b) of EPC) finds **plant and animal varieties not patentable,** decisions of patent eligibility have depended on the **interpretation of "variety."** The first application to the

EPO involving an animal was for the **Harvard OncoMouse**, the patent application for which was filed in the EPO in 1985, a year after it had been filed in the United States. Patent for the mouse was granted by the EPO 19 years later. Patent applications for the OncoMouse had also been filed in Canada, Japan, and several other countries. The treatment of the patent application in different jurisdictions is described in Box 14.4.

BOX 14.4 PATENTING OF THE HARVARD ONCOMOUSE

In the 1980s, Philip Leder and Timothy Stewart of Harvard Medical School isolated a gene, *"myc,"* that causes cancer in many mammals including humans and injected it into fertilized mouse eggs. The eggs were implanted into a female host mouse and the offspring tested for presence of the oncogene. The transgenic "founder" mice were bred with unmodified mice. In half the offspring, all cells express the oncogene and could be used as a model for studying cancer and testing therapies for cancer treatment. Patent applications were made by the President and Fellows of Harvard College for the "OncoMouse" both for the process by which the mice are produced, as well as the transgenic mice themselves. The first application was filed in the United States on June 22, 1984, and subsequently in Australia, Canada, United Kingdom, Ireland, Belgium, Denmark, Finland, France, Germany, Italy, Luxembourg, the Netherlands, Portugal, Spain, and Sweden. In the United States, due to the precedence of *Diamond v. Chakraborty*, patents were granted in 1988 without any objections (although it did take a period of 4 years). The patents were exclusively licensed to DuPont and "OncoMouse" is a registered trademark. The application did not receive the same treatment in other jurisdictions.

In Europe, where the application was filed in 1985, the EPO Patent Examining Division excluded the application in July 1989, on grounds that the basic requirements of Article 53(b) and 83 of the EPC were not met: an "animal variety" was contrary to Article 53(b). The Examination Division referred the application to the EPO Board of Appeals unable to decide whether it should also be excluded under 53(a) for violating public policy (contrary to "ordre public"). The EPO Board of Appeals returned the case to the Division with the instruction that the Division should decide whether the claim actually constituted an "animal variety." Also, the Board suggested the Division weigh the suffering of animals and possible risk to the environment against the usefulness to mankind of the invention (cited in Brashear, 2001). In accordance to the Board's orders, in determining whether the invention actually constituted an "animal variety," the Division compared it to "animal." A "variety" constitutes a "sub-unit of a species" and ranked lower than a species. The mouse (a rodent) fell within a taxonomic classification higher than a species; hence 53(b) did not exclude the OncoMouse. In deciding whether the mouse should also be excluded on the basis of 53(a), the Division noted that the mouse could be a useful tool to develop anticancer medicines and posed no threat to the environment. Also it could possibly decrease animal suffering since fewer animals would be required for research. The Patent was therefore granted in May 1992, making the OncoMouse the first animal to be patented in Europe. Several opposing applications were filed against the patent by animal rights groups, farming interest groups and green activists. The Opposition Division of the EPO dealt with many of the oppositions on the basis of the new EU Directive 98/44/EC and a final decision was given by the EPO Board of Appeals in July 2004, 19 years after the filing of the application. The final patent limited the patent to mice, whereas the original 1992 patent extended to all mammalian species with a transgenic oncogene (Park, 2004). This decision cannot be further appealed in the EPO, but can be appealed at the national level.

In Canada, although the Federal Court of Appeal ruled in favor of the patent, the Supreme Court of Canada rejected the patent in 2002 (Supreme Court Judgments, 2002) on the basis of animals not being patentable subject matter. However, after the patent application was amended to omit the "composition of matter" claims, a Canadian patent has been awarded to Harvard College in October 2003. Japan granted a patent for the OncoMouse in 1994.

References

Brashear, A. D. (2001). Evolving biotechnology patent laws in the United States and Europe: Are they inhibiting disease research? *Indiana International & Comparative Law Review, 12*(1), 183–218. Retrieved from https://mckinneylaw.iu.edu/iiclr/pdf/vol12p183.pdf.

Park, P. (2004, July 26) [Reporter] EPO restricts OncoMouse patent. *The Scientist*. Retrieved from http://www.the-scientist.com/?articles.view/articleNo/22980/title/EPO-restricts-OncoMouse-patent/.

Supreme Court Judgments (2002) Harvard College v. Canada (Commissioner of Patents) (2002-12-05). Retrieved from https://scc-csc.lexum.com/scc-csc/scc-csc/en/item/2019/index.do.

In the United States, the Patent and Trademark Office issued an announcement in 1987 that it would consider *non-naturally occurring, non-human multicellular living organisms, including animals, to be patentable subject matter within the scope of the statute* (USPTO, 2015a). This announcement was made subsequent the decision in ***Ex Parte Allen***—In 1984, Standish Allen and Sandra Downing of the University of Washington and Jonathan Chaiton of the Coast Oyster Company applied for a patent for triploid sterile Pacific oysters and the method to create them. The oyster had been genetically modified to be larger and edible through all stages of its growth. The patent examiners **rejection of the patent claims** was based on **subject matter (controlled by laws of nature)** and **obviousness** (polyploidy was known to increase size). The Board of Patent Appeals pointed out that the issue was not whether the invention was "controlled by laws of nature," but whether it **"was naturally occurring,"** and as polyploid oysters did not occur in nature, they **rejected the examiner's finding.** [The Board did however concur with the examiner's conclusion of obviousness, which decision was affirmed by the Federal Circuit, *In re Allen*, 1988 (Federal Circuit, 1988).] The announcement on patent eligibility of animals paved the way for granting patents for the OncoMouse in 1988 and several others subsequently.

KEY TAKEAWAYS

Patenting animals:

- The US PTO in 1987 announced that it would consider multicellular higher order living organisms patentable
- The EPO does not consider plant and animal varieties to be patentable. (The Oncomouse was however given a patent 19 years after the application was filed on the argument that the animal fell in a taxonomic classification higher than a "variety," and therefore patentable)
- Animals are not patent eligible in several jurisdictions such as India

14.6 PROTECTION OF PLANT VARIETIES

New plant varieties have been created by a number of different plant breeding techniques and have significant commercial impact in agriculture. There have however been glaring **ideological differences** in the need for and the manner in which new plant varieties are to be protected. In the United States for instance, plant varieties have been afforded protection through a number of IPR instruments including patents, whereas in Europe, protection of new plant varieties is as breeder's rights, plant varieties per se are not patent eligible. The concept of IPR for plant varieties in developing nations has been difficult to implement as seeds are considered a common resource, shared by all. Enforcing monopolistic rights on new varieties has been perceived as unjust to farmers and a denial of their rights to livelihood (Shiva, 2016).

14.6.1 PATENTS

In the **United States**, plant protection is available under the **Plant Patent Act (35 U.S.C. 161–164)** and the **Plant Variety Protection Act (7 U.S.C. 2321 et. seq.).** In addition, in a 1985 case *Ex parte Hibberd*, the PTO Board of Appeals held that the Plant Patent Act and the Plant Varietal Protection Act **do not narrow the scope of patentable subjects protected under the general utility statute** (USPTO, 2015a).

In **Europe**, patent protection for individual plant varieties is **excluded by Article 53(b) EPC**; however, the European Union (EU) has established a system similar to a patent called **Community Plant Variety Right** that is valid throughout the EU. It is implemented by the Community Plant Varietal Office in Angers (European Commission, 2016). It is also possible to obtain plant variety rights under the UPOV.

14.6.2 SUI GENERIS FORMS OF PROTECTION OF PLANT VARIETIES: INTERNATIONAL UNION FOR THE PROTECTION OF NEW VARIETIES OF PLANTS (*UNION INTERNATIONALE POUR LA PROTECTION DES OBTENTIONS VéGéTALES*, UPOV)

The UPOV (http://www.upov.int) is an international organization established by the **International Convention for the Protection of New Varieties of Plants** adopted in Paris in 1961, revised in 1972, 1978, and 1991. As of 2015, the convention has 74 members. The convention provides breeders of new plant varieties intellectual property right in the form of the breeder's right. A variety so protected requires authorization of the breeder to propagate the variety for commercial purposes. For a variety to be protected by the breeder's right, it must meet four criteria (abbreviated as **NDUS**):

1. *Novelty*: it should not have been marketed earlier in the country where rights are applied
2. *Distinctness*: must be distinct from other available varieties
3. *Uniformity*: the plants must display homogeneity
4. *Stability*: the trait(s) unique to the new variety should remain true to type after repeated cycles of propagation

The breeder's right is granted by each member state for a period of 20 years from the date of issue, except in the case of vines and trees, for which it is 25 years. There are explicit exemptions to the rights, known as "**breeder's exemption clause**" that makes it unnecessary to receive authorization from the breeder for private use or for use in further breeding efforts.

In harmonizing patent laws in different jurisdictions, the **TRIPS Agreement Article 27.3(b)** requires the **member countries to provide for protection of plant varieties either by a patent or an effective sui generis system or any combination thereof**. Many developing countries such as India, exercised the **option of framing legislation suiting their own system**. India has an agrarian economy with over 70% of its population subsisting on agriculture and related activities. Indian economists felt that input costs for seed would be beyond the means of farmers with small and marginal holdings, and any intellectual property system for crop plants would have to recognize a "**farmer's right**" to save and reuse farm saved seed. India therefore adopted the **Protection of Plant Varieties and Farmer's Rights (PPVFR) Act** in 2001(http://plantauthority.gov.in/). Under the Act, plant varieties are eligible for protection if they satisfy the **conditions of DUS**.

The term of the protection is 15 years from date of registration (or 18 years for trees and vines). The Act defines:

- *Breeders' rights*: exclusive rights to license, produce, sell, market, distribute, import or export the variety
- *Researchers' rights*: access to varieties for bona fide research purposes, and
- *Farmers' right*: farmers shall be deemed to be entitled to save, use, sow, resow, exchange, share, or sell farm produce including seed (unlabeled, brown bag sale) of the variety.

14.6.3 GEOGRAPHICAL INDICATIONS

Geographical Indications recognize the heritage of a geographical region in certain specialized goods and protect them from unfair trade practices (see Chapter 13: Relevance of Intellectual Property Rights in Biotechnology). The use of this instrument has been suggested for the protection of certain plant varieties which qualify for such protection in order to **prevent patent applications from claiming exclusive rights** on their use. An example is that of an aromatic rice variety, called Basmati, traditionally grown in areas of India and Pakistan. The variety was the cause of a dispute between the United States and India. In 1997, RiceTec, a Texas-based company, patented some lines and grains of Basmati which they had developed as "American Basmati" which was being sold in the international market under the trade names "Texmati" and "Kasmati." India took strong exception to the patents for several reasons including: Basmati had been grown for several centuries in India and the patent would affect its export market; it was seen as an exploitation of a developing countries rights (Mukherjee, 2008). The patent was officially challenged by the Indian government in 2000 on the grounds that the plant varieties and grains already exist as a staple in India, and should not have been granted in the first place. Upon reexamination, the USPTO canceled 15 of the 20 claims, allowing only the claims dealing with three varieties developed by the company. India has since actively pursued means of protecting its traditional plant varieties and traditional knowledge in a **defensive strategy** using instruments such as Geographical Indications.

KEY TAKEAWAYS

Protection of plant varieties:

- *Patents*: the USPTO permits plants to be patented under the Plant Patent Act and the Plant Varietal Protection Act
- *Sui Generis forms of protection: UPOV*: Plant varieties can be protected under UPOV if they satisfy conditions of *Novelty, Distinctness, Uniformity*, and *Stability*. Breeder's exemption clause allows new varieties to be created from a protected variety
- "Farmer's right" to use farm saved seed in addition to the "breeder's right" to use varieties for developing new varieties
- *Geographical Indications*:can be used to protect traditional varieties from being patented

14.7 SUMMARY

Modern biotechnologies have yielded processes and products which have commercial significance. Protecting intellectual property for these applications critical for commercial viability of innovator projects has been a challenge. This is primarily because living organisms and their components (proteins and nucleic acids) do not conform to conventional subject matter in the industrial sector. Establishing organisms or isolated nucleic acids as "compositions of matter" or "articles of manufacture" eligible for patent protection has been difficult. Also, living organisms have been viewed as "products of nature," free for all of mankind to enjoy and have raised ethical issues in conferring exclusive rights on their use. Patent protection for living organisms became possible subsequent to the award of a patent for a modified bacterium not found in nature (*Diamond v. Chakrabarty*, 1980), which was admittedly created through human ingenuity. Patents for plants and animals were not granted till the late 1980s, as they were not considered patentable subject matter, although in the United States new plant varieties could be protected by plant patents. In 1987, the USPTO announced that it would consider patents for *non-naturally occurring, non-human multicellular living organisms, including animals* resulting in the patenting of the OncoMouse in 1988. Patenting DNA sequences has been possible as sequences isolated from living organisms are considered compositions of matter and articles of manufacture. However, there is an increasing emphasis on sequences which demonstrate a "substantial, credible, and specific use" only being eligible for patenting.

REFERENCES

Cook-Deegan, R., & Heaney, C. (2010). Patents in genomics and human genetics. *Annual Review of Genomics and Human Genetics.*, *11*, 383−425. Retrieved from http://www.ncbi.nlm.nih.gov/pmc/articles/PMC2935940/.

European Commission (2016) *Plant varietal property rights* [Web page] Retrieved from http://ec.europa.eu/food/plant/plant_property_rights/index_en.htm.

EUR-Lex (1998, July 30) Directive 98/44/EC of the European Parliament and of the Council of 6 July 1998 on the legal protection of biotechnological inventions. Retrieved from http://eur-lex.europa.eu/legal-content/EN/TXT/?uri=CELEX:31998L0044.

Faure-Andre, G. (2015) Human embryonic stem cell patentability in Europe and the United States [Web article] Retrieved from http://www.regimbeau.eu/REGIMBEAU/GST/COM/PUBLICATIONS/2015-04-hESCs-GFA_EN.pdf.

Federal Circuit (1988) *In re Allen, 846 F.2d 77*. Retrieved from http://law.justia.com/cases/federal/appellate-courts/F2/846/77/397098/.

Federal Circuit (1993) *In re Graeme I. Bell, Leslie B. Rall and James P. Merryweather. 991F.2d 781, 61 USLW 2692, 26 U.S.P.Q.2d 1529*, United States Court of Appeals, Federal Circuit, No.92-1375. Retrieved from https://law.resource.org/pub/us/case/reporter/F2/991/991.F2d.781.92-1375.html.

Federal Circuit (1995) *In re Thomas F. Deuel, Yue-Sheng Li, Ned R. Siegel, and Peter G. Milner, 51 F.3d 1552* (Fed.Cir.1995). Retrieved from https://casetext.com/case/deuel-in-re.

Federal Circuit (2005) *In re Dane K. Fisher and Raghunath V. Lalgudi, 04-1465* (Serial No: 09/619,643). Retrieved from http://www.uspto.gov/web/offices/com/sol/fedcirdecision/04-1465.pdf.

Ginsberg, A.D. (2010) *The synthetic kingdom: Designing evolution [Blog] WordPress.com*. Retrieved from https://synthetickingdom.wordpress.com/2010/07/10/redesigning-the-tree-of-life/.

Gold, E. R., & Gallochet, A. (2001). The European Biotech Directive: past as prologue. *European Law Journal*, *7*, 331–366. Retrieved from https://www.researchgate.net/publication/227903993_The_European_Biotech_Directive_Past_as_Prologue.

IP Australia (2002) Australian patent manual of practice and procedure volume 2: National.

Martin-Rendon, E., & Blake, D. J. (2007). Patenting human genes and stem cells. *Recent Patents on DNA and Gene Sequences 2007*, *1*, 25–34.

Mukherjee, U. (2008) *A study of the Basmati case (India-US Basmati rice dispute): Geographical Indication perspective* [Web article] Retrieved from http://csbweb01.uncw.edu/people/eversp/classes/BLA361/Intl%20Law/Cases/Study%20of%20Basmati%20Rice%20Intl%20Case.ssrn.pdf.

NHGRI (2014) Intellectual Property and Genomics. [Web page]. Retrieved from https://www.genome.gov/19016590/intellectual-property/.

Nuffield Council on Bioethics (2002) *The ethics of patenting DNA- a discussion paper*. Nuffield Council on Bioethics, Enfield. Retrieved from http://nuffieldbioethics.org/wp-content/uploads/2014/07/The-ethics-of-patenting-DNA-a-discussion-paper.pdf.

Shiva, V. (2016, May 19) *Dr. Vandana Shiva: Seeds of suicide*. Retrieved from https://www.sott.net/article/318816-Dr-Vandana-Shiva-Seeds-of-suicide.

United States Supreme Court (1980) *Diamond v. Chakrabarty No.79-136*. Retrieved from http://caselaw.findlaw.com/us-supreme-court/447/303.html.

USPTO (2001) Utility examination guidelines. Retrieved from http://www.uspto.gov/sites/default/files/web/offices/com/sol/notices/utilexmguide.pdf.

USPTO (2015a) 2015 patentable subject matter—Living subject matter [R-07.2015] [Webpage] Retrieved from https://www.uspto.gov/web/offices/pac/mpep/s2105.html.

USPTO (2015b) 2014 Interim guidance on subject matter eligibility [Webpage] Retrieved from http://www.uspto.gov/patent/laws-and-regulations/examination-policy/2014-interim-guidance-subject-matter-eligibility-0.

Varma, A., & Abraham, D. (1996). DNA is different: Legal obviousness and the balance between biotech inventors and the market. *Harvard Journal of Law & Technology*, *9*(1), 53–85 . Retrieved from http://jolt.law.harvard.edu/articles/pdf/v09/09HarvJLTech053.pdf.

FURTHER READING

Justia (Website) *Funk Brothers Seed Co. v. Kalo Inoculant Co. 333U.S. 127*(1948). Retrieved from https://supreme.justia.com/cases/federal/us/333/127/case.html.

PATENTS IN BIOPHARMA

15

Intellectual-property rules are clearly necessary to spur innovation: if every invention could be stolen, or every new drug immediately copied, few people would invest in innovation. But too much protection can strangle competition and can limit what economists call "incremental innovation"—innovations that build, in some way, on others.

-James Surowiecki, "The Financial Page" columnist of *The New Yorker*.

CHAPTER OUTLINE

15.1 Introduction 330
15.2 Pharmaceuticals and Biopharmaceuticals.. 331
    15.2.1 Differences Between Biologics and Conventional Drugs........................... 331
    15.2.2 Biosimilars and "Interchangeable" Biologics... 333
15.3 The Need for Intellectual Property Protection of Biopharmaceuticals.............. 334
15.4 Patent Protection for Biologics 335
    15.4.1 Incremental Innovation 336
    15.4.2 "Evergreening"... 336
15.5 Patent Protection for Diagnostics 338
15.6 The Impact of Patent Protection in Genetic Testing 340
15.7 International Trade Agreements in Medicines 342
15.8 Summary 343
References 344

An Introduction to Ethical, Safety and Intellectual Property Rights Issues in Biotechnology.
DOI: http://dx.doi.org/10.1016/B978-0-12-809231-6.00015-6

Assorted dried plant and animal parts used in traditional Chinese medicines,
Clockwise from top left corner: dried Lingzhi (lit. "spirit mushrooms"), ginseng, Luo Han Guo, turtle shell underbelly (plastron), and dried curled snakes *(By User: Vberger (Personal picture) [Public domain], via Wikimedia Commons)*

15.1 INTRODUCTION

Biotechnology contributes to healthcare in two important ways: by the production of therapeutic products (biopharmaceuticals or biologics), and by means of genetic and proteomic testing for disease diagnosis and personalized medicine. Development of these healthcare products is expensive and time consuming, requires sophisticated laboratory and manufacturing facilities, and entails employing skilled labor. Although patents have been effectively used as a means of recovering research and development (R&D) costs, in the case of therapeutic products, patenting and licensing of diagnostic products appears to be limiting the use of genetic testing and further innovation. This chapter takes a closer look at biopharmaceuticals (biologics) and biosimilars (generic or "follow-on" biologics) in order to better understand the role of patents in the production and marketing of these unique classes of medicines, as well as the impact of patents on diagnostics (genetic and proteomic testing).

15.2 PHARMACEUTICALS AND BIOPHARMACEUTICALS

Pharmaceuticals have a significant economic potential, so much so that highly industrialized nations compete aggressively to attract investment in this area and create clusters of companies to act as innovation hubs to drive sustained economic growth. The majority of the large pharmaceutical ("Big Pharma") companies are located in the United States, east of the Mississippi and in New Jersey. Many others are in Western Europe, in United Kingdom, France, Germany, and Switzerland. The *biotech industry* is significantly smaller than the pharma industry and mostly clustered near prominent research universities, the largest concentration being in California, followed by Massachusetts. **Biopharmaceuticals are unique pharmaceutical substances produced by biotech companies** that have the potential to radically change the diagnosis and treatment of several diseases such as arthritis, cancer, Alzheimer's disease, and Parkinson's disease. The *biopharma industry* has grown over the years and in 2013 represented more than US$150 billion in global sales. **By 2020, biopharmaceuticals are predicted to generate US$290 billion in revenue and comprise 27% of the pharmaceutical market** (IMS Health, 2013).

By definition *biopharmaceuticals* (**also known as** *biologics* **or "biologic(al) medicinal products") are medicinal products manufactured by biotechnology methods from living organisms or their products and include all recombinant proteins, monoclonal antibodies, vaccines, blood/plasma-derived products, nonrecombinant culture-derived proteins, and cultured cells and tissues.** Commercialization of biopharmaceuticals typically entails **patenting of either the process or the product in order that companies may enjoy exclusive marketing rights** so as to (1) **recover the high costs of development** and labor costs of skilled personnel, and (2) **make a profit** in order to sustain further R&D efforts. Opponents of the patent system opine that this **inflates the cost of treatment or diagnosis and places healthcare out of the reach of ordinary citizens.** Proponents of the patent system argue that **without this system innovation will be hampered** as companies would **keep their technology secret** in order to retain their market edge. The issue has attained significance in the face of the patents of several biologics having expired or being on the verge of expiring. Due to the complexity of structure and manufacturing process, derivatives of biologics cannot be given the same treatment as generic chemically derived small molecule drugs. These *"follow-on biologics"* (*"biosimilars"*) therefore warrant special standards and guidelines for their market approval compared to generic drugs.

15.2.1 DIFFERENCES BETWEEN BIOLOGICS AND CONVENTIONAL DRUGS

Unlike conventional drugs that are synthesized by chemical processes and are homogenous and usually small in size, **biologics are complex and large**, and as they are **produced by living cells**, often **heterogeneous**. Many biologics are proteins and both production and activity is susceptible to physical factors such as heat, light, agitation, and pH. Being proteins, they often have the **potential to elicit immune reactions**. The process of manufacture of biologics is monitored at every step, which may not be necessary for chemically synthesized drugs. The safety considerations in the regulatory framework for market approval in the United States and EU for drugs are different from that for biologics and biosimilars (see Chapter 12: Recombinant DNA Safety Considerations in Large Scale Applications and Good Manufacturing Practice). Patents pertaining to the early biologics have expired and many more are due to expire in the near future (see Box 15.1). However, unlike the low-cost generics that enter the marker upon expiration of the patents of conventional "small molecule" drugs, due to complexities in the manufacture process, competing low-cost generic biologics (most regulatory authorities prefer to use the term biosimilars or follow-on

BOX 15.1 THE GROWING SIGNIFICANCE OF BIOSIMILARS

Market analysts are optimistic about the future of biosimilars for several reasons:

- Patents for several biologics have expired or on the verge of expiring
- Increasing global focus on healthcare
- National and international efforts to improve access to healthcare by lowering the cost of medicines

Currently, 48% of sales come from eleven biologics, for which the patents have expired or will expire over the next few years. These include:

Name	Generic Name	Patent Expiry in United States	Patent Expiry in European Union
Humira	Adalimumab	2016	2018
Embrel	Etanercept	2022	2015
Lantus	Insulin glargine	2014	2014
Rituxan	Rituximab	2016	2013
Remicade	Infliximab	2018	2015
Avastin	Bevacizumab	2019	2022
Herceptin	Trastuzumab	2019	2014
Gleevec	Imatinib	2015	2016
Neulasta	Pegfilgrastim	2015	2017
Copaxone	Glatiramer acetate	2014	2015
Revlimide	Lenalidomide	2019	2022

Reference: Winning with biosimilars: Opportunities in global markets *(Deloitte (2015)* Winning with biosimilars: Opportunities in global markets. *Retrieved from http://www2.deloitte.com/content/dam/Deloitte/us/Documents/life-sciences-health-care/us-lshc-biosimilars-whitepaper-final.pdf).*

The European Medicines Agency (EMA) was the first to approve biosimilars, having approved Sandoz's Omnitrope (somatropin) in 2006. As of May 2016, there are 22 biosimilars approved in Europe (GaBI, 2016). Australia gave its approval for biosimilars in 2008, while Japan and Canada authorized marketing of biosimilars in 2009. The first biosimilar approval in the United States was for filgrastim (Sandoz's Zarxio) in 2015 (Amgen, 2015). With the regulatory policies having been resolved in these countries and regions, pharmaceutical companies are gearing up to capitalize on business opportunities in developed as well as emerging markets (see Deloitte, 2015).

References

Amgen (2015, December 11) Biosimilar frenzy: Facts, fiction, and future. [Webinar] Retrieved from http://www.amgenbiosimilars.com/~/media/amgen/full/www-amgenbiosimilars-com/downloads/usa-bio-114753_biosimilar_frenzy_class_i.ashx?la=en.

Deloitte (2015) *Winning with biosimilars: Opportunities in global markets.* Retrieved from http://www2.deloitte.com/content/dam/Deloitte/us/Documents/life-sciences-health-care/us-lshc-biosimilars-whitepaper-final.pdf.

GaBI (2016, May 6) *Biosimilars approved in Europe.* Generics and biosimilars initiative (GaBI) [Web article] Retrieved from http://www.gabionline.net/Biosimilars/General/Biosimilars-approved-in-Europe.

BOX 15.2 EXPEDITING MARKET APPROVAL FOR GENERICS IN THE UNITED STATES

In the United States, a comprehensive health care reform bill, *Patient Protection and Affordable Care Act* (PPACA) was signed into law by President Obama on March 23, 2010. The Act sought to bring down the cost of health services by encouraging the manufacture of biosimilars. Under the Act, a new regulatory authority was established within the Food and Drug Administration (FDA) creating a licensure pathway for biosimilars. This is analogous to the FDA's authority for approving generic drugs. The *Drug Price Competition and Patent Term Restoration Act* of 1984 (also called the *Hatch-Waxman Act*), allows generic companies to manufacture a drug by establishing that it is chemically the same as the already approved innovator drug (termed as "bioequivalence"). By avoiding the expense of conducting clinical trials, as well as the R&D costs of the initial drug developer, the generic drug industry could achieve considerable cost saving, while making the drug available to the public at a lesser price than the brand-name manufacturer. The innovators are partly compensated by additional marketing exclusivity of up to five years over the twenty years granted by the award of patent. Unfortunately, due to the complex nature of biologics, establishing bioequivalence is difficult if not impossible.

Title VII of the PPACA incorporates the 2009 *Biologics Price Competition and Innovation Act* (BPCIA) which established a licensure pathway for competing versions of a previously marketed biologic and a regulatory regime for two types of follow-on biologics, the "biosimilars" and "interchangeable" biologics. The BPCIA created FDA administered periods of regulatory exclusivity for brand name and follow-on biologics and established a system of patent dispute resolution between manufactures of brand name and follow-on biologics.

The PPACA allows for a period of exclusive marketing for a biosimilar that is first to be established as interchangeable with the reference product. The Act also provides data exclusivity for 12 years for the reference product from the date on which the reference product was first approved. This serves as a period of exclusive marketing for the brand-name product. Provisions for resolution of patent infringement claims against an applicant or prospective applicant for a biosimilar product license are addressed by the PPACA.

Thomas J.R. (2014) Follow-on biologics: The law and intellectual property issues. *Congressional Research Service* Report for Congress. Retrieved from https://www.fas.org/sgp/crs/misc/R41483.pdf.

---

biologics) may not be available. **In order to make healthcare affordable, many countries are implementing policies to expedite market approval for biosimilars** (see Box 15.2 for the United States).

15.2.2 BIOSIMILARS AND "INTERCHANGEABLE" BIOLOGICS

By definition, a biosimilar is a product that is *highly similar, but not identical*, to the reference product. Minor difference may exist in clinically inactive components. It is essential however that there are *no clinically meaningful differences* between the biological product and the reference product in terms of safety, purity, and potency. For a biosimilar to be considered as an *interchangeable biologic*, additional criteria have to be met, and requires regulatory approval as biosimilar *and* designated as interchangeable (FDA, 2015). A proposed biosimilar product may be licensed in one or more additional conditions of use for which the reference product is licensed, if appropriate justification is provided and the patent landscape allows for it (termed as *extrapolation*). Copies of innovator drugs are often found in markets that do not have a dedicated regulatory system or with less stringent intellectual property laws and are known as *nonoriginal biologics*. *Bio-betters* are step-wise improvements on innovator molecules that follow the same regulatory pathway as the innovator drug. Bio-better strategies are often used by manufacturers of originator biologics to strengthen market positioning with an improved product (example, Roche's Gazyva is a bio-better of Rituxan).

KEY TAKEAWAYS

Chemical Drugs	Generics	Biologics	Biosimilars
Small ($M_W = \sim 180$ Da), well-defined, homogeneous, can be fully characterized, relatively stable	Bioequivalent and identical to reference product	Large (e.g., monoclonal antibody $M_W = \sim 150$ kDa), heterogeneous, often difficult to fully characterize, sensitive to storage and handling conditions	Similar to, but not identical to the reference product
Predictable chemical process	Identical copy can be made	Complex with many options for post-translational modifications	Manufactured using unique cell lines, expected to be similar but not identical
Low risk of immunogenicity	(Similar to drug)	High risk of immunogenicity	(Similar to biologic)
$2.6 billion (inclusive of failures and capital costs)	$1–5 million in development costs		$100–200 million in development costs
More than 10 years to develop	A 3–5-year development timeline	More than 10–15 years to develop	8–10 year development timeline
	Interchangeable with reference product		No interchangeability or automatic substitution
	80%–90% cheaper than reference product		20%–30% cheaper than reference product

15.3 THE NEED FOR INTELLECTUAL PROPERTY PROTECTION OF BIOPHARMACEUTICALS

Estimates in 2014 made by the Tufts Centre for the Study of Drug Development put the **average pretax industry cost per new prescription drug approval (inclusive of failures and capital costs) at US$ 2.6 billion** (DiMasi, Grabowski, & Hansen, 2016). Typically, only **one in around 5000–10,000 experimental molecules gains regulatory approval** as a medicine; the process takes an **average of more than 10 years,** whereas **less than 12% of drugs entering clinical trials result in an approved medicine** (PhRMA, 2015).

Patents and **data exclusivity protection** provide an opportunity for innovators to **recoup their investment** and **serve as an incentive for further development**. Data protection allows for a specified period of time following marketing approval during which **competing firms are denied use of the safety and efficacy data of the innovator to obtain marketing approval for a generic version of the drug.** It thus provides the innovator firm a period of marketing exclusivity and return on investment. Data exclusivity protection is **complementary to patents but is not an extension of patent rights** as it does not prevent a competitor from introducing a generic version of the pioneer drug provided the innovator's data is not used for getting the market approval. In the United States, the current period specified by law is **12 years from date of market approval** irrespective of the time taken to bring the drug to market (although President Obama's fiscal year 2016 budget proposes to reduce this to 7 years). The provision **forces competitors to generate independent data on the safety and efficacy** of the drug rather than use the innovator's data for market approval.

Pioneering biotech companies such as Genentech, Amgen, Chiron, relied on a **strong patent base to stay in business** (Grabowski and Vernon, 2000). Without intellectual property protection, competitors could copy and manufacture medicines as soon as they have been proven safe and effective, without the expense of R&D or testing. A strong patent regime is expected to promote innovation by incentivizing drug development.

KEY TAKEAWAYS

The need for IPR protection for pharmaceuticals:

- Typically, only one in around 5000–10,000 experimental molecules gains regulatory approval as a medicine.
- Less than 12% of drugs entering clinical trials result in an approved medicine
- Average pretax industry cost per new prescription drug approval (inclusive of failures and capital costs)—US$ 2.6 billion.

15.4 PATENT PROTECTION FOR BIOLOGICS

For an effective intellectual property system, it is necessary that the system not only provides fair and effective incentives and certainty for innovators but also allow patent holders the means for enforcing and for defending infringed patents. Patent protection in biologics industry differs from that in any other industry on several counts:

- *Time taken for development*: Biologics typically take 10–15 years for development from discovery, to preclinical and clinical trials, review by the regulatory authority and finally, market approval. This is longer than the patent process in any other industry.
- *A significant portion of the patent term is lost even before the drug enters the market*: The term of the patent in most countries is **20 years from date of filing**. Most companies do not wait for market approval before filing for the patent; hence, the **effective patent life is shorter** (estimated to be 8 years in the early 1980s, and 10.8 years in late 1980s by Grabowski and

Vernon, 2000). This is because the **grant of a patent** by a patent office and the **award of marketing approval** by the regulatory body are **distinct events** that depend on different criteria. Generic companies can challenge patents as soon as 4 years of the brand drug entering the market.

- *Cost to the innovator*: A large portion of the cost of bringing a drug from discovery stage to marketing stage is the expense involved with the **conduct of preclinical and clinical trials to demonstrate safety and efficacy of a new drug for market approval**. Without adequate patent protection this data could be used by generic companies to bring out cheaper versions of the brand drug.

The problem with patents is that brand drugs are more expensive for the consumer and in the absence of generic versions, could put healthcare out of the reach of the common man. In the event of the new drug being deemed crucial to the health of its citizens, countries address this issue by a procedure known as *compulsory licensing*. This is a legal process that allows **governments to ignore eligible and issued patent rights of an inventor** and to produce the drug through **contractors or third parties without the license or approval of the patent holder**. This is usually resorted to in the face of a **national emergency** such as a threat to national security or medical emergency. Such a provision has been used by several countries against rising costs of drugs against HIV/AIDS. In the United States, the threat of compulsory licensing was used in 2001 to negotiate with Bayer a 70% reduction in the price of antibiotic Ciprofloxacin (Cipro) used to treat anthrax. This allowed the US government to stockpile the antibiotic against a possible anthrax threat (Reichman, 2009). Compulsory licensure provisions in patent laws have thus allowed many countries, especially developing countries, to control the prices of many life-saving (including anti-cancer) drugs (see Box 15.3).

15.4.1 INCREMENTAL INNOVATION

Incremental improvements in drugs often represent **advances in safety and efficacy, new formulations, and dosing options** that provide doctors with multiple therapeutic possibilities. Incremental innovations of a patented drug may be released prior to its patent expiration thereby possibly diminishing losses experienced by the innovator company when a patent expires. Typically, generic companies are prepared to enter a market immediately after the innovator company's patent expires, resulting in a loss of **about 40% of the market for the innovator company**. Innovators generally file multiple patents for the same product based on the product's active ingredient, formulations, or methods of use. Some of the multiple patents may also have different expiration times, often as part of an "evergreening" strategy. In cases where the invention is made up of several components, each patented by different parties, it often results in what has been termed a "**patent thicket**." Such patent thickets generally delay or prevent development of technologies. Suggested solution to the patent thicket is the creation of a patent pool or a clearinghouse.

15.4.2 "EVERGREENING"

The tactic used by pharmaceutical companies to preempt their own patent expirations by incremental innovation is referred to as "evergreening." Critics maintain that "evergreening" has a negative

BOX 15.3 COMPULSORY LICENSING

One of the contentious issues discussed in the Uruguay Round of the World Trade Organizations that led to the Trade Related aspects of Intellectual Property rights (TRIPS) Agreement in 1995 was the form of patents to be used for protecting innovations in the pharma sector. India in its Patent Act of 1970, had limited patents in medicines to process patents. This allowed pharma companies to invent around patented drugs so as to keep drug prices low. This patent policy was adopted primarily to keep healthcare affordable to the majority of Indians, but it had the happy outcome of positioning India as a global supplier of low-cost medicines to other developing nations. In keeping with the TRIPS Agreement, India had to introduce product patents, which was strongly opposed by the domestic pharma sector. In the discussions in the Uruguay Round, India was able to insist on the insertion of a clause (Article 31) in the TRIPS Agreement that allowed member countries to retain the right to issue compulsory license to domestic firms for a patented medicine if the patent holder did not provide the medicine at an affordable price. This clause was put to test in 2012 when an Indian company, Natco Pharma (http://natcopharma.co.in/), applied to the Controller General of Patents in India for a license to manufacture an anticancer drug used to treat liver and kidney cancer. The patented drug Naxevar (sorafenib tosylate) was being sold by a German multinational company, Bayer Corporation, at the selling price of $5420 (Rs. 280,428/-) for a monthly dose. Sorafenib fights cancers by targeting vascular endothelial growth factor receptors (VEGFR) 2 and 3 as well as platelet derived growth factor receptor (PDGFR) β active against melanoma, renal cell carcinoma or hepatocellular carcinoma. Natco in its license application offered to sell it at $170 (Rs. 8800/-). P.H. Kurian, the then Controller General of Patents, on the basis of arguments presented by both companies and on investigation, in a landmark decision granted a compulsory license to Natco on March 9, 2012, to manufacture Naxevar (Panagariya, 2012).

Other examples of compulsory licensing being used include (WHO, 2006):

- May 2002—Zimbabwe declared a period of emergency over its AIDS epidemic authorizing the government to override patents to permit local production of Anti RetroViral (ARV) medicine
- October 2003—Malaysia allowed import of generic didanosine, zidovudine, and lamivudine/zidovudine combination from India to supply to public hospitals
- March 2004—Mozambique granted compulsory license for local manufacture of first in line triple combination ARV
- September 2004—Zambia issued compulsory license to permit local production of first-line ARV therapy
- 2004—Indonesia authorized government use of patents to enable local production of nevirapine and lamivudine
- April 2010—Ecuador granted compulsory license to supply lopinavir/ ritonavir.

References

Panagariya A. (2012, May 30) (Reporter) *How PH Kurien took on global patents to make very costly drug affordable for poor.* The Economic Times. Retrieved from http://articles.economictimes.indiatimes.com/2012-05-30/news/31900229_1_patent-act-product-patent-patent-law.

WHO (2006). Access to AIDS medicines stumbles on trade rules. *Bulletin of the World Health Organization, 84*(5), 337−424. Retrieved from http://www.who.int/bulletin/volumes/84/5/news10506/en/.

impact upon public health through higher prices and reduced access to medicines. Advocates however maintain that it is an effective intellectual property policy that provides for patent protection for technological advancements. The crux of the problem is that many of the incremental innovations claimed are often "minor" changes and may not have a significant effect on the properties or efficacy of the innovator drug (see Box 15.4).

KEY TAKEAWAYS

- **Incremental innovation:** the period of patent protection for a drug may be increased by advances in its safety or efficacy by new formulations or dosing options.

BOX 15.4 INCREMENTAL INNOVATION AND "EVERGREENING"

A much-publicized case that exemplifies "evergreening" is *Novartis AG v. Union of India & Others*, tried in the Supreme Court of India, decided on April 1, 2013. Novartis AG in 1993 patented *imatinib* with salts vaguely specified in several countries and marketed the same under the trade-name *Glivec* in Europe, or *Gleevec* in the United States. The drug was hailed as a magic bullet against cancer, being effective against newly diagnosed patients with Philadelphia chromosome-positive chronic myeloid leukemia and gastrointestinal stromal tumors (GIST). Since product patents were not allowed in India at that time, Novartis made a "mail-box" patent application in 1998. Subsequent to the patent laws in India becoming TRIPS compliant, the application was examined in 2005 for patent eligibility. Novartis had applied for Exclusive Marketing Rights in India which had been granted by the Indian Patent Office in 2003. The price set by Novartis was US$2666 (Rs. 1.2 lakh) per month for a patient. Meanwhile generic versions of the drug were available in the market for US$177 to 266 (~Rs. 8000) per month. The 1998 patent application made by Novartis differed from the 1993 patent in that a specific salt, the beta crystalline form of *imatinib mesylate*, was claimed as Gleevec with enhanced bioavailability. The patent application was rejected by the patent office in 2006 and the appeal board in 2009 on the grounds that new uses for known drugs or modifications of known drugs are patentable only if *they differ significantly in properties with regard to efficacy* under Section 3(d) of the Indian Patents (Amendment) Act, 2005. On appeal the case was taken up by the Supreme Court. The Court tried the case *de novo*, but the decision of the two-judge bench upheld the rejection of the patent application since the incremental invention did not qualify the test of Section 3(d) of the Act. The decision was hailed as being in the interest of patients, providing affordable medicine for the poor. The ruling also was appreciated for having *"set a precedent that would prevent international pharmaceutical companies from obtaining fresh patents in India on updated versions of existing drugs"* (BBC News, 2013).

Reference

BBC News (2013, April 1). Novartis: India rejects patent plea for cancer drug Glivec. Retrieved from http://www.bbc.com/news/business-21991179.

- **Compulsory licensing**: a legal process that allows governments to ignore patent rights of an inventor and to produce the drug through third parties without license from patent holder. Usually resorted to in the face of a national emergency.

15.5 PATENT PROTECTION FOR DIAGNOSTICS

Diagnostic and personalized medicine technologies rely on "gene patents" for commercial success. Whether gene sequences are patentable subject matter has itself been controversial and challenged in courts in the United States and in Australia (see Chapter 14: Patenting of Life Forms). In most jurisdictions, exemptions to patent subject matter eligibility include laws of nature and natural phenomenon, fundamental principles, and abstract ideas. Patents challenges in US courts on applications in disease diagnosis and personalized medicine have revolved around whether the invention is **an abstract idea** (*Bilski et al. v. Kappos*, see Box 15.5), **a law of nature** (*Mayo Collaborative Services v. Prometheus Laboratories, Inc.*, see Box 15.6), **or a natural phenomenon** (*Association for Molecular Pathology v. Myriad Genetics, Inc.*, see Box-14.1 in Chapter 14: Patenting of Life Forms). The decisions arrived at in these court cases have prompted the USPTO to implement major changes to the evaluation by patent examiners of **claims that recite a judicial exemption**. The first of the examination guidance documents was issued by the USPTO in March 2014: *"Guidance for*

BOX 15.5 BILSKI ET AL. V. KAPPOS

Bilski et al. v. Kappos, 561 U.S., 130S.Ct. 3218, 3231 (2010)

This case is not related to genomics but for a claimed invention in the business world. Its significance lies in that it established that the "machine or transformation test" is insufficient for determining the patent eligibility of a process and emphasized that an abstract idea cannot be patented.

Bernard Bilski and Rand Warsaw sought patent protection for a claimed invention that protects buyers and sellers of commodities in the energy market against the risk of price changes. Claim 1 in the patent application described a series of steps instructing how to hedge risk, whereas claim 4 put the claim in a simple mathematical formula. The remaining claims in the application explained how claims 1 and 4 can be applied. The application was rejected by the patent examiner on the grounds that it merely manipulates an abstract idea and solves a mathematical equation "without any limitation to a practical application." The Board of Patent Appeals and Interferences affirmed, as did the United States Court of Appeals for the Federal Circuit. Applying the machine-or-transformation test, the court held that the application was not patent eligible. The United States Supreme Court on writ of certiorari to the United States Court of Appeals for the Federal Circuit in 2010, rejected the patent application on the unpatentability of abstract ideas.

Supreme Court of United States No.08-964: 561 U.S. (2010): Bernard L. Bilski and Rand A. Warsaw, Petitioners v. David J. Kappos, Under Secretary of Commerce for Intellectual Property and Director, Patent and Trademark Office on writ of certiorari to the United States Court of Appeals for the Federal Circuit. Retrieved from https://supreme.justia.com/cases/federal/us/561/08-964/opinion.html.

BOX 15.6 MAYO COLLABORATIVE SERVICES V. PROMETHEUS LABORATORIES, INC.

Mayo Collaborative Services v. Prometheus Laboratories, Inc., 566 U.S., 132S.Ct. 1289, 101 USPQ2d 1961 (2012)

This case argued in the U.S. Supreme Court in December 2011 concerned the measurement of the metabolites of a drug in order to estimate the effective dosage to be given to a patient, an example of a diagnostic test used in personalized medicine. In the verdict delivered in March 2012, the Court unanimously ruled that the claims involved observing a natural correlation (law of nature) and was not eligible subject matter under 35 U.S.C. § 101.

Two patents owned by Hospital Sainte-Justine in Montreal concern the use of thiopurine drugs used in the treatment of autoimmune diseases. These drugs are metabolized differently by different people, so doctors have to empirically determine the dose for each patient as high doses produce adverse side effects and too low doses have no effect. Scientists at the hospital discovered that estimating the level of 6-thioguanine, a metabolite of the drug, in patient blood samples could help deciding the drug dose to be used. The patents were exclusively licensed to Prometheus Laboratories Inc. (http://www.prometheuslabs.com/) that sells diagnostic kits based on them. Until 2004, Mayo Collaborative Services (http://www.mayomedicallaboratories.com/) and Mayo Clinic Rochester bought and used diagnostic kits from Prometheus. But in 2004, Mayo announced that it intended to sell its own somewhat different diagnostic test. Prometheus sued Mayo for patent infringement. The District Court in 2008 found the patent infringed, but held the claims invalid under Section 101. Prometheus appealed, and in 2009, the Federal Circuit reversed the District Court verdict and held the patents valid on the basis of sufficient transformation of the natural law. Mayo appealed to the Supreme Court which granted certiorari, vacated the Federal Circuit verdict, and remanded the case to the Federal Circuit for reconsideration. The Federal Circuit did not change its verdict, and Mayo again appealed to the Supreme Court. The Court called the correlation between the naturally produced metabolites and the therapeutic efficacy and toxicity an unpatentable "natural law" and held that Prometheus' process is not patent eligible.

The significance of the case is that the decision had rendered any form of medical diagnostic testing as patent ineligible. The inventive concept described in such patent applications lies in the discovery of the relationship of a biomarker with a certain medical condition. Once such an inventive concept is removed from the claim, the technology would not be sufficient to transform the claim into a patent eligible application.

Supreme Court of the United States: Mayo Collaborative Services, dba Mayo Medical Laboratories et al. v Prometheus Laboratories Inc. Certiorari to the United States Court of Appeals for the Federal Circuit. No. 10-1150. Argued December 7, 2011- Decided March 20, 2012. Retrieved from https://www.supremecourt.gov/opinions/11pdf/10-1150.pdf.

Determining Subject Matter Eligibility of Claims Reciting or Involving Laws of Nature, Natural Phenomena, & Natural Products." This was replaced in December 2014 by the *"2014 Interim Guidance on Patent Subject Matter Eligibility."* Patent examination is to be a three step process:

1. Is the claimed invention directed to a composition of matter, a process, defined as an act or a series of acts or steps, a manufacture or machine?
2. Does the claim recite a judicial exemption: focus on use of a law of nature, a natural phenomenon or naturally occurring relation or correlation?
3. Does the claim amount to more than the natural principle itself: does it include additional elements/steps or combination thereof that integrate the natural principle into the claimed invention?

The guidance has been further updated in July 2015 based on discussions on various issues raised by the public comments (USPTO, 2015). The guidance establishes that for claims that recite a natural product or natural law to be patent eligible, there has to be *significantly more* than a mere monopolization of the naturally occurring principle (*Ariosa Diagnostics, Inc. v. Sequenom, Inc.*, see Box 14.2 in Chapter 14: Patenting of Life Forms). Isolated biomolecules such as peptides or oligonucleotides with sequence similarity to portions of a naturally occurring larger molecule and having same or similar functions as the larger molecule are not patent eligible. This rationale could probably be extended to RNA, antibodies, and proteins where the claims on therapeutic applications derive from the naturally occurring function of the biomolecule. The patent eligibility of DNA primers and probes is not currently clear: they are probably not eligible if identical in sequence to naturally occurring DNA, but eligible if conjugated to a nonnaturally occurring label so that it is now structurally different from the naturally occurring molecule.

In **Australia**, isolated nucleic acids are not patent eligible as established by the High Court of Australia ruling of October 7, 2015, in the *D'Arcy v Myriad Genetics Inc & Anor [2015] HCA35* (see Box 14.1 in Chapter 14: Patenting of Life Forms). In **Canada**, the Children's Hospital of Eastern Ontario (CHEO) has challenged patents covering the testing of a genetic disorder Long QT Syndrome (a heart rhythm disorder that can cause seizures and sudden death) held by University of Utah Research Foundation, Genzyme Genetics, and Yale University. The patents prevented CHEO from conducting on-site genetic analysis, and the mandatory outside testing at the University of Utah more than doubled the cost and time taken for diagnosis. CHEO commenced patent proceedings against University of Utah in November 2014 asking for a declaration of noninfringement and/or invalidation of claims on isolated nucleic acids and methods of assessing risk to Long QT Syndrome (DeLuca and Bonter, 2015).

15.6 THE IMPACT OF PATENT PROTECTION IN GENETIC TESTING

Patent challenges in the area of diagnostics in the United States and elsewhere have prompted examination of the **impact of gene patents and licensing practices on access to genetic testing**. To quote Richard Gold and Julia Carbone, *"From the late 1980s, a storm surrounding the wisdom, ethics, and economics of human gene patents has been brewing. The winds of concern in this storm touched on the impact of gene patents on the basic and clinical research, health care delivery and ability of public health care systems to provide equal access when faced with costly* *patented genetic diagnostic tests"* (Gold and Carbone, 2010). In a report submitted by the **Secretary's Advisory Committee on Genetics, Health and Society (SACGHS) in April 2010** titled *"Gene Patents and Licensing Practices and Their Impact on Patient Access to Genetic Tests,"* based on a thorough literature review and case studies of genetic testing for 10 clinical conditions, the Committee concluded that it *"did not appear the patents were necessary for either basic genetic research or the development of available genetic tests"* (SACGHS, 2010). Moreover, patents were being used *"to narrow or clear the market of existing tests, thereby limiting rather than promoting availability of testing"* (see Box 15.7). In the period following the report, the Supreme Court and the Federal Circuit in the United States, as well as courts in Australia and Canada, interpreted various provisions of the patent law making it more difficult for diagnostic

---

BOX 15.7 GENE PATENTS AND LICENSING PRACTICES AND THEIR IMPACT ON PATIENT ACCESS TO GENETIC TESTS

The report submitted to the Secretary of Health and Human Services in the United States by the Secretary's Advisory Committee on Genetics, Health, and Society (SACGHS) on March 21, 2010, considered three major issues regarding gene patents and licensing practices and their impact on access to genetic tests, and made the following observations:

1 Effect of Patents and licenses on genetic research and test development:
   a. Prospect of patent protection of a genetic research discovery does not play a significant role in motivating scientists
   b. Patents can harm genetic research; evidence suggests that patents on genes discourage follow-on research
   c. Possession of exclusive rights was not necessary for development of particular genetic tests
   d. Exclusive rights do not result in faster test development
   e. Development of future testing methods such as multiplex tests, parallel sequencing, and whole genome sequencing could be inhibited due to the need to acquire multiple licenses to all relevant patents (patent thicket). Negotiations could be complicated and licenses expensive, making products expensive/ unmarketable.
2 *Effects of patents and licensing practices on patient access to existing tests:*
       When patents and licensing practices have created sole provider of a genetic test, patients have suffered because they are as follows:
   a. Unable to obtain insurance cover
   b. Unable to obtain second opinion testing from independent laboratory
3 *Effects of patents and licensing practices on test quality:*
   a. Since neither sample sharing nor competition possible due to exclusive rights, significant concerns about quality arise.

       The observations were based on eight case studies of genetic testing for ten clinical conditions:

1. Inherited susceptibility to breast/ovarian cancer and colon cancer
2. Hearing loss
3. Cystic fibrosis
4. Inherited susceptibility to Alzheimer disease
5. Hereditary hemochromatosis
6. Spinocerebellar ataxias
7. Familial long QT syndrome, and
8. Canavan disease and Tay-Sachs disease.

SACGHS (2010) *Gene patents and licensing practices and their impact on patient access to genetic tests.* Report of the Secretary's Advisory Committee on Genetics, Health and Society. Retrieved from http://osp.od.nih.gov/sites/default/files/SACGHS_patents_report_2010.pdf.

methods to be patented and for patents to be enforced (discussed in the previous section). There appear to be many obstacles to patenting diagnostic methods causing **serious concerns for the future of innovation in diagnostic testing** (Eisenberg, 2015). Rachel Sachs argues *"ensuring that academic researchers and diagnostic testing companies have sufficient resources and incentives to develop those tests is critically important"* (Sachs, 2016). In Sachs opinion, in the case of health technologies, patent law alone may be insufficient to incentivize innovations. As the Food and Drug Administration regulations dictate the conditions under which these technologies enter the market, and the Centers for Medicare and Medicaid Services considers whether to pay for a given technology as part of its role as insurer, an array of policy choices involving each of the three systems may be necessary (Sachs, 2016).

KEY TAKEAWAYS

Patent protection for diagnostics:

- Whether gene sequences crucial to diagnostics are patentable subject matter has been challenged in courts in the United States and elsewhere.
- The USPTO 2015 guidance on Patent Subject Matter Eligibility establishes that for *claims that recite a natural product or natural law to be patent eligible, there has to be "significantly more" than a mere monopolization of the naturally occurring principle.*
- The SACGHS report of 2010 on gene patents and licensing practices concluded that patents were not necessary for either basic research or development of genetic tests, and were *limiting rather than promoting availability of testing.*

15.7 INTERNATIONAL TRADE AGREEMENTS IN MEDICINES

Increasingly international trade agreements are including clauses on patent protection and pricing of products including medicines. This could have far reaching consequences on healthcare policies and affordability of healthcare across nations. One such agreement that has had an impact on the development of biotechnology patent policy is the Agreement on Trade Related Aspects of Intellectual Property Rights (TRIPS). Articles 27 to 34 of the TRIPS Agreement deals with the obligations of the WTO members with respect to patent law and establishes common minimum standards regardless of technology. **Article 27(3) deals specifically with biotechnology**. As a result of this agreement, except in extreme cases governed by morality, countries will not be able to craft their patent laws to take into account the specific and different ethical and social concerns that arise out of biotechnology. An illustrative example is the impact that the TRIPS Agreement has had on the pharmaceutical industry in India.

Prior to India joining the WTO in 1975, the **Indian Patent Act, 1970 vide** Section 5, **allowed process patents and not product patents for agrochemicals and pharmaceuticals**. **Process patents encouraged the development of generic version of even on-patent drugs, keeping drug prices low** and enabling access to medical treatment to all. With a strong base in chemical engineering and the capability to reverse engineer drugs, the domestic pharmaceutical industry developed into a world class generics industry. As a supplier of low-cost medicines and drugs India has been called "**the pharmacy of the developing world**" (Medecins sans Frontiers, n.d.). For example, in 2012, India provided the largest amount ($558 million worth) of services and supplies to the UNICEF (2012). However, in keeping with its obligations under the TRIPS Agreement, the **Patents (Amendment) Act, 2002 introduced product patents in India**. The pharmaceutical industry has since been forced to either increase funding for domestic R&D efforts, or produce generic versions of off-patent drugs.

Another trade agreement speculated to have an impact on the pharma sector is the Trans Pacific Partnership Agreement (TPP). The TPP is a trade agreement between 12 Pacific rim countries: Australia, Brunei, Canada, Chile, Japan, Malaysia, Mexico, New Zealand, Peru, Singapore, the United States, and Vietnam. Negotiations on various issues such as agriculture, intellectual property, environment, services and investments began in 2008 and an agreement was finally reached on October 5, 2015. The treaty apparently proposes to require all countries to use "**competitive market-derived prices**" or to establish the price of the drug on benchmarks that "appropriately recognize the value" of the drug.

KEY TAKEAWAYS

International agreements such as the TRIPS Agreement and the TPP trade agreement have raised concerns of increasing health care costs by imposing patent obligations on developing nations and inhibiting the development of low-cost generics.

15.8 SUMMARY

A major component of the appeal of biotechnology is the impact that it could potentially have on human healthcare. Genomics and recombinant DNA technology have made it possible to isolate and manipulate genes of interest. Genes for production of therapeutic proteins such as hormones (insulin, growth hormone, erythropoietin) have made treatment of medical conditions easier. Gene sequences can be used in diagnostic genetic testing for several inherited disorders. Patents have played a significant role in the nascent biotech industry helping companies recover cost of R&D and sustain future efforts. With the patents of many of the first-generation products having expired or on the verge of expiry, several nations are focused on patenting and marketing of biosimilars and making healthcare affordable to all. Gene patents and patents for diagnostics however is becoming increasingly difficult due to reasons that are technical (whether gene sequences, or correlations between a marker and a disease condition revealed by the test is indeed patentable subject matter) and ethical (whether premium rates charged due to exclusive licensing puts testing out of the reach of the common man). Proponents of the patent system insist that a strong patent policy is required as without it innovators would not be able to sustain R&D activities. Critics however are of the opinion that patents make healthcare expensive and unaffordable to most citizens.

REFERENCES

DeLuca, C. & Bonter K. (2015, March 9) Personalized medicine: Patenting controversies. [Blog] Biotechnology Focus.ca. Retrieved from http://biotechnologyfocus.ca/personalized-medicine-patenting-controversies/.

DiMasi, J. A., Grabowski, H. G., & Hansen, R. W. (2016). Innovation in the pharmaceutical industry: New estimates of R&D costs. *Journal of Health Economics*, *47*, 20–33.

Eisenberg, R. S. (2015). Diagnostics need not apply. *Boston University Journal of Science & Technology Law*, *21*(2). Retrieved from https://www.bu.edu/jostl/files/2015/12/EISENBERG_ART_FINAL-web.pdf.

FDA (2015) Guidance for industry: Scientific considerations in demonstrating biosimilarity to a reference product. Retrieved from http://www.fda.gov/downloads/DrugsGuidanceComplianceRegulatoryInformation/Guidances/UCM291128.pdf.

Gold, R. E., & Carbone, J. (2010). Myriad genetics: In the eye of the policy storm. *Genetics in Medicine*, *12*(4 Suppl.), S39–S70. Retrieved from http://www.ncbi.nlm.nih.gov/pmc/articles/PMC3037261/.

Grabowski, H. G., & Vernon, J. M. (2000). Effective patent life in pharmaceuticals. *International Journal of Technology Management*, *19*(1/2), 98–120. Retrieved from https://fds.duke.edu/db/attachment/182.

IMS Health (2013) Searching for terra firma in the biosimilars and non-original biologics market: Insights for the coming decade of change. Cited in Deloitte (2015) *Winning with biosimilars: Opportunities in global markets.*

Medecins sans Frontieres (n.d.) Examples of the importance of India as the "pharmacy for the developing world". *Campaign for Access to Essential Medicines*. Retrieved from http://www.msfaccess.org/sites/default/files/MSF_assets/Access/Docs/ACCESS_briefing_PharmacyForDevelopingWorld_India_ENG_2007.pdf.

PhRMA (2015) *2015 Profile Biopharmaceutical Research Industry*. Pharmaceutical Research and Manufacturers of America. Retrieved from http://www.phrma.org/sites/default/files/pdf/2015_phrma_profile.pdf.

Reichman, J. H. (2009). Compulsory licensing of patented pharmaceutical inventions: Evaluating the options. *The Journal of Law, Medicine & Ethics*, *37*(2), 247–263 . Retrieved from http://www.ncbi.nlm.nih.gov/pmc/articles/PMC2893582/#FN40.

SACGHS (2010) *Gene patents and licensing practices and their impact on patient access to genetic tests.* Report of the Secretary's Advisory Committee on Genetics, Health and Society. Retrieved from http://osp.od.nih.gov/sites/default/files/SACGHS_patents_report_2010.pdf.

Sachs, R. (2016). Innovation law and policy: Preserving the future of personalized medicine. . *U.C. Davis L. Rev. 49 (forthcoming)*. Draft retrieved from http://papers.ssrn.com/sol3/papers.cfm?abstract_id=2596875.

UNICEF (2012) *2012 supply annual report*. Retrieved from http://www.unicef.org/supply/files/UNICEF_Supply_Annual_Report_2012_web.pdf.

USPTO (2015) 2014 Interim Guidance on Subject Matter Eligibility [Webpage] Retrieved from http://www.uspto.gov/patent/laws-and-regulations/examination-policy/2014-interim-guidance-subject-matter-eligibility-0.

PROTECTION OF TRADITIONAL KNOWLEDGE ASSOCIATED WITH GENETIC RESOURCES

16

"In a society where all are related, simple decisions require the approval of nearly everyone in that society. It is society as a whole, not merely a part of it, that must survive. This is the indigenous understanding. It is the understanding in a global sense. We are all indigenous people on this planet, and we have to reorganize to get along."

-Rebecca Adamson, Founder, First Peoples Worldwide and First Nations Development Institute.

CHAPTER OUTLINE

16.1 Introduction 346
16.2 The Importance and Need for Protection 346
    16.2.1 The Convention on Biological Diversity and Traditional Knowledge Associated With Genetic Resources 347
    16.2.2 The World Intellectual Property Organization and Traditional Knowledge 348
    16.2.3 The World Trade Organization and Traditional Knowledge 350
    16.2.4 The Food and Agriculture Organization and Traditional Knowledge 352
16.3 Legal Protection for Traditional Knowledge 352
    16.3.1 Positive Protection Strategies 353
    16.3.2 Defensive Protection Strategies 353
    16.3.3 Problems With Implementation 354
16.4 Summary 354
References 356

An Introduction to Ethical, Safety and Intellectual Property Rights Issues in Biotechnology.
DOI: http://dx.doi.org/10.1016/B978-0-12-809231-6.00016-8

Turmeric (Curcuma longa): fresh rhizome and powder
{By Simon A. Eugster (Own work) [GFDL (http://www.gnu.org/copyleft/fdl.html) or CC BY-SA] https://commons.wikimedia.org/wiki/File:Curcuma_longa_roots.jpg}

16.1 INTRODUCTION

Traditional knowledge (TK) is an intrinsic part of the cultural heritage identifying and defining a community. It is a functional knowledge system developed through observation and interaction with a habitat that has been preserved and refined by generations of people. While TK extends to all aspects of a community including sustainable livelihood, art, craft, and literature, biotechnologists have been especially interested in TK in the context of conservation and sustainable use of biodiversity and genetic resources (GRs), the use of traditional medicines in healthcare, and knowledge of traditional farming practices. As many communities exist around biodiversity hotspots, TK held by these peoples are increasingly being accessed for bioprospecting. Concerns of misappropriation however arise when industrial or commercial benefits derived from this TK fail to recognize and respect the role of the holders of the knowledge. Access and benefit sharing (ABS) has been integrated into many international treaties such as the Convention for Biodiversity (see Chapter 8: Biodiversity and Sharing of Biological Resources). This chapter looks at mechanisms that allow for the protection of intellectual property (IP) with respect to TK and GRs.

16.2 THE IMPORTANCE AND NEED FOR PROTECTION

Variously referred to as Indigenous Knowledge, Traditional Environmental Knowledge, and Local Knowledge, TK has helped to **meet the livelihood needs** of a large proportion of the **global population dependent on biological resources**. For instance, the knowledge of medicinal herbs: The World Health Organization estimates that 80% of the world's population (including urban populations for many ailments) **depends on traditional medicine for primary health needs**. Increasing urbanization and changing lifestyles in developing nations has however reduced this knowledge to indigenous tribes and fewer members of local communities. Preserving TK for future generations has been the objective of several conservation efforts. The task is challenging since **TK exists both as undocumented knowledge** (largely oral traditions often part of a larger cultural collective knowledge), **as well as documented knowledge** (in treatises, scriptures, and folklore). This chapter discusses protection strategies for TK linked to GR which the **United Nations Convention on Biological Diversity (CBD)** refers to as "TK associated with GRs." In addition to the CBD, other international bodies involved in devising strategies for protection of TK and GR include the **World Intellectual Property Organization (WIPO)**, the **World Trade Organization (WTO)**, and the **Food and Agriculture Organization (FAO)**. The UNESCO and the United Nations Declaration on the Rights of Indigenous Peoples are working for the protection of Traditional Cultural Expressions (TCE), a discussion on which is beyond the scope of this book.

Protection of TK has two major objectives:

1. **to prevent misappropriation** by others, to ensure that the TK (and GR) are not used inappropriately, and to establish a method by which the **benefits arising from a commercial application of the resource is shared with the holders of the resource,** and
2. **to prevent erosion of the TK and to document it for posterity.** The concept of "protection" of TK is complex, partly because of difficulties in defining in legal terms what makes a TK "traditional," and what is the manner and purpose of protection that is to be accorded to it. In treating it as a form of IP several nations have modified existing national laws to prevent misuse or misappropriation of TK. Classic IP instruments such as patents and breeder's rights are however not considered suitable for protection of TK or GR for reasons that may be both **technical** (cannot be considered "novel," GR are "not creations of human mind") and **ideological** (exclusionary, monopolistic rights which do not reflect the collective aspects of the resource).

KEY TAKEAWAYS

TK is a "functional knowledge system developed through observation and interaction with a habitat that has been preserved and refined by generations of people."

Purpose of protection of TK:

- To prevent misappropriation
- To preserve for posterity

16.2.1 THE CONVENTION ON BIOLOGICAL DIVERSITY AND TRADITIONAL KNOWLEDGE ASSOCIATED WITH GENETIC RESOURCES

The United Nations CBD signed in 1992 during the Rio "Earth Summit" (see Chapter 8: Biodiversity and Sharing of Biological Resources) explicitly recognized the role of indigenous communities in conservation efforts and the importance of TK in sustainable use of biodiversity. This is articulated in **Article 8(j)** of the CBD. In **Article 10**, the CBD encourages the use of traditional cultural practices that are compatible with conservation or sustainable use requirements. Although the CBD does not specifically address the rights of indigenous people over their knowledge, it established the following:

- The sovereignty of the state over genetic and biological resources (Articles 3, 15)
- Prior informed consent (PIC) of both the state as well as resource providers/ holders of TK for access to genetic and biological resources (Article 15) and exchange of information including indigenous and TK (Article 17)
- Fair and equitable sharing, on mutually agreed terms (MAT), of benefits arising from use of GRs and/or the TK associated with it (Article 19).

The **Working Group on Article 8(j)** was established in 1998 by the COP to the CBD to help further the implementation of the CBD's provisions relating to TK (https://www.cbd.int/traditional). Recognition of TK was encouraged in CBD discussions on ABS and incorporated in the Bonn Guidelines adopted in 2002 and more specifically in the **Nagoya Protocol** adopted in 2010 (see Chapter: Biodiversity and Sharing of Biological Resources). The Nagoya Protocol further builds on Article 8(j) of the CBD; access to and use of TK should be subject to the PIC of indigenous people and MAT (**Article 7**). **Article 12** of the Nagoya Protocol specifies conditions for use of TK associated with GR, while **Article 16** mandates compliance with domestic legislation or regulatory requirements on ABS for TK associated with GR.

16.2.2 THE WORLD INTELLECTUAL PROPERTY ORGANIZATION AND TRADITIONAL KNOWLEDGE

Bioprospecting, especially for discovering new medicinal products from biological resources, has relied heavily on the knowledge of local healers of the unique properties of various organisms (particularly herbs) in their medicine chest. With many new products and applications based on TK being patented by innovators, often without the knowledge of the TK holders, and many of these patents being subsequently challenged (see Box 16.1), policy makers are increasingly aware of a need for an IP system suitable for protecting TK and associated GR. Some of the salient issues are as below: (paraphrased from WIPO Booklet No. 2, n.d.):

1. Whether the existing IP system which privilege individual rights over collective interests of the community is compatible with the values and interests of the community.
2. Whether an IP system could give a community greater recognition for their TK and control the use and management of their TK.
3. Whether an IP system could help communities safeguard their interests, and ensure a fair and equitable share in the profits.

The WIPO began work on TK in 1998 and at its twenty-sixth session held in Geneva in 2000 established the **Intergovernmental Committee on Intellectual Property and Genetic Resources, Traditional Knowledge and Folklore (IGC)** to primarily look into IP issues that arise in the context of:

1. Access to GRs and benefit sharing
2. Protection of TK, whether or not associated with those resources
3. The protection of folklore, including handicrafts (WIPO, 2001)

International cooperation and coordination in the protection of TK is necessary as utilization of TK is not confined to nations and international partnerships are often crucial to research and development

BOX 16.1 SOME EXAMPLES OF "BIOPIRACY"

By definition biopiracy is the use of indigenous knowledge of natural resources for commercial applications without the consent and with little or no compensation or recognition to the community from which it originates. There are numerous examples in recent literature, some well-known examples are discussed here:

1. **Turmeric patent (*Curcuma longa L.*):**
   Turmeric is a spice with numerous medicinal properties popular in India and South-East Asian countries. It forms an important part of traditional medicine in India and is part of every household's medicinal chest used for curing a variety of ailments. Turmeric became the center of a patent storm when two expatriate Indians, Suman K. Das and HariHar P. Kohly, at the University of Mississippi Medical Centre were granted a U.S. patent in 1995 for the use of turmeric for wound healing. In the ensuing furore, the Council of Scientific and Industrial Research (CSIR) filed a re-examination case with the USPTO challenging the patent on the grounds that it was not "novel" and known to Indians for centuries. Documentary evidence was provided from ancient Sanskrit texts and a 1953 paper published by the Indian Medical Association. The U.S. patent was revoked in 1997.

2. **Neem patent (*Azadirachta indica A. Juss*):**
   The neem tree, common in India and South-East Asia, is considered as the "Village Pharmacy" as every part of the tree can be used as medicine, and the leaves and seed-oil as insecticide and fungicide for storing grain and protecting crop plants. In 1994, the European Patent Office granted a patent to W.R. Grace and Company and the United States Department of Agriculture for the use of a hydrophobic substance extracted from neem oil for controlling fungi on plants. Legal opposition to the patent was mounted by a group of NGOs and farmers from India in 1995. Based on the evidence presented, the patent was revoked in May 2000.

3. **Hoodia (*Hoodia gordonii*):**
   Hoodia is a cactus that has been used for thousands of years by African bushmen of the San tribe for suppressing hunger and thirst on long hunting trips. In 1995, the South African Council of Scientific and Industrial Research (CSIR) patented Hoodia's appetite suppressing element, and in 1997 licensed it to a Phytopharm, a British biotech company. Pfizer in 1998 acquired the rights from Phytopharm for $32 million to develop and market the ingredient as a potential cure for obesity, a market worth several billions. The San soon got to know of it and in June 2001, launched legal action against the South African CSIR and pharma industry. Phytopharm conducted extensive enquiries but were unable to identify the knowledge holders. The San are a nomadic tribe spread over four countries. In trying to enter into negotiations with the tribe, it was unclear as to who should be benefited: the person who shared the information, his family, the tribe, or the entire country. In 2002 the issue had been resolved with the San tribe to receive a share of future royalties from the patent holder, the South African CSIR. This would be a fraction of the royalties paid by Pfizer to Phytopharm and the South African CSIR.

4. **Ayahuasca (*Banisteriopsis caapi Mort.*):**
   Ayahuasca is a ceremonial drink made from the processed bark of a liana of the same name, used by the Shamans (medicine men) of indigenous tribes throughout the Amazon valley to diagnose and treat diseases, as well as meet with spirits and divine the future. In 1986, Loren Miller obtained a U.S. Patent for a variety of ayahuasca which he had collected from a domestic garden in Amazon which had flowers of a different color qualifying it as a new and distinct variety. The Coordinating Body of Indigenous Organizations of the Amazon Basin (COICA) which represents more than 400 indigenous tribes in the Amazon region protested as the ayahuasca was known for several generations and they felt this was an appropriation of their traditional knowledge. The patent was revoked in 1999, but as the inventor was able to convince the USPTO of its novelty, the patent was restored in 2001 with the original claims.

Biopiracy of Traditional Knowledge. Retrieved from http://www.tkdl.res.in/tkdl/langdefault/common/Biopiracy.asp?GL=Eng.

efforts in the utilization of the TK. The IGC currently provides the main forum in WIPO for an international discussion on various aspects of protection of TK and has since its formation generated considerable information toward understanding and resolving the issues associated with access to GRs and benefit sharing as well as protection of TK (available at http://www.wipo.int/tk/en/tk/). Participants in the IGC include Member States, indigenous and local communities, representatives from business and civil societies, and Non-Governmental Organizations (NGOs). Since 2009, the IGC has concentrated on text-based negotiations with the objective of reaching an agreement on a text of an international legal instrument(s) which will ensure effective protection of TK and GRs. A *"Draft objectives and principles relating to intellectual property and genetic resources"* was discussed in the twentieth session of the IGC in Geneva in February 2012. By 2014, three revisions of the text had been made, but the recommendations could not be agreed upon to transmit to the Assembly (Saez, 2014). At the General Assembly, no decision could be reached on the draft of the IGC, and with the mandate of the IGC coming to an end in 2015, it needed to be renewed if work on policy solutions to provide international protection for TK, GR and TCE was to be continued (Saez, 2015). The United States attributed the collapse of the discussions to lack of consensus on fundamental issues, such as what should be protected, who should be the beneficiaries of the protection, and the exceptions to the proposal. The United States suggested that the IGC should be replaced by an ad-hoc expert working group which by seminars and studies would be able to better understand the issues at stake. However, other countries, Switzerland, Norway, New Zealand, Holy See, Kenya, and Mozambique, proposed that the mandate be renewed to continue negotiations on text-based solution (Saez, 2015). The fifty-fifth session of the Assemblies of Member States of WIPO held in October 2015, extended the mandate of the IGC during the next budgetary biennium 2016/2017 to continue to expedite its work (WIPO, 2015).

16.2.2.1 Traditional knowledge

Two types of IP protection are being sought (WIPO, n.d.):

- *Positive protection*: granting of rights that empower communities to promote their TK, control its uses and benefit from its commercial exploitation.
- *Defensive protection*: aims to stop people outside the community from acquiring IP rights over TK

16.2.2.2 Genetic resources

GRs themselves are not IP as they are not creations of the human mind, hence cannot be protected as IP. However, inventions based on GR may be patentable, or capable of protection under plant breeder's rights. Issues being discussed at WIPO include:

- *Disclosure requirements*: Several countries have enacted domestic legislation putting into effect the CBD obligations that necessitate access to the country's GR to be on the basis of PIC and mutually agreeable understanding of fair and equitable sharing of the benefits. WIPO is considering whether it should be incorporated into international legal instruments.
- *Defensive protection of genetic resources*: prevents patents from being granted over GR and associated TK, and possible disqualification of patents that do not comply with CBD obligations of PIC and MAT, and disclosure of origin. Aims to prevent biopiracy.

16.2.3 THE WORLD TRADE ORGANIZATION AND TRADITIONAL KNOWLEDGE

IP issues in trade are dealt with by the TRIPS Agreement of the WTO. The TRIPS Agreement does not have any provision to protect TK and GR through patents, but it does create opportunities for protection by other measures. One method is the use of **Article 27.3(b)** which allows governments to exclude some kinds of inventions from patenting, for example, biological processes, animals, and plants (although plant *varieties* are to be protected by plant patents or other means created specifically for the purpose). A review of Article 27.3 began in 1999 with the TRIPS Council discussing how the existing TRIPs provisions were to be applied on patent protection for animals, plants and plant varieties including how to handle moral and ethical issues. Also under discussion was how to ensure that the TRIPS Agreement and the CBD support each other and how to deal with commercial applications of TK and GR. In 2001, the Doha Declaration (in paragraph 9) emphasized that the TRIPS Council should examine the relation between the TRIPS Agreement and the CBD, and discuss proposals on disclosing the source of biological material and associated TK (WTO, 2016 a,b). A **"GI extension"** (extending to other products the higher level of protection for Geographical Indications beyond wines and spirits) was another option discussed for protection of TK and GR. In 2002, the developing nations Brazil, China, Cuba, Dominican Republic, Ecuador, India, Pakistan, Thailand, Venezuela, Zambia, and Zimbabwe gave a submission to the TRIPS Council for an amendment in the TRIPS Agreement to include a provision that members shall require a patent applicant making use of TK or GRs in the invention to provide, as a condition to acquiring patent rights:

1. "disclosure of the source of the biological resource and/or the TK used in the invention
2. evidence of PIC through approval of authorities under the relevant national regimes where applicable; and
3. evidence of fair and equitable benefit sharing under the national regime of the country of origin where applicable" (WTO, 2002).

The underlying reason for the submission was because most of the developing nations which were biodiversity rich did not have adequate national legal systems to prevent biopiracy and were hoping that **a link between the CBD obligations and TRIPS Agreement might help them protect their resources**. While the request was supported by the European Commission, it was opposed by the United States and Japan on grounds that it would make patent laws cumbersome, and that the CBD alone should be sufficient protection.

In 2008, proponents of the TRIPS related issues under the Doha Work Programme agreed to include the GI Register, TRIPS disclosure requirement, and GI Extension, as part of the horizontal process in order to have the modality texts for negotiating the final legal draft with respect to each of the issues as part of a single undertaking.

- *"GI Register*: The members agree to establish a register open to geographical indications for wines and spirits protected by any of the WTO Members as per TRIPS
- *TRIPS/CBD disclosure*: Members agree to amend the TRIPS Agreement to include a mandatory requirement for the disclosure of the country providing/source of GR, and/or associated TK for which a definition will be agreed in patent applications. Patent applications will not be processed without completion of the disclosure requirement. Members agree to define the nature and extent of a reference to PIC and ABS
- *GI-Extension*: Members agree to the extension of the protection of Article 23 of the TRIPS Agreement to geographical indications for all products, including extension of the Register" (excerpted from WTO, 2008).

16.2.4 THE FOOD AND AGRICULTURE ORGANIZATION AND TRADITIONAL KNOWLEDGE

The FAO addresses concerns of access and fair use of plant GRs and the role of IP rights with respect to biological resources by means of the legally binding International Treaty on Plant Genetic Resources for Food and Agriculture (ITPGRFA) of 2001 (see Chapter 8: Biodiversity and Sharing of Biological Resources).

KEY TAKEAWAYS

Conservation of TK, and GRs associated with TK addressed by:

- CBD:
  - Article 8(j)—recognizes the role of TK and indigenous communities in conservation efforts
  - 1998—*Working Group on Article 8(j)* was established by the COP to the CBD to help implementation of CBD's provisions relating to TK
  - 2010—*Nagoya Protocol to the CBD*—Articles 7, 12, 16, 21, and 22 specify aspects of PIC and MAT for ABS of TK
- WIPO:
  - 2000—established the *Intergovernmental Committee on Intellectual Property and Genetic Resources, Traditional Knowledge and Folklore* (IGC) to primarily look into IP issues that arise in the context of access to GRs and benefit sharing
  - Since 2009, working on a text-based negotiation for establishing international legal instrument(s) for effective protection of TK and GRs
- WTO:
  - Article 27.3(b) of TRIPS Agreement—allows government to exclude some kinds of inventions from patenting
  - 2008—Doha Work Programme agreed to include the GI Register, TRIPS disclosure requirement, and GI extension for protection of TK and GR
- FAO:
  - 2001—ITPGRFA incorporates aspects of ABS to GR and TK associated with GR

16.3 LEGAL PROTECTION FOR TRADITIONAL KNOWLEDGE

Protection of TK by laws is complicated by the **nature of the knowledge**: it could be produced by individuals, small groups of individuals, or by indigenous communities. The knowledge could be retained as confidential [known only to the originator(s) of the knowledge and his/her (their) descendants, or limited to certain members of the community often with restricted access], or widely disseminated in the community as well as with outsiders and may be considered to be in the public domain. The framework of protection therefore would **have to be case specific in order to be effective**. Several nations and regions have adopted diverse mechanisms to address this issue. National laws implementing the recommendations of the CBD for ABS have been used in Bhutan,

India, South Africa, and Vietnam, while *sui generis* laws to protect specific aspects of TK have been adopted by Brazil, Costa Rica, Peru, Panama, the Philippines, Portugal, Thailand, and the United States of America (see http://www.wipo.int/tk/en/databases/tklaws/). **Increasingly, countries are amending laws conferring intellectual property rights to include TK.** While the overarching purpose of the IP-form of protection is to prevent third parties from commercially exploiting the TK, like other protection of IP, **protection of TK cannot be considered an end in itself.** The protection of TK should:

- Encourage a greater value and respect for the TK system
- Support and empower the TK holders
- Encourage tradition based creativity and innovation
- Promote equitable benefit sharing and discourage misappropriation and unfair uses
- Be responsive to the actual needs of the TK holders

Two types of legal protection have been devised:

1. *Positive protection*: This recognizes the right of the TK holders and empowers them to prevent misappropriation of the TK by third parties.
2. *Defensive protection*: This safeguards against illegitimate IP rights taken out by third parties over TK subject matter.

16.3.1 POSITIVE PROTECTION STRATEGIES

Due to the diversity of the nature of TK, several options have been investigated for protecting the interests of TK holders. These include **non-IP options** such as:

- laws ensuring fair trade practices and preventing unfair competition,
- labeling laws,
- regulation of access to GRs and associated TK,
- laws of civil liability, or
- *sui generis* systems adapted to protect TK.

IP options include:

- patents, for example, in 2001, China granted 3300 patents for innovations in Traditional Chinese Medicines (TCMs) (WIPO Booklet No. 2, n.d.),
- distinctive signs, such as, trademarks, collective marks, certification marks and geographical indications, and
- law of confidentiality and trade secrets

16.3.2 DEFENSIVE PROTECTION STRATEGIES

The purpose of "defensive" protection strategies is to thwart attempts by third parties to obtain IP rights, mainly patents, over the TK. Since an invention cannot be patented if it is available in "prior art," the method has been to systematically document TK in databases accessible to patent examiners. An illustrative example is the *Traditional Knowledge Digital Library (TKDL*, http://www.tkdl.res.in/), a project initiated by the Indian government, that aims to collate all information on TK available on traditional medical practices in India such as Ayurveda, Unani, and Siddha. Compiled in digital format, the TKDL is available in five international languages, English, French, German, Spanish, and Japanese.

A similar effort has been made by the State Intellectual Property Office of the People's Republic of China to establish a Traditional Chinese Medicine, **TCM Patent Database**, created to meet the need of patent examination. The Chinese version of the database covers TCM related patent applications published from 1985 in China and contains over 19,000 bibliographic records and over 40,000 TCM formulas (Liu & Sun, 2004).

16.3.3 PROBLEMS WITH IMPLEMENTATION

The originators and custodians of TK are mostly local and indigenous communities. Fair use of their knowledge often does not happen because of lack of education and poverty making them vulnerable to exploitation. Even when mentored by government institutions and NGOs, the communities sometimes lack the ability to convert their resource into a sustainable endeavor benefiting the community. An instance of this is the commercialization of an herb with medicinal properties called "*Arogyapacha*" used by the "kani" tribes in India (see Box 16.2). Capacity building to help local communities understand the nuances of ABS and empowering communities to negotiate mutually acceptable terms and sustainable use of their resource has been emphasized in the CBD and is addressed in Articles 21 and 22 of the Nagoya Protocol.

KEY TAKEAWAYS

Two types of legal protection strategies exist for protection of TK:

- Positive protection strategies:
  - Non-IP options: laws for fair trade practices, labeling, regulation of access, civil liability
  - IP options: patents, distinctive signs (such as trademarks, collective marks, geographical indications), trade secrets
- Defensive protection strategies:
  - Documenting the TK (as for example a digital library/ database)
  Problems of implementation:
- Identifying the owners of TK (collective nature of the knowledge)
- Unequal negotiations (TK holders often uneducated and poor, vulnerable to exploitation)
- TK holders often lack the ability to convert their resource into a sustainable endeavor benefiting the community

---

16.4 SUMMARY

While the need to protect TK and knowledge associated with GRs is universally accepted, what is not obvious is the manner and the purpose of protection that is to be accorded. TK is held by communities and is a dynamic, evolving aspect of the group, defining it in legal terms is challenging.

BOX 16.2 THE *AROGYAPACHA* STORY

The story began in the 1980s when a team of scientists led by P. Pushpangadan and S. Rajasekharan, working at the CSIR Regional Research Laboratory, Jammu, were guided by three members of the *kani* tribe, Mathen Kani, Mallan Kani, and Eachen Kani, on an expedition to the Agasthya hills. The scientists noticed that the tribal members ate the fruit of a plant they referred to as *Arogyapacha* and showed no signs of fatigue during the journey. The scientists persuaded them to share their information on medicinal plants for validation through modern scientific methods. In 1987, scientists at the Jawaharlal Nehru Tropical Botanical Garden and Research Centre (JNTBGRI), Palode, headed by P. Pushpangadan, (who had by then become the Director), identified the plant named *Arogyapacha* as *Trichopus zeylanicus ssp. travancoricus* and found rejuvenating properties in the leaves, fruits and seeds. In a project spanning eight years, a compound drug named *Jeevani* with immuno-enhancing, antistress, liver-protective, and antifatigue properties was developed. In September 1995, the JNTBGRI entered into a technology transfer agreement with the Arya Vaidya Pharmacy, Coimbatore for commercialization of the drug. The Institute was instrumental in evolving a mechanism to share 50% of the commercial profits with the *kani* tribe. The money was to be used to assist development activities in twenty-seven tribal settlements. In order to facilitate the sharing of benefits, the Kerala Kani Samudaya Kshema Trust was formed by the tribals. The Trust received half of the license fee of Rs.10 lakh and half of the royalty of two percent. Of that Mathen Kani and Mallan Kani received Rs. 20,000 each, and Eachen Kani Rs. 10,000 in one-time payment. Mathen Kani and Mallan Kani were also employed as "consultants" in TBGRI drawing Rs. 1500 a month for a few years. The tribe earned a few lakh rupees collecting the plant from the Agasthya peak and supplying the same to Arya Vaidya Pharmacy. This however led to depletion of the plants in its natural habitat and efforts were made to cultivate it. Due to procurement problems, in 2008, Arya Vaidya Pharmacy walked out on the deal. Unfortunately, the Trust mismanaged its funds, dipping into its fixed deposits to buy a jeep and then to maintain it, although it had to sold when revenues ceased. Disputes regarding the sharing led to a new team of office bearers in the trust, who however could not sustain the enterprise. Meanwhile, the process patent for Jeevani that had been obtained by JNTBGRI expired and no patent was taken for the product. Illegal collection of the plants could not be checked (Mathew, 2012).

This initiative was recognized by the U.N. Environment Programme at the World Trade Organization as a global model in benefit-sharing and recognition of the intellectual property rights of tribal communities. Being the first of its kind, and preceding the Convention on Biological Diversity it unfortunately had a few drawbacks. For one, although efforts were made for capacity building through training and counseling, the tribe was clearly unable to manage their resource. For another, as only oral consent was obtained from the *kanis* for commercialization of their traditional knowledge, protection for the knowledge was inadequate.

In 2012, the project was presented before the panel on Indigenous and Local Communities at the 20th session of the Intergovernmental Committee on Intellectual Property and Genetic Resources, Traditional Knowledge and Folklore held at Geneva under the auspices of the World Intellectual Property Organization (Nandakumar, 2012).

References

Nandakumar, T. (2012). *Kerala project to take center stage at WIPO meet. The Hindu.* February 13, Retrieved from http://www.thehindu.com/todays-paper/tp-national/kerala-project-to-take-centre-stage-at-wipo-meet/article2887369.ece.

Mathew, R. (2012). *A benefit-sharing model that did not yield desired results. The Hindu.* October 18, Retrieved from http://www.thehindu.com/news/national/A-benefit-sharing-model-that-did-not-yield-desired-results/article12561312.ece.

There are basically two reasons why TK should be protected: preserving it for future generations (conservation), and ensuring that the knowledge is not misappropriated (benefit sharing). With commercial applications emerging from the use of GRs, it is only fair that the holders of the GR and the TK associated with the resource are respected, and adequately compensated. Appropriate ABS arrangements with PIC and mutually agreeable terms for sharing of the benefits are mandated by international agreements such as the United Nations' Convention on Biodiversity, and the Food and Agricultural Organization's ITPGRFA. Organizations such as the WIPO and WTO are working toward establishing IP options for protection of TK and GR. Two strategies that are emerging are: defensive protection strategies (such as setting up of public databases documenting the TK) and positive protection strategies (using IP instruments such as Trade Secrets, labeling/marking, and Geographical Indications).

REFERENCES

Liu, Y., & Sun, Y. (2004). China traditional Chinese medicine (TCM) patent database. *World Patent Information, 26*(1), 91–96.

Saez, C. (2014) *WIPO meeting on TK protection ends with no agreement, draft texts heading to Assembly.* Intellectual property watch. Retrieved from http://www.ip-watch.org/2014/07/09/wipo-meeting-on-tk-protection-ends-with-no-agreement-draft-texts-heading-to-assembly/.

Saez, C. (2015) *US proposes suspension of WIPO TK Committee; Switzerland and others counter.* Intellectual property watch. Retrieved from http://www.ip-watch.org/2015/09/11/us-proposes-suspension-of-wipo-tk-committee-switzerland-and-others-counter/.

WIPO (n.d.) *Traditional knowledge and intellectual property—background brief.* [Website] Retrieved from http://www.wipo.int/pressroom/en/briefs/tk_ip.html.

WIPO (2001) Matters concerning intellectual property and genetic resources, traditional knowledge and folklore—an overview. WIPO/GRTKF/IC/1/3. Intergovernmental committee on intellectual property on intellectual property and genetic resources, traditional knowledge and folklore. Retrieved from https://www.google.co.in/url?sa = t&rct = j&q = &esrc = s&source = web&cd = 1&cad = rja&uact = 8&ved = 0ahUKEwiQ_7ynz-HSAhWILo8KHVihAGkQFggZMAA&url = http%3A%2F%2Fwww.wipo.int%2Fedocs%2Fmdocs%2Ftk%2Fen%2Fwipo_grtkf_ic_1%2Fwipo_grtkf_ic_1_3.doc&usg = AFQjCNEK1UTkjAixvVc5ZLEDcesatIQTfg&sig2 = rSIUWlawHcjOMY0D26gUCA&bvm = bv.149760088,d.c2I.

WIPO (2015) Decision on agenda item 17, matters concerning the Intergovernmental Committee on intellectual property and genetic resources, traditional knowledge and folklore. Assemblies of member states of WIPO, fifty-fifth session, October 5 to 14, 2015. Retrieved from http://www.wipo.int/export/sites/www/tk/en/igc/pdf/igc_mandate_1617.pdf.

WTO (2002) Council for trade-related aspects of intellectual property rights, the relationship between the TRIPS agreement and the convention on biological diversity and the protection of traditional knowledge: Submission by Brazil on behalf of the delegations of Brazil, China, Cuba, Dominican Republic, Ecuador, India, Pakistan, Thailand, Venezuela, Zambia and Zimbabwe, IP/C/W/356, June 2002. Retrieved from https://www.google.co.in/url?sa = t&rct = j&q = &esrc = s&source = web&cd = 1&cad = rja&uact = 8&ved = 0ahUKEwjNuLTlzuHSAhUHpI8KHV3yBi4QFggZMAA&url = https%3A%2F%2Fdocsonline.wto.org%2Fdol2fe%2FPages%2FSS%2FDirectDoc.aspx%3Ffilename%3Dt%253A%252Fip%252Fc%252Fw356.doc%26&usg = AFQjCNGTnW2neWUjXazQ0wfgYepHtjEiNg&sig2 = S4AcLpKChUx5hVlPHT9oIg.

WTO (2008) Draft modalities for TRIPS related issues: Communication from Albania, Brazil, China, Columbia, Ecuador, the European Communities, Iceland, India, Indonesia, the Kyrgyz Republic, Liechtenstein, the Former Yugoslav Republic of Macedonia, Pakistan, Peru, Sri Lanka, Switzerland, Thailand, Turkey, the ACP Group and the African Group, TN/C/W/52, 19 July 2008. Retrieved from http://trade.ec.europa.eu/doclib/docs/2008/september/tradoc_140562.pdf.

WTO (2016a) *TRIPS: Reviews, Article27.3(b) and related issues: Background and current situation.* [Website] Retrieved from https://www.wto.org/english/tratop_e/trips_e/art27_3b_background_e.htm.

WTO (2016b) *Article 27.3, traditional knowledge, biodiversity* [Website] Retrieved from https://www.wto.org/english/tratop_e/trips_e/art27_3b_e.htm.

WIPO Booklet No. 2 (n.d.) Intellectual Property and Traditional Knowledge. WIPO Publication No. 920(E) Retrieved from http://www.wipo.int/edocs/pubdocs/en/tk/920/wipo_pub_920.pdf.

Index

Note: Page numbers followed by "*f*," "*t*," and "*b*" refer to figures, tables, and boxes, respectively.

A

ABS. *See* Access and benefit sharing (ABS)
Access and benefit sharing (ABS), 190–191, 346
  countries
    with a biodiversity or environmental law with provisions, 206
    with national law specifically devoted, 206–209
    with no national laws specifically devoted, 205
  frameworks, 204–209
  national and regional laws, 207*t*
  national parks system, 205
  provisions, 204–205
ADA. *See* Americans with Disability Act (ADA)
Adenosine deaminase (ADA) gene, 17–18
Ad Hoc Group of States Parties, 154–155
Ad Hoc Working Group of Technical and Legal experts, 195–196
Administrative Order No. 20, 206
Adult stem cells, 57–60
  transdifferentiation, 59–60
Advance informed agreement (AIA), 218–219
Advisory Committee on Health Research report, Genomics and World Health, 179–180
Advocacy, rights based on, 180–182
  Charter of Fundamental Rights (EU), 182
  human rights, 181
  International Declaration on Human Genetic Data, 181
  Universal Declaration on Bioethics and Human Rights, 181–182
  Universal Declaration on the Human Genome and Human Rights, 181
Aedes aegypti mosquitoes, 110*b*
Agreement on Trade-Related Aspects of Intellectual Property Rights (TRIPS). *See* TRIPS Agreement
AgResearch Limited, 40–41
Allergenicity, GM crops, 103
Alternate splicing, 5
Alzheimer's Mouse, 106
American Association for the Advancement of Science (AAAS), 48
Americans with Disability Act (ADA), 176*b*, 183–184
American Type Culture Collection (ATCC), 306
Animals
  cloning, 26–41
    applications, 30–36
    Dolly, 28–29, 28*f*
    ethical issues, 36–37
    laws and public policy, 38–41

    nuclear transfer, 26–27, 30
    progress, 29–30
    SCNT, 26, 27*f*
  patenting, 321–323
  physical and biological containment, 267–269, 267*b*
  rights, 134–137
  safety considerations for field release, 282–283
  transgenic, 105–111
    creation of, 106*f*
    disease models, 106
    food production, 106–107
    hypoallergenic pets, 111
    for molecular farming, 100
    pets, 109–110
    proteins, 108
    vector control, 109
Anthrax, 150
  September 11 terrorist attacks, 152–153
  World War I, 152–153
  World War II, 152–153
Anthropocentric ethical theories, 129–130
Antibiotic resistance, GM crops, 104
Antibiotic resistance marker (ARM) genes, 248–249
Antibodies, 91–97
  Aga2p fusion protein, 96–97
  binding to antigen target, 97
  daughter phages, 96–97
  immunization
    active, 91–96
    passive, 91–96
  phage display libraries, 96–97
  therapeutic monoclonal, 98*t*
Anticancer compounds, 194
Antiterrorism, Crime and Securities Act (ATCSA) of 2001, 155–156
Anti-Terrorism and Effective Death Penalty Act of 1996, 157–158
AquaBounty Technologies, 107*b*
AquAdvantage Salmon, 107*b*
Argentina, animal cloning, 40
Ariosa Diagnostics, Inc. v. Sequenom, Inc., 2015, 319, 320*b*
ARM. *See* Antibiotic resistance marker (ARM) genes
Arogyapacha, 354, 355*b*
Artemisinin, 191–194
Article 12 of the Nagoya Protocol, 348
Article 16 of the Nagoya Protocol, 348
Article 27(3) of TRIPS Agreement, 342
Articles 27 to 34 of TRIPS Agreement, 342

Arya Vaidya Pharmacy, 355*b*
Ashkenazi Jews, genetic testing and, 173—174
Asilomar Conference Centre in California, 212—213
Assisted reproductive technologies (ART), 30
Association for Molecular Pathology v. Myriad Genetics, Inc.,
    317*b*
Australia, 206
  animal cloning, 40
  human cloning, 48—49
  IP Australia, 313—314
  patent protection for diagnostics, 340
  stem cell research, 74
Australia Group, 156*t*
Autologous stem cells, 60
Avidin (egg protein), 99

B
Babies
  designer, 43*b*
  three parents, 43*b*
Bacillus anthracis (anthrax), 152—153
Bacillus thuringiensis (Bt), 105
Bacteria, weaponizing, 151—152
Banana Xanthomonas Wilt (BXW), 114—117
BARDA. *See* Biomedical Advanced Research and
    Development Authority (BARDA)
Belmont Report, 143—144, 144*b*
Bentham, Jeremy, 129—130
Berg Committee, 212—213
Bermuda Principles, 301—302
Bial clinical trial, 140
Bilski et al. v. Kappos, 339*b*
BioBricks, 18
Biodefense for the 21st Century, 162
Biodefense programs, 162—163
Biodiversity
  CBD. *See* Convention on Biological Diversity (CBD)
  concept, 191—195
  functions, 191—194
  GM corps and loss of, 104
  hotspots, 191, 192*f*
  overview, 190—191
  rDNA technology and, 112—113
Biodiversity Act (South Africa), 206
Bioethics, 143—146
  defined, 129
Bioethics Defense Fund, 48
Bioethics Programme (UNESCO), 181
Biological containment, 263
  for animals, 267—269
  for plants, 263—266
Biological weapons, 151
  BWC of 1972, 153—155

development of, 152
  history, 152—153
  misconception, 151
Biological Weapons Convention (BWC) of 1972, 153—155
Biologics, 91
  clinical trials of, 285—286
  conventional drugs *vs.*, 331—333
  defined, 331
  intellectual property protection, 334—335
  interchangeable, 333—334
  patenting, 331
  patent protection for, 335—338
    evergreening, 336—338
    incremental innovation, 336
  safety considerations for market approval, 284—286
Biologics License Application (BLA), 285
Biomedical Advanced Research and Development Authority
    (BARDA), 161
Biopharmaceuticals. *See* Biologics
Biopiracy, 195
Biopolitics, GMO and, 117—121
Biopreparat, 152—153
Bioprospecting, 194
  concerns regarding, 194—195
  defined, 194
  phases, 194
  politics, 195
  scale back activities, 194
  traditional knowledge (TK), 195
Bioprospecting and Access and Benefit-Sharing Regulations
    (South Africa), 206
Bioprospecting Guidelines (South Africa), 206
Biosafety
  Commission on Genetic Resources for Food and
    Agriculture, 221—222
  Informal Working Group on Biosafety, 216—217
  International Plant Protection Convention, 221
Biosafety levels (BL), 256—263
  biological containment, 263
    for animals, 267—269
    for plants, 263—266
  physical containment, 257—262
    for animals, 267—269, 267*b*
    for plants, 263—266, 264*b*
    for standard laboratory experiments, 258*b*
  relation of risk groups to, 256*t*
  requirements, 257*t*
Biosecurity, 155—157
  defined, 155—156
  international initiatives, 156*t*
Bioserfdom, 113, 134
BioShield, 162—163
Biosimilars
  defined, 286

interchangeable biologics, 333–334
safety considerations for market approval, 286
significance of, 332*b*
BioSteel, 108
Biotech industry, 91–100, 274
good developmental principles, 278–279
good industrial large-scale practice, 275–278
safety considerations for, 274–275
Biotechnology
ethical issues, 131–140
animal rights, 134–137
environmental ethics, 131–132
human clinical trials, 138–140
plant biotechnology, 132–134
IPR issues, 297–308
enforcement, 302–305
ethical issues, 307–308
ownership, 302
patentability, 298–302
sharing of costs and benefits, 306–307
overview, 212–213
Bioterrorism
biosecurity measures preventing, 155–157, 156*t*
defined, 151
European Union (EU) on, 161–162
United States on, 157–161
weaponizing microbes, 151–152, 151*t*
Bioterrorism Preparedness Act, 158
Biothreat agents, 155–156
Biotoxins, 151–152
Blastomeres, nuclear transfer of cells from, 30
B-lymphocytes, 91–96
Bovine Spongiform Encephalopathy, 118
Bowman v. Monsanto Co. et al., 305*b*
Brazil, animal cloning, 40
BRCA genes, 173*b*
Breast cancer, 173–174
genetic testing, 173*b*
mutations, 173*b*
British Ministry of Agriculture, 118
Brotherhood of Maintenance of Way Employees (BMWE),
176*b*
Bt brinjal, 118
in India, 118, 119*b*
Bt corn, 248–249
Bt cotton, in India, 119*b*
Bt maize, 104
Budapest Treaty, 296
Budapest Treaty on the International Recognition of the
Deposit of Microorganisms for the Purposes of Patent
Procedure, 298
Burlington Northern Santa Fe (BNSF) Railroad employees,
175–176, 176*b*
BWC. *See* Biological Weapons Convention (BWC) of 1972

C
Canada, 205
animal cloning, 40
human cloning, 48
patent protection for diagnostics, 340
stem cell research, 74
Canadian Environment Protection Act, 40
The Cancer Genome Atlas, 16
Cancer treatment, stem cells and, 66
Cartagena Protocol on Biosafety, 217–221
advance informed agreement, 219
assessment and review, 220
capacity building, 220
compliance, 220
international negotiations and, 217*b*
liability and redressal, 220–221
Categorical imperative, 129–130
CBD. *See* Convention on Biological Diversity (CBD)
CCFL. *See* Codex Committee on Food Labeling (CCFL)
CDER. *See* Center for Drug Evaluation and Research (CDER)
CDH 13, 20
CEER. *See* Centers of Excellence in ELSI Research (CEER)
Cellular differentiation, 64
Center for Drug Evaluation and Research (CDER), 243*b*
Centers for Disease Control and Prevention (CDC), 152–153
Centers of Excellence in ELSI Research (CEER), 179
"Central Dogma of Biology" (Crick), 3
Centre for Biologics Evaluation and Research, 48
Centre for Veterinary Medicine (CVM), 32–33, 243*b*
Charter of Fundamental Rights (EU), 182
human cloning, 47
CHDI Foundation Inc., 177
Children's Hospital of Eastern Ontario (CHEO), 340
Chromosomal aberrations, 172–173
Chromosomes, 2–3
Clinical trials
biologics, 285–286
human, 138–140
safety assessment, 249–250
stem cells, 71
Clone, defined, 26
Cloning, 85*f*
animals, 26–41
applications, 30–36
Dolly, 28–29, 28*f*
ethical issues, 36–37
laws and public policy, 38–41
nuclear transfer, 26–27, 30
progress, 29–30
SCNT, 26, 27*f*
human, 41–49
ethical considerations, 42–44
laws and public policy, 45–49

Cloning (*Continued*)
overview, 26
CMX001, 162—163
Code of Federal Regulations (U.S.), 205
Codex Alimentarius Commission, 247—249
Codex Committee on Food Labeling (CCFL), 222
Coding region, 5
Colorectal cancer, 173—174
Combined DNA Index System, 16—17
Commission on GR for Food and Agriculture, 202, 221—222
Committee for Medicinal Products for Human (CHMP), 285—286
Commodification of life, 44
Common morality, 130
Comparative genomics, 9
Competitive market-derived prices, 343
Compulsory licensing, 337*b*
Conference of the Parties (COP), 196—198
Consequentialist ethical theories, 129—130
Consolidated Appropriations Act, 2014, 161
Consultative Document on Ethical Guidelines for Biomedical Research on Human Subjects, 49—50
Contamination of food, 103—104
Conventional drugs *vs*. biologics, 331—333
Convention on Biological Diversity (CBD), 195—201, 346—347
    Article 6, 197*b*
    Article 8(j), 197*b*, 347—348
    Article 10, 347—348
    Article 19, 196
    Article 23, 197*b*
    Articles 7 to 14, 197*b*
    Articles 16 and 19, 197*b*
    Articles 20 and 21, 197*b*
    Articles 23, 197*b*
    Articles 24 to 42, 197*b*
    Cartagena Protocol on Biosafety, 217—221, 217*b*
    Conference of the Parties (COP), 196—198
    implementation, 197*b*
    objectives, 196
    TK with GR, 347—348
    Working Group on Article 8(j), 348
    working groups, 196—197
Cooperative Research and Development Agreement (CRADA), 306
COP. *See* Conference of the Parties (COP)
Cord blood stem cells, 59
Council for Responsible Genetics, 175—176
Council of Europe Convention on Human Rights and Biomedicine, 46. *See also* Oviedo Convention
CRADA. *See* Cooperative Research and Development Agreement (CRADA)
Creutzfeldt-Jakob disease, 118
Crimes against humanity, 143

CRISPR/Cas9, 90
Crohn's disease, 173—174
Cry1Ac gene, 119*b*
CRYO-CELL International, USA, 59
Cultivated plants, centers of origin of, 191, 193*b*
Cystic fibrosis, 17—18

D

D'Arcy v Myriad Genetics Inc & Anor, 317*b*
Data exclusivity protection, 335
Declaration of Helsinki, 140, 141*b*
deCODE Genetics Inc., 174*t*
    ELSI concerns, 178*b*
De-extinction, 31*b*
Deontological ethics, duty based on, 129—130
Department of Environment and Natural Resources, 206
Designer babies, 43*b*
Diagnostics, 13—16
    patent protection for, 338—340
    sequences used for, 319
Diamond v. Chakrabarty, 300—301
Dickey-Wicker Amendment, 74—75
Dideoxy sequencing technique, 11*b*
Differentiation, 26
Dihybrid crosses, 4*f*
Direct-to-customer genetic testing kits, 174—175, 174*t*
Discovery *vs*. invention, 313—314
Disease progression, stem cells and, 65
DNA, 86
    cloning. *See* Cloning
    double helical model for, 3, 6*f*
    stability of, 3
DNA evidence, 16—17
DNA profiles, 16—17
DNA sequences, 9, 318—321
    for diagnostic testing, 319
    as research tools, 319—321
Dolly, 28—29, 28*f*. *See also* Animals: cloning
Down's syndrome, 172—173
Drug targets, identifying, stem cells, 67
DsRed2, 110*b*
Dual-use research of concern (DURC), 150, 163—165
    life scientists responsibilities, 165
    NSABB, 163—164
Duty based on deontological ethics, 129—130

E

Earth Summit, 195—196
Ecological Equilibrium and Environmental Protection General Act (Mexico), 206
Eight ball, 152—153
Elelyso, 99

Embryonic Stem Cell Research Oversight (ESCRO) Committees, 75–76
Embryonic stem cells, 34, 57–58
  nuclear transfer, 58
Enforcement issues, of IPR, 302–305
Environment, rDNA technology and, 112
Environmental concerns, GM crops, 104–105
  emergence of resistance, 105
  genetic pollution, 104
  loss of biodiversity, 104
Environmental Law Programme (IUCN), 190–191
Environmental Protection Agency (EPA), 100–101
Environment Programme (UNEP), 195–196, 216–217
Environment Protection and Biodiversity Conservation Act (Australia), 206
Environment Protection and Biodiversity Conservation Regulations (Australia), 206
Enzymes, 99
Epigenomics, 9
Equal Employment Opportunity Commission (EEOC), 176b, 183–184
Escherichia coli, 213
Ethical, Legal and Social Implication (ELSI), 172, 177–180
  deCODE Genetics Inc., 178b
  HGP, 172, 177–179
  impact of, 179
  NIH Revitalization Act of 1993, 19
  overview, 19
  programs on, 179–180, 180t
Ethical issues
  animals cloning, 36–37
  biotechnology, 131–140
    animal rights, 134–137
    environmental ethics, 131–132
    human clinical trials, 138–140
    plant biotechnology, 132–134
  intellectual property, 307–308
  stem cells
    research, 68–70
    translation, 71–73
Ethics
  critical, 129
  defined, 129
  descriptive, 129
  morals, 129
  normative, 129
  principles, 128
  theories, 129–130
    common morality, 130
    consequentialist, 129–130
  values, 129
Eugenics, 44
Eukaryotes, 5, 7f
European Convention on Human Rights, 182–183

European Food Safety Authority (EFSA), 38
European Group on Ethics, 38
European Huntington's Disease Network, 177
European Medicines Agency (EMA), 332b
European Patent Convention (EPC), 313
European Patent Office (EPO), 313
European Union
  animal cloning, 38–39
  anti-GMO wave, 117–118
  Charter of Fundamental Rights, 47, 182
  GM crop, 117–118
Evergreening, 336–338, 338b
Executive Order 247 of 1995, 206
Executive order 2001, 74–75
Ex parte Hibberd, 323–324
Expression of genes, 7–8, 20
  check points, 7–8
  operons, 7–8

F
Familial dysautonomia, 173–174
Farmers, socioeconomic status, 113
Fetal stem cells, 59
Fibroblast cells, 61b
Field release, safety considerations for
  GM animals, 282–283
  GM crops, 279–282
FlavrSavr, 248–249
Food and Agriculture Organization (FAO), 202, 346–347
  biosafety, 221–222
  Commission on GR for Food and Agriculture, 202, 221–222
  International Plant Protection Convention, 221
Food safety
  CCFL, 222
  Codex Alimentarius Commission, 222
Forensics, 16–17
Francisella tularensis, 152–153
Functional genomics, 9

G
GATT. *See* General Agreement on Tariffs and Trade (GATT)
Gaucher's disease, 173–174
GCP. *See* Good clinical practices (GCP)
GE animals. *See also* Animals
  FDA on, 243b
  US regulation of, 243b
Gelsinger, Jesse, 89–90
Gene editing therapy, 90–91
Genentech, 91
General Agreement on Tariffs and Trade (GATT), 294–295
Generics, market approval for, 333b

Genes
  cloning. *See* Cloning
  concept, 2−8
  expression, 7−8, 20
  housekeeping, 8
  luxury, 8
  patenting, 301−302
  patents, 341*b*
  regions, 5
  role of, 20
Gene Technology Act 2000, 48−49
Gene Technology Act of 1986, 223
Gene therapy, 17−18, 86−90
  application, 88−89
  cancer cells, killing, 88
  challenges, 89−90
  defective gene, 87
    inhibiting, 87, 87*f*
    replacing, 87, 87*f*
  delivery systems, 88*t*
  germ-line, 86
  overview, 84, 86
  SCID, 89
  somatic, 86
    ex vivo, 87
    in vivo, 87
  technique, 88
  vector, 88
Genetically engineered microorganisms (GEM), 238−240
Genetically modified (GM) crops, 100−105
  commercial cultivation of, 100−101
  critics of, 102
  environmental concerns, 104−105
    emergence of resistance, 105
    genetic pollution, 104
    loss of biodiversity, 104
  ethical and socioeconomic concerns, 105
  for food and feed, 101−102
  health and safety concerns, 103−104
    allergenicity, 103
    antibiotic resistance, 104
    contamination of food, 103−104
    nutritional composition, 104
    toxicity potential, 103
  OECD BioTrack Product Database, 100−101
  patent infringement, 305*b*
  safety considerations for field release, 279−282
  trends in, 102
Genetically modified organisms (GMO), 2
  biopolitics and, 117−121
  challenges in applications, 112−113
    biodiversity, 112−113
    environment, 112
    sociocultural norms, 113

    socioeconomic status of farmers, 113
  safety considerations for field/market release, 279−286
  traceability and labeling of, 284
Genetic code, 5−6
Genetic counselling, 173
Genetic discrimination, 175−177
  by employers and insurance companies, 175−176, 176*b*
  fear of, 177
  preventing, 180−185
    legislation, 183−185, 185*t*
    rights. *See* Rights
Genetic engineering (GE). *See* Recombinant DNA (rDNA)
    technology
Genetic exceptionalism, 175
Genetic information, 175
Genetic Information Non-discrimination Act (GINA), 184
Genetic pollution, 104
Genetic reductionism, 19−20
Genetic rescue, 31*b*
Genetic Technologies Limited v. Merial LLC, 319, 320*b*
Genetic testing, 172−175
  BNSF Railroad employees, 175−176, 176*b*
  BRCA genes, 173*b*
  direct-to-customer kits, 174−175, 174*t*
  overview, 13−16
  patent protection, 340−342
  reasons, 173
  select populations, 173−174
  treatment regimen, 173
Genetic use restriction technologies (GURT), 105
Geneva Protocol of 1925, 153−154
Genographic Project, 14*b*
Genome
  DNA, 8−9
  overview, 8−9
  sequencing, 9
Genome editing, 13−19
Genomics
  applications of, 13−19
  comparative, 9
  epigenomics, 9
  functional, 9
  HGP. *See* Human Genome Project (HGP)
  overview, 9
  pharmacogenomics, 9
  structural, 9
Geographical Indications, 325
Germany, 152−154
  World War II, 128
Gert, Bernard, 130
GINA. *See* Genetic Information Non-discrimination Act
    (GINA)
Global Health Security Initiative (GHSI), 156*t*
GloFish, 109

Glybera, 88–89
GM foods, 112
    safety considerations for marketing of, 283–284
    sociocultural norms, 113
GM-free labels, 284
GM Salmon, 106
Golden Rice, 114–117, 115*b*
Good clinical practices (GCP), 141–142, 285
    Investigator's Brochure, 142
    Standard Operating Procedures, 142
Good industrial large-scale practice, 275–278
Good laboratory practice, 270–271
Good manufacturing practices (GMP), 286–288, 287*b*
Good microbiological techniques (GMT), 254
Great Barrier Reef Marine Park Act 1975, 206
Great Barrier Reef Marine Park Regulations 1983, 206
Green Paper on Bio-preparedness, 162
Greenpeace, 115*b*
Growing meat in vitro, 66
Growth hormone genes, 106
GTC Biotherapeutics, Inc., 243*b*
Guidelines for Human Embryonic Stem Cell Research, 75–76
Guillain-Barre Syndrome, 19

H

Hague Agreement, 296
Hairy cell leukemia, 91
Health and safety concerns, GM crops, 103–104
    allergenicity, 103
    antibiotic resistance, 104
    contamination of food, 103–104
    nutritional composition, 104
    toxicity potential, 103
Health Insurance Portability and Accountability Act (HIPAA), 184
    National Standards to Protect Patients' Personal Medical Records, 184
Health Security Committee, 161–162
HeLa cell line, 304*b*
Helicobacter pylori, 65
HFEA. *See* Human Fertilization and Embryology Authority (HFEA)
HGN. *See* Human Genetics (HGN)
HGP. *See* Human Genome Project (HGP)
Highly repetitive sequences, 9
HIV infection
    gene editing therapy, 90
Housekeeping genes, 8
htt gene, 177
Human clinical trials, 138–140
Human cloning, 41–49
    ethical considerations, 42–44
    laws and public policy, 45–49, 49*b*

Human Cloning and Human Dignity: An Ethical Inquiry, 48
Human Fertilization and Embryology Authority (HFEA), 43*b*
Human gastric diseases, 65
Human Genetics (HGN), 179–180
Human Genome Project (HGP), 2, 10–13, 172
    DNA sequencing, 10–13, 11*b*
    model organisms, 10–13
    objective, 10
Human Papilloma Virus (HPV), 304*b*
Human pharmaceutical proteins, 99
Human Reproductive Cloning Act 2001, 47
Human rights, 181
    defined, 181
    Universal Declaration on Bioethics and Human Rights, 181–182
    Universal Declaration on the Human Genome and Human Rights, 181
Huntington's Disease (HD), 177
Huntington Study Group, 177
Hypoallergenic pets, 111

I

IBM, 14*b*
ICCP. *See* Intergovernmental Committee for the Cartagena Protocol on Biosafety (ICCP)
Immunization
    active, 91–96
    passive, 91–96
Incremental innovation, 336, 338*b*
India
    Bt brinjal, 119*b*
    Bt cotton, 119*b*
    Health Department, 174–175
    human cloning, 49
    low-cost medicines and drugs in, 342–343
    National Guidelines for Stem Cell Research, 74
    new-borns screening, 174–175
    stem cell research, 74
    TRIPS Agreement, 342–343
Indian Council of Medical Research, 49–50
Indian Patent Act, 1970, 342–343
Induced pluripotent stem cells (iPSC), 60, 61*b*
Industrial applicability/utility requirement, 316
Innovations, restricting access to, 307–308
In re Bell, 315
In re Deuel, 315
Institute for Genomic Research, Maryland, 11*b*
Institutional Committee for Stem Cell Research (IC-SCR), 74
Institutional Review Board/Independent Ethics Committee (IRB/IEC), 142
Intellectual property rights (IPR), 120–121, 292
    biotechnology, 297–308
    enforcement, 302–305

Intellectual property rights (IPR) (*Continued*)
    ethical issues, 307–308
    ownership, 302
    patentability, 298–302
    sharing of costs and benefits, 306–307
    international conventions and treaties, 295–297
        Paris Convention, 296
        WIPO, 296–297
    overview, 293
    salient features of, 294*b*
    trade, 293–295
Interferon alpha-2a, 91
Intergovernmental Committee for the Cartagena Protocol on
    Biosafety (ICCP), 217–219
Interleukin 4 (IL-4) gene, 163
Interleukin Genetics, 174*t*
International Bioethics Committee (IBC), 45
International Conference on Harmonization (ICH), 285
International Congress on Recombinant DNA Molecules,
    212–213
International Convention for the Protection of Industrial
    Property, 293
International Convention for the Protection of New Varieties
    of Plants, 324
International Declaration on Human Genetic Data, 181
International Genetically Engineered Machine, 18
International Human Genome Sequencing Consortium, 13
International Plant Protection Convention, 221
International trade agreements, in medicines, 342–343
International Treaty on Plant Genetic Resources for Food and
    Agriculture (ITPGRFA), 198, 202–204, 221–222
    ABS obligations, 203
    Article 9 of, 203
    conservation centers, 203–204
    description, 202
    Farmer's Rights, 203
    intellectual property, 203
    International Undertaking on Plant GR, 202
International Undertaking on Plant GR, 202
International Union for Conservation of Nature (IUCN),
    190–191
Invention *vs.* discovery, 313–314
Inventive step/non-obviousness step, 315
Investigational New Drug (IND) Application, 285
Investigator's Brochure, 142
IP Australia, 313–314
IPR. *See* Intellectual property rights (IPR)

J

Japanese Army Unit 731, 152–153
Jawaharlal Nehru Tropical Botanical Garden and Research
    Centre (JNTBGRI), 355*b*
J. Craig Venter Institute (JCVI), 19

Jeevani, 355*b*
JNTBGRI. *See* Jawaharlal Nehru Tropical Botanical Garden
    and Research Centre (JNTBGRI)
Johns Hopkins University, Baltimore, 84–85

K

Kani tribes, 354, 355*b*
Kant, Immanuel, 129–130
Kaposi's sarcoma, 173–174
Kerala Kani Samudaya Kshema Trust, 355*b*

L

Lawrence Berkeley Laboratory, 176*b*
Lazarus Project, 31*b*
Legal/social status of a clone, 44
Legislations, on genetic discrimination, 183–185, 185*t*
    ADA, 183–184
    GINA, 184
    HIPAA, 184
        National Standards to Protect Patients' Personal Medical
        Records, 184
Licensing
    compulsory, 337*b*
    genetic tests, 341*b*
LifeCell, 59
Life scientists, responsibilities related to DURC,
    165
Lisbon Agreement, 296
Living Modified Organisms (LMO), 217–219
Lou Gehrig disease, 42

M

Madrid Agreement and Madrid Protocol, 296
Maharashtra Hybrid Seed Company, 119*b*
Mapping sequences, 11*b*
Market approvals
    for generics, 333*b*
    safety considerations for
        of biopharmaceuticals, 284–286
        of biosimilars, 286
Market Authorization Application (MAA), 285–286
Massachusetts Institute of Technology (MIT), 18
Mayo Collaborative Services v. Prometheus Laboratories,
    Inc., 339*b*
Medicines. *See also* Biologics
    competitive market-derived prices, 343
    international trade agreements in, 342–343
        TPP, 343
        TRIPS Agreement, 342–343
    pain, 191–194
    regenerative, 65
    substances used in, 191–194

Meganucleases, 90
Mendel, Gregor, 2–3
    laws, 4f
messenger RNA (mRNA), 5–6
    mRNA reading frame, 5–6
Mexico, 206
Microbial enzymes, 191
Microorganisms, risk categories of, 255–256
Mill, John Stuart, 129–130
Millions Against Monsanto, 120–121
Minimal genome, 111–112
Missiplicity project, 35b
MND. *See* Motor neuron disease (MND)
Moderately repeated sequences, 9
Molecular farming, 97–100
    transgenic animals, 100
    transgenic plants, 99
MON810, 117–118
Monoamine oxidase (MAO) A, 20
monohybrid crosses, 4f
Monsanto Canada Inc. v. Schmeiser, 305b
Monsanto Company, 120–121
Moore v. Regents of California, 302, 303b
Morality of patenting life forms, 307
Morals
    defined, 129
    socially acceptable behaviors, 129
Mosquitoes, 110b
Motor neuron disease (MND), 42
Mousepox virus, 163
mRNA. *See* Messenger RNA (mRNA)
Multipotent adult progenitor cells, 59–60
Mutations
    cystic fibrosis, 17–18
    in several genes, 172–173
    in single gene, 172–173
Mutually acceptable data, 271
Mutually agreed terms (MAT), 196
Mycoplasma genitalium, 19, 111–112

N

Nagoya Protocol, 199–201, 348
    ABSCH, 201
    access obligations, 199
    Article 8(j) of the CBD on the role of TK, 200–201
    Article 14, 201
    benefit sharing obligations, 200
    compliance obligations, 200–201
    description, 199
    parties to, 199, 199f
Nanog, 61b
National Academies
    ESCRO Committees, 75–76

Guidelines for Human Embryonic Stem Cell Research,
    75–76
National Academies of Sciences' National Research Council
    (NAS/NRC), 39
National Apex Committee for Stem Cell Research and
    Therapy (NAC-SCRT), 74
National Bioethics Advisory Committee, 48
National Center for Biotechnology Center (NCBI), 10–13
National Commission for the Protection of Human Subjects of
    Biomedical and Behavioral Research, 143–144
National Geographic Society, 14b
National Guidelines for Stem Cell Research (India), 74
National Human Genome Research Institute (NHGRI),
    178–179
National Institutes of Health (NIH)
    HGP. *See* Human Genome Project (HGP)
    NSABB, 163–164
    research at, 162–163
    stem cell research, 74–75
National Parks Omnibus Management Act of 1998, 205
National parks system, ABS and, 205
National Research Act, 143–144
National Science Advisory Board for Biosecurity (NSABB),
    163–164
    code of conduct, 164–165
    responsible conduct of research (RCR), 164
    Transformational Medical Technologies Initiative,
        163–164
National Strategy for the Conservation of Australia's
    Biological Diversity, 206
Native Americans, 152–153
Natural Product Repository, 194
Nature, 62b
Nazi War Crime Tribunals, 143
New Drug Application (NDA), 285
New Zealand, 205
    animal cloning, 40–41
Nexia Biotechnologies, 108
Next Gen Sequencers, 11b
NHGRI. *See* National Human Genome Research Institute
    (NHGRI)
NIH Revitalization Act of 1993, 19
Nonallergenic gene, 109
Noncoding stretches, 5
Non-governmental organizations (NGOs), 119b
Noninvasive Prenatal Testing (NIPT), 320b
Nontoxic gene, 109
Norman-Bloodsaw v. Lawrence Berkeley Laboratory, 176b
Novelty, 314–315
NSABB. *See* National Science Advisory Board for
    Biosecurity (NSABB)
Nuclear transfer, 26–27
    embryonic stem cells, 58
    limitations of, 30

Nucleic acids, chemical structure of, 3
Nuremberg Code, 143, 143*b*
Nutritional composition, GM crops, 104

O

OECD. *See* Organization for Economic Cooperation and Development (OECD)
Omics, 9, 10*t*. *See also* Genomics
On the Origin of Species (Darwin), 312
Open-ended Ad Hoc Intergovernmental Committee for the Nagoya Protocol on ABS, 197
Open-ended Ad-Hoc Working Group on Biosafety, 217–219
Operator region of operons, 7–8
Operons, 5, 7–8
    operator region, 7–8
    promoter region, 7–8
Organization for Economic Cooperation and Development (OECD), 156*t*
    GLP, 270
Organ transplantation, stem cells and, 66
Oviedo Convention, 182–183
Oxitec Limited, 110*b*

P

Pain medicine, 191–194
Pandemic and All-Hazards Preparedness Reauthorization Act of 2013, 161
Paris Convention, 296
Parkinson's Fly, 106
Passive immunization, 91–96
Patentability
    essential requirements for, 314–316
        industrial applicability/utility requirement, 316
        inventive step/nonobviousness step, 315
        novelty, 314–315
    prerequisites for, 313–314
Patent Cooperation, 301–302
Patent Cooperation Treaty (PCT), 296
Patent infringement, GM crops, 305*b*
Patenting animals, 321–323
Patenting cells and cell lines, 316–318
Patenting genes, 301–302, 318–321
Patenting life forms, morality of, 307
Patenting organisms, 298–301
Patent protection
    for biologics, 335–338
        evergreening, 336–338
        incremental innovation, 336
    for diagnostics, 338–340
    genetic testing, 340–342
Patents, 299*b*
    criteria for award of, 312–316
    protection of, 323–325

Patents Act (South Africa), 206
Patents (Amendment) Act, 2002, 342–343
Pathway Genomics, 174*t*
PATRIOT Act of 2001, 158
Pets, 109–110. *See also* Transgenic animals
    hypoallergenic, 111
Pharmaceuticals, patents, 331–334
Pharmacogenomics, 9, 16, 172
Pharmacovigilance plan, 286
PhiX174, synthetic version of, 19
Phylogenetic analysis, 13
Phylogenetic trees, 13
Physical containment, 257–262
    for animals, 267–269, 267*b*
    for plants, 263–266, 264*b*
    for standard laboratory experiments, 258*b*
Plague corpses, 152–153
Plant molecular farming (PMF), 99
Plants
    physical and biological containment, 263–266, 264*b*
    transgenic
        creation, 100
        for molecular farming, 99
Plasmids, 84–85
Playing God, 42–44
Pluripotent, 26
Policy and tools, rights based on
    Oviedo Convention, 182–183
    Protocol on Genetic Testing for Health Purposes, 183
    Title VII of the Civil Rights Act of 1964, 183
Polymerase chain reaction (PCR), 306
Precautionary Principle, 250–251
Precision Medicine Initiative, 16
President's Council on Bioethics, 48
Prior informed consent (PIC), 196
Prohibition of Human Cloning Act 2002, 48–49
Project BioShield Act, 158, 159*b*, 161
Prokaryotes, 6
    genetic elements in, 7*f*
    rapid response to environmental cues, 5
Promoter gene, 5
Promoter region, 5
    operons, 7–8
Proposed Framework for the Oversight of Dual-Use Life Sciences Research: Strategies for Minimizing the Potential Misuse of Research Information, 164
Protection of Plant Varieties and Farmer's Rights (PPVFR) Act, 324–325
Protein-induced pluripotent stem cells (piPSCs), 61*b*
Proteins
    human pharmaceutical, 99, 99*t*
    recombinant, 91, 92*t*
Protein synthesis, 5–6

Protocol for the Prohibition of the Use in War of
  Asphyxiating, Poisonous or Other Gases and of
  Bacteriological Methods of Warfare. *See* Geneva
  Protocol of 1925
Protocol on Genetic Testing for Health Purposes, 183
Pyrosequencing, 11*b*

R
Recombinant DNA Advisory Committee (RDAC), 212–213
Recombinant DNA (rDNA) technology
  antibodies, 91–97
  Berg Committee, 212–213
  challenges in applications, 112–113
    biodiversity, 112–113
    environment, 112
    sociocultural norms, 113
    socioeconomic status of farmers, 113
  good industrial large-scale practice, 275–278
  industrial applications, safety considerations for, 274–275
  objections to, 114–117
    proportionality, 114
    slippery slope, 114
    subsidiarity, 114–117
  overview, 17, 84–85, 212–213
  proteins, 91
    pharmaceutical applications, 92*t*
  safety issues, 213–215
    biopharmaceuticals, 214–215
    food produced by GMOs, 214
    gene therapy, 215
    GMO's impact on environment, 213–214
Regenerative medicine, 65
Registry of Standard Biological Parts, 18
Regulation (EC) No. 1830/2003, 284
Release of Insects with Dominant Lethal (RIDL) genes, 109,
  110*b*
Removing Barriers to Responsible Scientific Research
  Involving Human Stem Cells, 74–75
Research Involving Human Embryos Act 2002, 48–49
Research tools, sequences used as, 319–321
Respecting Assisted Human Reproduction and Related
  Research, 48
Responsible conduct of research (RCR), 164
Restricting access to innovations, 307–308
Restriction endonucleases, 84–85
RiceTec, 325–326
Rickettsia, 151–152
RIDL. *See* Release of Insects with Dominant Lethal (RIDL)
  genes
Rights
  advocacy, 180–182
    Charter of Fundamental Rights (EU), 182
    human rights, 181

International Declaration on Human Genetic Data, 181
  Universal Declaration on Bioethics and Human Rights,
    181–182
  Universal Declaration on the Human Genome and
    Human Rights, 181
  policy and tools
    Oviedo Convention, 182–183
    Protocol on Genetic Testing for Health Purposes, 183
    Title VII of the Civil Rights Act of 1964, 183
RIKEN Centre for Developmental Biology, Kobe, Japan, 60,
  62*b*
Risk assessment
  defined, 234–235
  GEM, 238–240
  GM crops, 240–242
  prospective, 234–235
  retrospective, 234–235
  science-based, 235–236
  six-step process, 236–237, 236*f*
  transgenic animals, 243–247
Risk management, 237
  objective, 237
  post-release monitoring and documentation, 237
RNA
  intermediate, 5–6
  modifications, 6
Roslin Institute of Edinburg, Scotland, 26–27, 40
Royal Institute of Technology, Stockholm, 11*b*
Rural Advancement Foundation International, 194–195

S
Safety assessment
  clinical trials, 249–250
  foods derived from GMO, 247–249
Safety considerations
  for biotech industry, 274–275
  for field release
    GM animals, 282–283
    GM crops, 279–282
  industrial applications of rDNA technology, 274–275
  for market approval
    of biopharmaceuticals, 284–286
    of biosimilars, 286
  for marketing of GM foods, 283–284
Sangamo BioSciences, Richmond, California, 90
Scientific Fraud and Stimulus Triggered Acquisition of
  Pluripotent Stem Cells, 62*b*
SCNT. *See* Somatic cell nuclear transfer (SCNT)
Secretary's Advisory Committee on Genetics, Health and
  Society (SACGHS), 340–342
Section 101 of Title 35 U.S.C., 300–301
Select Agent Rule, 157–158
Self-limiting technique, 109

September 11, 2001, terrorist attacks, 150
  anthrax, 152—153
Sequencing genome, 9
Severe Combined Immune Deficiency (SCID), 17—18
  gene therapy, 84, 89
Shotgun sequencing, 11*b*
single stranded DNA (ssDNA), 11*b*
SIT. *See* Sterile Insect Technique (SIT)
Smallpox virus, 152—153
Social benefits, of stem cells, 60—67
  cancer treatment, 66
  cellular differentiation, 64
  identifying drug targets, 67
  organ transplantation, 66
  regenerative medicine, 65
  studying disease progression, 65
  tissue engineering, 66
  toxicity testing, 67
Sociocultural norms, rDNA technology and, 113
Socioeconomic status of farmers, 113
Somatic cell nuclear transfer (SCNT), 26, 27*f*
Somatostatin, 91
Sooam Biotech Research Foundation, 35*b*
South Africa, 206
Spider silk, 108
Splicing, 5
State Intellectual Property Office of the People's Republic of
  China, 354
Stem Cell Research Enhancement Act (H.R. 810),
  74—75
Stem cells
  adult, 57—60
    multipotent adult progenitor cells, 59—60
    transdifferentiation, 59—60
  benefit to society, 60—67
    cancer treatment, 66
    cellular differentiation, 64
    identifying drug targets, 67
    organ transplantation, 66
    regenerative medicine, 65
    studying disease progression, 65
    tissue engineering, 66
    toxicity testing, 67
  clinical trials, 71
  cord blood, 59
  embryonic, 57—58
    nuclear transfer, 58
  ethical and scientific misconduct in research, 62*b*
  ethical issues
    research, 68—70
    translation, 71—73
  fetal, 59
  guidelines for clinical translation, 71—73
  overview, 56

  politics/public opinion shaping policy on, 76—77
  sources, 57—60
Sterile Insect Technique (SIT), 109
Stimulus-triggered acquisition of pluripotency (STAP) cells,
  60, 62*b*
Structural genes, 5
Structural genomics, 9
Superweeds, 104
Sustainable Forestry Development Act (Mexico), 206
Symptomatic genetic disabilities, 183—184
Synthetic biology, 18—19
Synthetic organisms, 111—112

T
TALEN (transcription activator like effector nucleases), 90
Taq polymerase, 306
Tartars, 152—153
Tasmanian tiger, 31*b*
Taxol, 191—194
Tay-Sachs disease, 173—174
Technical Report 986 (WHO), 287
TeGenero clinical trial, 139
Terminator region, 5
Terminator technology, 133
Terrorism. *See also* Bioterrorism
  defined, 151
Thalidomide disaster, 139
Theories of ethics, 129—130
  common morality, 130
  consequentialist, 129—130
Therapeutic cloning, 34
Thermus aquaticus (Taq), 306
*Threat Awareness, Prevention and Protection, Surveillance
  and Detection, and Response and Recovery*, 162
Three 'R's of animal experimentation, 134—135
Tissue engineering, 66
Tissue-plasmogen activator, 91
Title VII of the Civil Rights Act of 1964, 176*b*, 183
Title 35 United States Code (U.S.C.), 298
Totipotency, 26
Toxicity potential, GM crops, 103
Toxicity testing, 67
Toxins, cone snail, 191—194
TPP. *See* Trans Pacific Partnership Agreement (TPP)
Trade, IPR in, 293—295
Traditional Cultural Expressions (TCE), 346—347
Traditional knowledge (TK)
  concept, 346, 347*b*
  as documented/undocumented knowledge, 346—347
  livelihood needs, 346—347
  overview, 346
  preserving, 346—347
  protection, 346—347
    CBD, 347—348

concept, 347
defensive strategies, 353–354
FAO, 352
importance and need, 346–352
legal, 352–354
objectives, 347, 347b
positive strategies, 353
WIPO, 348–350
WTO, 350–351
Traditional Knowledge Digital Library (TKDL), 353–354
Traitor technology, 133
Transcription, 5–6
Transdifferentiation, adult stem cells, 59–60
Transduction, 88. *See also* Gene therapy
Transfection, 88. *See also* Gene therapy
Transformational Medical Technologies Initiative, 163–164
Transgenic animals, 105–111
creation of, 106f
disease models, 106
food production, 106–107
for molecular farming, 100
pets, 109–110
hypoallergenic, 111
proteins, 108
vector control, 109
Transgenic livestock
for nutraceuticals, 33–34
for pharmaceuticals, 33
Transgenic plants
creation, 100
for molecular farming, 99
Translation, 5–6
Trans Pacific Partnership Agreement (TPP), 343
Treaty of Lisbon, 182
Triplet codons, 5–6
TRIPS Agreement, 295, 342
Tuskegee syphilis trials, 138
T-VEC, 88
23andMe, 174t

U

UCART19, 90
UCB. *See* Umbilical cord blood (UCB)
Umbilical cord blood (UCB), 59
UCB-banking, 59
Unique sequences, 9
United Kingdom, 152–153
ban on export of British beef and culling of animals, 118
human cloning, 47
Human Reproductive Cloning Act 2001, 47
stem cell research, 74
United Nations
CBD. *See* Convention on Biological Diversity (CBD)
Environment Programme, 195–196, 216–217

United Nations Conference on Environment and Development
(UNCED), 216–217
United Nations Convention on the Law of the Sea
(UNCLOS), 201
maritime zones, 201
parties, 201, 202f
United Nations Declaration on Human
Cloning, 45, 46b
United Nations Declaration on the Rights of Indigenous
Peoples, 346–347
United Nations Educational, Scientific and Cultural
Organization (UNESCO), 179–180, 346–347
Bioethics Programme, 181
General Conference of, 181–182
stem cell research, 73–74
United Nations Industrial Development Organization
(UNIDO), 216–217
United Nations Office for Disarmament Affairs (UNODA),
154
United States
animal cloning, 39
approach to bioterrorism, 157–161
Bioterrorism Preparedness Act, 158
PATRIOT Act of 2001, 158
Project BioShield Act, 158, 159b, 161
biological arsenal, 152–153
GM food crop, 117
human cloning, 47–48
legislations preventing genetic discrimination, 183–185,
185t
ADA, 183–184
GINA, 184
HIPAA, 184
National Standards to Protect Patients' Personal Medical
Records (HIPAA), 184
stem cell research, 74–75
United States Army Medical Research Institute of Infectious
Diseases, 152–153
United States Patent and Trademark Office
(USPTO), 298
Universal Declaration on Bioethics and Human Rights,
144–146, 181–182
Universal Declaration on Human Rights,
181
Universal Declaration on the Human Genome and Human
Rights (UNESCO), 45, 181
University of California, San Francisco, 84–85
UNODA. *See* United Nations Office for Disarmament Affairs
(UNODA)
UPOV, 296, 324–325
US Department of Energy (DOE), 10
US National Cancer Institute (NCI), 194
US National Institutes of Health, 194
Utilitarianism, 129–130

V

Vaccine antigens, 99
Values
  defined, 129
  intrinsic, 129
Viruses, 151—152
Voluntary moratorium on rDNA research, 212—213

W

Wildlife General Act (Mexico), 206
WIPO. *See* World Intellectual Property Organization (WIPO)
Woolly mammoth, 31*b*
Working Group on ABS, 196
Working Group on Article 8(j), 196
Working Group on Protected Areas, 196
Working Group on the Review of Implementation of the
  Convention, 197
World Federation for Culture Collections, 156—157
World Health Organization (WHO)
  HGN, 179—180
  Laboratory Biosafety Manual, 155—156, 254
  Technical Report 986, 287
  on traditional medicine, 346—347
World Intellectual Property Organization (WIPO), 296—297,
  346—347
World Medical Association (WMA), 140, 141*b*
World Trade Centre, terrorist attacks on, 150
World Trade Organization (WTO), 294—295, 346—347
World War I, 152—153

Y

Yamanaka factors, 60, 62*b*
Yellowstone National Park, 306
Yorktown Technologies, L.P., Austin, Texas, 109
Y-shaped glycoproteins, 91—96

Z

ZFN (zinc finger nucleases), 90
Zika virus, 19